11th Workshop on Biomedical Natural Language Processing 2012

(BioNLP 2012)

Held at the 2012 Conference of the North American Chapter of the Association for Computational Linguistics: Human Language Technologies (NAACL-HLT 2012)

Montreal, Quebec, Canada
3-8 June 2012

ISBN: 978-1-62276-552-2

Printed from e-media with permission by:

Curran Associates, Inc.
57 Morehouse Lane
Red Hook, NY 12571

Some format issues inherent in the e-media version may also appear in this print version.

Printed by Curran Associates, Inc. (2013)

For permission requests, please contact the Association for Computational Linguistics at the address below.

Association for Computational Linguistics
209 N. Eighth Street
Stroudsburg, Pennsylvania 18360

Phone: 1-570-476-8006
Fax: 1-570-476-0860

acl@aclweb.org

Additional copies of this publication are available from:

Curran Associates, Inc.
57 Morehouse Lane
Red Hook, NY 12571 USA
Phone: 845-758-0400
Fax: 845-758-2634
Email: curran@proceedings.com
Web: www.proceedings.com

This page intentionally left blank.

Introduction

BioNLP 2012 received 31 submissions exceeding even the traditionally high quality of the preceding eleven years of BioNLP. Due to uniformly positive reviews, eleven submissions were accepted as full papers and 19 as poster presentations.

The themes in this year's papers and posters continue reflecting researchers' growing interest in clinical text processing, while maintaining a steady mature work in biological language processing. This year presents a wide range of innovative methods applied to interesting problems in both domains.

Acknowledgments

We are profoundly grateful to the authors who chose BioNLP as venue for presenting their innovative research.

The authors' willingness to share their work through BioNLP consistently makes the workshop not only noteworthy and stimulating, but also one of the largest, and some years the largest workshop, at ACL/NAACL.

We are equally indebted to the program committee members (listed elsewhere in this volume) who produced three thorough reviews per paper on a tight review schedule and with an admirable level of insight.

Organizers:

Kevin Bretonnel Cohen, University of Colorado School of Medicine
Dina Demner-Fushman, US National Library of Medicine
Sophia Ananiadou, University of Manchester and National Centre for Text Mining, UK
John Pestian, Computational Medical Center, University of Cincinnati,
Cincinnati Children's Hospital Medical Center
Jun'ichi Tsujii, University of Tokyo
and Microsoft Research Asia
Bonnie Webber,University of Edinburgh, UK

Program Committee:

Sophia Ananiadou
Galia Angelova
Emilia Apostolova
Alan Aronson
Olivier Bodenreider
Wendy Chapman
Kevin Cohen
Nigel Collier
Dina Demner-Fushman
Noemie Elhadad
Marcelo Fiszman
Filip Ginter
Su Jian
Jin-Dong Kim
Zhiyong Lu
Aurelie Neveol
Jon Patrick
John Pestian
Sampo Pyysalo
Bastien Rance
Fabio Rinaldi
Thomas Rindflesch
Brian Roark
Andrey Rzhetsky
Daniel Rubin
Guergana Savova
Hagit Shatkay
Matthew Simpson
Pontus Stenetorp
Yuka Tateisi

Jun'ichi Tsujii
Yoshimasa Tsuruoka
Ozlem Uzuner
Karin Verspoor
Bonnie Webber
Peter White
W. John Wilbur
Limsoon Wong
Antonio Yepes
Guodong Zhou
Pierre Zweigenbaum

Invited Speaker:

Wendy W. Chapman, University of California San Diego
Challenges and Opportunities in Clinical Text Annotation

Table of Contents

Conference Program

Friday, June 8, 2012

8:40–8:50	Opening Remarks
	Session 1: Alignment, similarity, classification
8:50–9:10	*Graph-based alignment of narratives for automated neurological assessment* Emily Prud'hommeaux and Brian Roark
9:10–9:30	*Bootstrapping Biomedical Ontologies for Scientific Text using NELL* Dana Movshovitz-Attias and William W. Cohen
9:30–9:50	*Semantic distance and terminology structuring methods for the detection of semantically close terms* Marie Dupuch, Laëtitia Dupuch, Thierry Hamon and Natalia Grabar
9:50–10:10	*Temporal Classification of Medical Events* Preethi Raghavan, Eric Fosler-Lussier and Albert Lai
10:10–10:30	*Analyzing Patient Records to Establish If and When a Patient Suffered from a Medical Condition* James Cogley, Nicola Stokes, Joe Carthy and John Dunnion
10:30–11:00	Morning coffee break
11:00–12:10	Invited Talk by Wendy Chapman
12:10–12:30	*Alignment-HMM-based Extraction of Abbreviations from Biomedical Text* Dana Movshovitz-Attias and William W. Cohen
12:30–14:00	Lunch break
14:00–14:20	*Medical diagnosis lost in translation – Analysis of uncertainty and negation expressions in English and Swedish clinical texts* Danielle L. Mowery, Sumithra Velupillai and Wendy W. Chapman
14:20–14:40	*A Hybrid Stepwise Approach for De-identifying Person Names in Clinical Documents* Oscar Ferrandez, Brett South, Shuying Shen and Stephane Meystre

Friday, June 8, 2012 (continued)

14:40–15:00 *Active Learning for Coreference Resolution*
Timothy Miller, Dmitriy Dligach and Guergana Savova

15:00–15:20 *PubMed-Scale Event Extraction for Post-Translational Modifications, Epigenetics and Protein Structural Relations*
Jari Björne, Sofie Van Landeghem, Sampo Pyysalo, Tomoko Ohta, Filip Ginter, Yves Van de Peer, Sophia Ananiadou and Tapio Salakoski

15:20–15:40 *An improved corpus of disease mentions in PubMed citations*
Rezarta Islamaj Dogan and Zhiyong Lu

15:30–16:00 Afternoon coffee break

Poster Session (16:00–18:00)

New Resources and Perspectives for Biomedical Event Extraction
Sampo Pyysalo, Pontus Stenetorp, Tomoko Ohta, Jin-Dong Kim and Sophia Ananiadou

Combining Compositionality and Pagerank for the Identification of Semantic Relations between Biomedical Words
Thierry Hamon, Christopher Engström, Mounira Manser, Zina Badji, Natalia Grabar and Sergei Silvestrov

Domain Adaptation of Coreference Resolution for Radiology Reports
Emilia Apostolova, Noriko Tomuro, Pattanasak Mongkolwat and Dina Demner-Fushman

What can NLP tell us about BioNLP?
Attapol Thamrongrattanarit, Michael Shafir, Michael Crivaro, Bensiin Borukhov and Marie Meteer

A Prototype Tool Set to Support Machine-Assisted Annotation
Brett South, Shuying Shen, Jianwei Leng, Tyler Forbush, Scott DuVall and Wendy Chapman

MedLingMap: A growing resource mapping the Bio-Medical NLP field
Marie Meteer, Borukhov Bensiin, Mike Crivaro, Michael Shafir and Attapol Thamrongrattanarit

Exploring Label Dependency in Active Learning for Phenotype Mapping
Shefali Sharma, Leslie Lange, Jose Luis Ambite, Yigal Arens and Chun-Nan Hsu

Evaluating Joint Modeling of Yeast Biology Literature and Protein-Protein Interaction Networks
Ramnath Balasubramanyan, Kathryn Rivard, William W. Cohen, Jelena Jakovljevic and John L. Woolford

Friday, June 8, 2012 (continued)

Graph-based alignment of narratives for automated neurological assessment

Emily T. Prud'hommeaux and Brian Roark
Center for Spoken Language Understanding
Oregon Health & Science University
{emilypx, roarkbr}@gmail.com

Abstract

Narrative recall tasks are widely used in neuropsychological evaluation protocols in order to detect symptoms of disorders such as autism, language impairment, and dementia. In this paper, we propose a graph-based method commonly used in information retrieval to improve word-level alignments in order to align a source narrative to narrative retellings elicited in a clinical setting. From these alignments, we automatically extract narrative recall scores which can then be used for diagnostic screening. The significant reduction in alignment error rate (AER) afforded by the graph-based method results in improved automatic scoring and diagnostic classification. The approach described here is general enough to be applied to almost any narrative recall scenario, and the reductions in AER achieved in this work attest to the potential utility of this graph-based method for enhancing multilingual word alignment and alignment of comparable corpora for more standard NLP tasks.

1 Introduction

Much of the work in biomedical natural language processing has focused on mining information from electronic health records, clinical notes, and medical literature, but NLP is also very well suited for analyzing patient language data, in terms of both content and linguistic features, for neurological evaluation. NLP-driven analysis of clinical language data has been used to assess language development (Sagae et al., 2005), language impairment (Gabani et al., 2009) and cognitive status (Roark et al., 2007; Roark et al., 2011). These approaches rely on the extraction of syntactic features from spoken language transcripts in order to identify characteristics of language use associated with a particular disorder. In this paper, rather than focusing on linguistic features, we instead propose an NLP-based method for automating the standard manual method for scoring the Wechsler Logical Memory (WLM) subtest of the Wechsler Memory Scale (Wechsler, 1997) with the eventual goal of developing a screening tool for Mild Cognitive Impairment (MCI), the earliest observable precursor to dementia. During standard administration of the WLM, the examiner reads a brief narrative to the subject, who then retells the story to the examiner, once immediately upon hearing the story and a second time after a 30-minute delay. The examiner scores the retelling in real time by counting the number of recalled *story elements*, each of which corresponds to a word or short phrase in the source narrative. Our method for automatically extracting the score from a retelling relies on an alignment between substrings in the retelling and substrings in the original narrative. The scores thus extracted can then be used for diagnostic classification.

Previous approaches to alignment-based narrative analysis (Prud'hommeaux and Roark, 2011a; Prud'hommeaux and Roark, 2011b) have relied exclusively on modified versions of standard word alignment algorithms typically applied to large bilingual parallel corpora for building machine translation models (Liang et al., 2006; Och et al., 2000). Scores extracted from the alignments produced using these algorithms achieved fairly high classification accuracy, but the somewhat weak alignment quality limited performance. In this paper, we compare these word alignment approaches to a new approach that uses traditionally-derived word alignments between retellings as the input for graph-based exploration of the alignment space in order to improve alignment accuracy. Using both earlier approaches and our novel method for word alignment, we then evaluate the accuracy of automated scoring and diagnostic classification for MCI.

Although the alignment error rates for our data might be considered high in the context of building phrase tables for machine translation, the alignments produced using the graph-based method are remarkably accurate given the small size of our training corpus. In addition, these more accurate alignments lead to gains in scoring accuracy and to classification performance approaching that of manually derived scores. This method for word alignment and score extraction is general enough to be easily adapted to other tests used in neuropsychological evaluation, including not only those related to narrative recall, such as the NEPSY Narrative Memory subtest (Korkman et al., 1998) but also picture description tasks, such as the Cookie Theft picture description task of the Boston Diagnostic Aphasia Examination (Goodglass et al., 2001) or the Renfrew Bus Story (Glasgow and Cowley, 1994). In addition, this technique has the potential to improve word alignment for more general NLP tasks that rely on small corpora, such as multilingual word alignment or word alignment of comparable corpora.

2 Background

The act of retelling or producing a narrative taps into a wide array of cognitive functions, not only memory but also language comprehension, language production, executive function, and theory of mind. The inability to coherently produce or recall a narrative is therefore associated with many different cognitive and developmental disorders, including dementia, autism (Tager-Flusberg, 1995), and language impairment (Dodwell and Bavin, 2008; Botting, 2002). Narrative tasks are widely used in neuropsychological assessment, and many commonly used instruments and diagnostic protocols include a task involving narrative recall or production (Korkman et al., 1998; Wechsler, 1997; Lord et al., 2002).

In this paper, we focus on evaluating narrative recall within the context of Mild Cognitive Impairment (MCI), the earliest clinically significant precursor of dementia. The cognitive and memory problems associated with MCI do not necessarily interfere with daily living activities (Ritchie and Touchon, 2000) and can therefore be difficult to diagnose using standard dementia screening tools, such as the Mini-Mental State Exam (Folstein et al., 1975). A definitive diagnosis of MCI requires an extensive interview with the patient and a family member or caregiver. Because of the effort required for diagnosis and the insensitivity of the standard screening tools, MCI frequently goes undiagnosed, delaying the introduction of appropriate treatment and remediation. Early and unobtrusive detection will become increasingly important as the elderly population grows and as research advances in delaying and potentially stopping the progression of MCI into moderate and severe dementia.

Narrative recall tasks, such as the test used in research presented here, the Wechsler Logical Memory subtest (WLM), are often used in conjunction with other cognitive measures in attempts to identify MCI and dementia. Multiple studies have demonstrated a significant difference in performance on the WLM between subjects with MCI and typically aging controls, particularly in combination with tests of verbal fluency and memory (Storandt and Hill, 1989; Peterson et al., 1999; Nordlund et al., 2005). The WLM can also serve as a cognitive indicator of physiological characteristics associated with symptomatic Alzheimers disease, even in the absence of previously reported dementia (Schmitt et al., 2000; Bennett et al., 2006).

Some previous work on automated analysis of the WLM has focused on using the retellings as a source of linguistic data for extracting syntactic and phonetic features that can distinguish subjects with MCI from typically aging controls (Roark et al., 2011). There has been some work on automating scoring of other narrative recall tasks using unigram overlap (Hakkani-Tur et al., 2010), but Dunn et al. (2002) are among the only researchers to apply automated methods to scoring the WLM for the purpose of identifying dementia, using latent semantic analysis to measure the semantic distance between a retelling

Dx	n	Age	Education
MCI	72	88.7	14.9 yr
Non-MCI	163	87.3	15.1 yr

Table 1: Subject demographic data.

and the source narrative. Although scoring automation is not typically used in a clinical setting, the objectivity offered by automated measures is particularly important for tests like the WLM, which are often administered by practitioners working in a community setting and serving a diverse population.

Researchers working on NLP tasks such as paraphrase extraction (Barzilay and McKeown, 2001), word-sense disambiguation (Diab and Resnik, 2002), and bilingual lexicon induction (Sahlgren and Karlgren, 2005), often rely on aligned parallel or comparable corpora. Recasting the automated scoring of a neuropsychological test as another NLP task involving the analysis of parallel texts, however, is a relatively new idea. We hope that the methods presented here will both highlight the flexibility of techniques originally developed for standard NLP tasks and attract attention to the wide variety of biomedical data sources and potential clinical applications for these techniques.

3 Data

3.1 Subjects

The data examined in this study was collected from participants in a longitudinal study on brain aging at the Layton Aging and Alzheimers Disease Center at the Oregon Health and Science University (OHSU), including 72 subjects with MCI and 163 typically aging seniors roughly matched for age and years of education. Table 1 shows the mean age and mean years of education for the two diagnostic groups. There were no significant between-group differences in either measure.

Following (Shankle et al., 2005), we assign a diagnosis of MCI according to the Clinical Dementia Rating (CDR) (Morris, 1993). A CDR of 0.5 corresponds to MCI (Ritchie and Touchon, 2000), while a CDR of zero indicates the absence of MCI or any dementia. The CDR is measured via the Neurobehavioral Cognitive Status Examination (Kiernan et al., 1987) and a semi-structured interview with the patient and a family member or caregiver that allows the examiner to assess the subject in several key areas of cognitive function, such as memory, orientation, problem solving, and personal care. The CDR has high inter-annotator reliability (Morris, 1993) when conducted by trained experts. It is crucial to note that the calculation of CDR is completely independent of the neuropsychological test investigated in this paper, the Wechsler Logical Memory subtest of the Wechsler Memory Scale. We refer readers to the above cited papers for a further details.

3.2 Wechsler Logical Memory Test

The Wechsler Logical Memory subtest (WLM) is part of the Wechsler Memory Scale (Wechsler, 1997), a diagnostic instrument used to assess memory and cognition in adults. In the WLM, the subject listens to the examiner read a brief narrative, shown in Figure 1. The subject then retells the narrative to the examiner twice: once immediately upon hearing it (Logical Memory I, LM-I) and again after a 30-minute delay (Logical Memory II, LM-II). The narrative is divided into 25 *story elements*. In Figure 1, the boundaries between story elements are denoted by slashes. The examiner notes in real time which story elements the subject uses. The score that is reported under standard administration of the task is a summary score, which is simply the raw number of story elements recalled. Story elements do not need to be recalled verbatim or in the correct temporal order. The published scoring guidelines describe the permissible substitutions for each story element. The first story element, *Anna*, can be replaced in the retelling with *Annie* or *Ann*, while the 16th story element, *fifty-six dollars*, can be replaced with any number of dollars between fifty and sixty.

An example LM-I retelling is shown in Figure 2. According to the published scoring guidelines, this retelling receives a score of 12, since it contains the following 12 elements: *Anna, employed, Boston, as a cook, was robbed of, she had four, small children, reported, station, touched by the woman's story, took up a collection,* and *for her*.

3.3 Word alignment data

The Wechsler Logical Memory immediate and delayed retellings for all of the 235 experimental subjects were transcribed at the word level. We sup-

Anna / Thompson / of South / Boston / employed / as a cook / in a school / cafeteria / reported / at the police / station / that she had been held up / on State Street / the night before / and robbed of / fifty-six dollars. / She had four / small children / the rent was due / and they hadn't eaten / for two days. / The police / touched by the woman's story / took up a collection / for her.

Figure 1: Text of WLM narrative segmented into 25 story elements.

Ann Taylor worked in Boston as a cook. And she was robbed of sixty-seven dollars. Is that right? And she had four children and reported at the some kind of station. The fellow was sympathetic and made a collection for her so that she can feed the children.

Figure 2: Sample retelling of the Wechsler narrative.

plemented the data collected from our experimental subjects with transcriptions of retellings from 26 additional individuals whose diagnosis had not been confirmed at the time of publication or who did not meet the eligibility criteria for this study. Partial words, punctuation, and pause-fillers were excluded from all transcriptions used for this study. The retellings were manually scored according to published guidelines. In addition, we manually produced word-level alignments between each retelling and the source narrative presented in Figure 1.

Word alignment for phrase-based machine translation typically takes as input a sentence-aligned parallel corpus or bi-text, in which a sentence on one side of the corpus is a translation of the sentence in that same position on the other side of the corpus. Since we are interested in learning how to align words in the source narrative to words in the retellings, our primary parallel corpus must consist of source narrative text on one side and retelling text on the other. Because the retellings contain omissions, reorderings, and embellishments, we are obliged to consider the full text of the source narrative and of each retelling to be a "sentence" in the parallel corpus.

We compiled three parallel corpora to be used for the word alignment experiments:

- **Corpus 1:** A roughly 500-line source-to-retelling corpus consisting of the source narrative on one side and each retelling on the other.
- **Corpus 2:** A roughly 250,000-line pairwise retelling-to-retelling corpus, consisting of every possible pairwise combination of retellings.
- **Corpus 3:** A roughly 900-line word identity corpus, consisting of every word that appears in every retelling and the source narrative.

The explicit parallel alignments of word identities that compose Corpus 3 are included in order to encourage the alignment of a word in a retelling to that same word in the source, if it exists.

The word alignment techniques that we use are entirely unsupervised. Therefore, as in the case with most experiments involving word alignment, we build a model for the data we wish to evaluate using that same data. We do, however, use the retellings from the 26 individuals who were not experimental subjects as a development set for tuning the various parameters of our system, which is described below.

4 Word Alignment

4.1 Baseline alignment

We begin by building two word alignment models using the Berkeley aligner (Liang et al., 2006), a state-of-the-art word alignment package that relies on IBM mixture models 1 and 2 (Brown et al., 1993) and an HMM. We chose to use the Berkeley aligner, rather than the more widely used Giza++ alignment package, for this task because its joint training and posterior decoding algorithms yield lower alignment error rates on most data sets and because it offers functionality for testing an existing model on new data and for outputting posterior probabilities. The smaller of our two Berkeley-generated models is trained on Corpus 1 (the source-to-retelling parallel corpus described above) and ten copies of Corpus 3 (the word identity corpus). The larger model is trained on Corpus 1, Corpus 2 (the pairwise retelling corpus), and 100 copies of Corpus 3. Both models are then tested on the 470 retellings from our 235 experimental subjects. In addition, we use both models to align every retelling to every other retelling so that we will have all pairwise alignments available for use in the graph-based model.

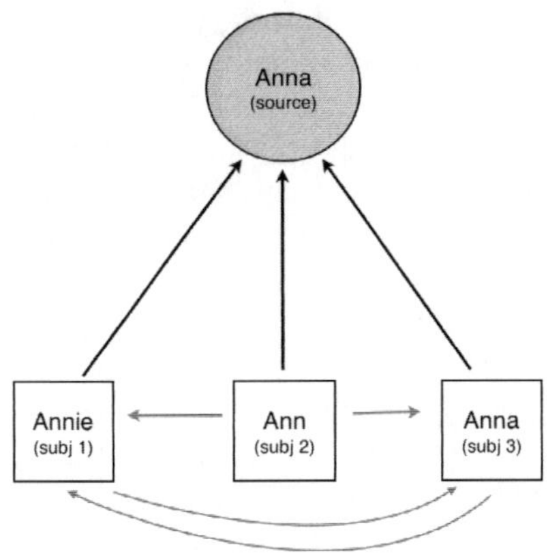

Figure 3: Depiction of word graph.

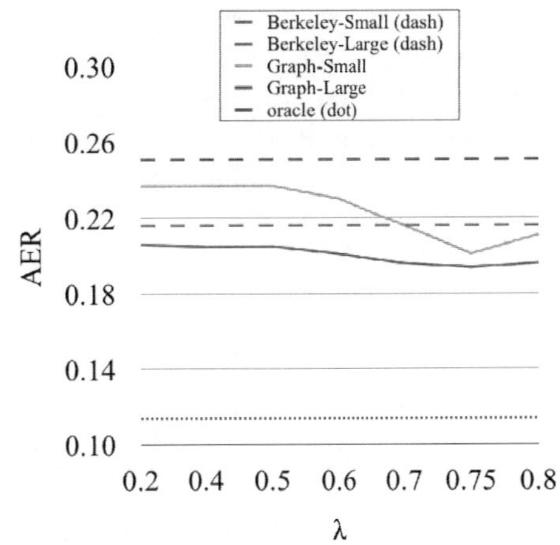

Figure 4: Changes in AER as λ increases.

The first two rows of Table 2 show the precision, recall, F-measure, and alignment error rate (AER) (Och and Ney, 2003) for these two Berkeley aligner models. We note that although AER for the larger model is lower, the time required to train the model is significantly larger. The alignments generated by the Berkeley aligner serve not only as a baseline for comparison but also as a springboard for the novel graph-based method of alignment we will now discuss.

4.2 Graph-based refinement

Graph-based methods, in which paths or *random walks* are traced through an interconnected graph of nodes in order to learn more about the nodes themselves, have been used for various NLP tasks in information extraction and retrieval, including webpage ranking (PageRank (Page et al., 1999)) and extractive summarization (LexRank (Erkan and Radev, 2004; Otterbacher et al., 2009)). In the PageRank algorithm, the nodes of the graph are web pages and the edges connecting the nodes are the hyperlinks leading from those pages to other pages. The nodes in the LexRank algorithm are sentences in a document and the edges are the similarity scores between those sentences. The likelihood of a random walk through the graph starting at a particular node and ending at another node provides information about the relationship between those two nodes and the importance of the starting node.

In the case of our graph-based method for word alignment, each node represents a word in one of the retellings or in the source narrative. The edges are the normalized posterior-weighted alignments that the Berkeley aligner proposes between each word and (1) words in the source narrative, and (2) words in the other retellings, as depicted in Figure 3. Starting at a particular node (i.e., a word in one of the retellings), our algorithm can either walk from that node to another node in the graph or to a word in the source narrative. At each step in the walk, there is a set probability λ that determines the likelihood of transitioning to another retelling word versus a word in the source narrative. When transitioning to a retelling word, the destination word is chosen according to the posterior probability assigned by the Berkeley aligner to that alignment. When the walk arrives at a source narrative word, that word is the new proposed alignment for the starting word.

For each word in each retelling, we perform 1000 of these random walks, thereby generating a distribution for each retelling word over all of the words in the source narrative. The new alignment for the word is the source word with the highest frequency in that distribution.

We build two graphs on which to carry out these random walks: one graph is built using the alignments generated by the smaller Berkeley alignment model, and the other is built from the alignments generated by the larger Berkeley alignment model. Alignments with posterior probabilities of 0.5 or greater are included as edges within the graph, since this is the default posterior threshold used by the Berkeley aligner. The value of λ, the probability of walking to a retelling word node rather than a source word, is tuned to the development set of retellings,

Model	P	R	F	AER
Berkeley-Small	72.1	79.6	75.6	24.5
Berkeley-Large	78.6	80.5	79.5	20.5
Graph-Small	77.9	81.2	79.5	20.6
Graph-Large	85.4	76.9	81.0	18.9

Table 2: Aligner performance comparison.

discussed in Section 3.3. Figure 4 shows how AER varies according to the value of λ for the two graph-based approaches.

Each of these four alignment models produces, for each retelling, a set of word pairs containing one word from the original narrative and one word from the retelling. The manual gold alignments for the 235 experimental subjects were evaluated against the alignments produced by each of the four models. Table 2 shows the accuracy of word alignment using these two graph-based models in terms of precision, accuracy, F-measure, and alignment error rate, alongside the same measures for the two Berkeley models. We see that each of the graph-based models outperforms the Berkeley model of the same size. The performance of the small graph-based model is especially remarkable since it an AER comparable to the large Berkeley model while requiring significantly fewer computing resources. The difference in processing time between the two approaches was especially remarkable: the graph-based model completed in only a few minutes, while the large Berkeley model required 14 hours of training.

Figures 5 and 6 show the results of aligning the retelling presented in Figure 2 using the small Berkeley model and the large graph-based model, respectively. Comparing these two alignments, we see that the latter model yields more precise alignments with very little loss of recall, as is borne out by the overall statistics shown in Table 2.

5 Scoring

The published scoring guidelines for the WLM specify the source words that compose each story element. Figure 7 displays the source narrative with the element IDs ($A - Y$) and word IDs ($1 - 65$) explicitly labeled. Element Q, for instance, consists of the words 39 and 40, *small children*. Using this information, we extract scores from the alignments as follows: for each word in the original narrative, if

[A anna$_{1}$] [B thompson$_{2}$] [C of$_{3}$ south$_{4}$] [D boston$_{5}$] [E employed$_{6}$] [F as$_{7}$ a$_{8}$ cook$_{9}$] [G in$_{10}$ a$_{11}$ school$_{12}$] [H cafeteria$_{13}$] [I reported$_{14}$] [J at$_{15}$ the$_{16}$ police$_{17}$] [K station$_{18}$] [L that$_{19}$ she$_{20}$ had$_{21}$ been$_{22}$ held$_{23}$ up$_{24}$] [M on$_{25}$ state$_{26}$ street$_{27}$] [N the$_{28}$ night$_{29}$ before$_{30}$] [O and$_{31}$ robbed$_{32}$ of$_{33}$] [P fifty-six$_{34}$ dollars$_{35}$] [Q she$_{36}$ had$_{37}$ four$_{38}$] [R small$_{39}$ children$_{40}$] [S the$_{41}$ rent$_{42}$ was$_{43}$ due$_{44}$] [T and$_{45}$ they$_{46}$ had$_{47}$ n't$_{48}$ eaten$_{49}$] [U for$_{50}$ two$_{51}$ days$_{52}$] [V the$_{53}$ police$_{54}$] [W touched$_{55}$ by$_{56}$ the$_{57}$ woman's$_{58}$ story$_{59}$] [X took$_{60}$ up$_{61}$ a$_{62}$ collection$_{63}$] [Y for$_{64}$ her$_{65}$]

Figure 7: Text of Wechsler Logical Memory narrative with story-element labeled bracketing and word IDs.

anna(1) : A	robbed(32) : O	station(18) : K
thompson(2) : B	fifty-six(34) : P	took(60) : X
employed(6) : E	four(38) : Q	collection(63) : X
boston(5) : D	children(40) : R	for(64) : Y
cook(9) : F	reported(14) : I	her(65) : Y

Figure 8: Source content words from the alignment in Figure 6 with corresponding story element IDs.

that word is aligned to a word in the retelling, the story element that it is associated with is considered to be recalled. Figure 8 shows the story elements extracted from the word alignments in Figure 6.

When we convert alignments to scores in this way, any alignment can be mapped to an element, even an alignment between function words such as *the* and *of*, which would be unlikely to indicate that the story element had been recalled. To avoid such scoring errors, we disregard any word-alignment pair containing a source function word. The two exceptions to this rule are the final two words, *for her*, which are not content words but together make a single story element.

The element-level scores induced from the four word alignments for all 235 experimental subjects were evaluated against the manual per-element scores. We report the precision, recall, and f-measure for all four alignment models in Table 3. In addition, report Cohen's kappa as a measure of reliability between our automated scores and the manually assigned scores. We see that as AER improves, scoring accuracy also improves, with the large graph-based model outperforming all other models in terms of precision, f-measure, and inter-

ann(1) : anna(1)
worked(3) : employed(6)
in(4) : in(10)
boston(5) : boston(5)
as(6) : as(7)
a(7) : a(8)
cook(8) : cook(9)
and(9) : and(31)
robbed(12) : robbed(32)
of(13) : of(33)
dollars(15) : dollars(35)
is(16) : was(43)
that(17) : that(19)
and(19) : and(45)
she(20) : she(36)
had(21) : had(37)
four(22) : four(38)
children(23) : children(40)
reported(25) : reported(14)
at(26) : at(15)
the(27) : the(16)
some(28) : police(17)
station(31) : station(18)
made(37) : up(61)
made(37) : took(60)
a(38) : a(62)
collection(39) : collection(63)
for(40) : for(64)
her(41) : her(65)
so(42) : woman's(58)
she(44) : she(20)

Figure 5: Word alignment generated by the small Berkeley alignment model with retelling words italicized.

ann(1) : anna(1)
taylor(2) : thompson(2)
worked(3) : employed(6)
in(4) : in(10)
boston(5) : boston(5)
as(6) : as(7)
a(7) : a(8)
cook(8) : cook(9)
robbed(12) : robbed(32)
of(13) : of(33)
sixty-seven(14) : fifty-six(34)
dollars(15) : dollars(35)
she(20) : she(36)
had(21) : had(37)
four(22) : four(38)
children(23) : children(40)
reported(25) : reported(14)
at(26) : at(15)
the(27) : the(16)
station(31) : station(18)
made(37) : took(60)
a(38) : a(62)
collection(39) : collection(63)
for(40) : for(64)
her(41) : her(65)

Figure 6: Word alignment generated by the large graph-based model with retelling words italicized.

Model	P	R	F	κ
Berkeley-Small	87.2	88.9	88.0	76.1
Berkeley-Large	86.8	90.7	88.7	77.1
Graph-Small	84.7	93.6	88.9	76.9
Graph-Big	88.8	89.3	89.1	78.3

Table 3: Scoring accuracy results.

rater reliability. The scoring accuracy levels reported here are comparable to the levels of inter-rater agreement typically reported for the WLM, and reliability between our automated scores and the manual scores, as measured by Cohen's kappa, is well within the ranges reported in the literature (Johnson et al., 2003). As will be shown in the following section, scoring accuracy is very important for achieving high diagnostic classification accuracy, which is the ultimate goal of this work.

6 Diagnostic Classification

As discussed in Section 2, poor performance on the Wechsler Logical Memory test is associated with Mild Cognitive Impairment. We now use the scores we have extracted from the word alignments as features with a support vector machine (SVM) to perform diagnostic classification for distinguishing subjects with MCI from those without. For each of the 235 experimental subjects, we generate 2 summary scores: one for the immediate retelling and one for the delayed retelling. The summary score ranges from 0, indicating that no elements were recalled, to 25, indicating that all elements were recalled. In addition to the summary score, we also provide the SVM with a vector of 50 per-element scores: for each of the 25 element in each of the two retellings per subject, there is a vector element with the value of 0 if the element was not recalled, or 1 if the element was recalled. Since previous work has indicated that certain elements may be more powerful in their ability to predict the presence of MCI, we expect that giving the SVM these per-elements scores may improve classification performance. To train and test our classifiers, we use the WEKA API (Hall et al., 2009) and LibSVM (Chang and Lin, 2011), with a second-order polynomial kernel and default parameter settings.

We evaluate the performance of the SVMs using a leave-pair-out validation scheme (Cortes et al., 2007; Pahikkala et al., 2008). In the leave-pair-out technique, every pairing between a negative example and a positive example is tested using a classifier trained on all of the remaining examples. The resulting pairs of scores can be used to calculate the area under the receiver operating characteristic (ROC) curve (Egan, 1975), which is a plot of the false positive rate of a classifier against its true positive rate. The area under this curve (AUC) has a

Model	Summ. (s.d.)	Elem. (s.d.)
Manual Scores	73.3 (3.76)	81.3 (3.32)
Berkeley-Small	73.7 (3.74)	77.9 (3.52)
Berkeley-Big	75.1 (3.67)	79.2 (3.45)
Graph-Small	74.2 (3.71)	78.9 (3.47)
Graph-Big	74.8 (3.69)	78.6 (3.49)

Table 4: Classification accuracy results (AUC).

value of 0.5 when the classifier performs at chance and a value 1.0 when perfect classification accuracy is achieved.

Table 4 shows the classification results for the scores derived from the four alignment models along with the classification results using the examiner-assigned manual scores. It appears that, in all cases, the per-element scores are more effective than the summary scores in classifying the two diagnostic groups. In addition, we see that our automated scores have classificatory power comparable to that of the manual gold scores, and that as scoring accuracy increases from the small Berkeley model to the graph-based models and bigger models, classification accuracy improves. This suggests both that accurate scores are crucial for accurate classification and that pursuing even further improvements in word alignment is likely to result in improved diagnostic differentiation. We note that although the large Berkeley model achieved the highest classification accuracy, this very slight margin of difference may not justify its significantly greater computational requirements.

7 Conclusions and Future Work

The work presented here demonstrates the utility of adapting techniques drawn from a diverse set of NLP research areas to tasks in biomedicine. In particular, the approach we describe for automatically analyzing clinically elicited language data shows promise as part of a pipeline for a screening tool for Mild Cognitive Impairment. Our novel graph-based approach to word alignment resulted in large reductions in alignment error rate. These reductions in error rate in turn led to human-level scoring accuracy and improved diagnostic classification.

As we have mentioned, the methods outlined here are general enough to be used for other episodic recall and description scenarios. Although the results are quite robust, several enhancements and improvements should be made before we apply the system to other tasks. First, although we were able to achieve decent word alignment accuracy, especially with our graph-based approach, many alignment errors remain. As shown in Figure 4, the graph-based alignment technique could potentially result in an AER of as low as 11%. We expect that our decision to select as a new alignment the most frequent source word over the distribution of source words at the end of 1000 walks could be improved, since it does not allow for one-to-many mappings. In addition, it would be worthwhile to experiment with several posterior thresholds, both during the decoding step of the Berkeley aligner and in the graph edges.

In order to produce a viable clinical screening tool, it is crucial that we incorporate speech recognition in the pipeline. Our very preliminary investigation into using ASR to generate transcripts for alignment seems promising and surprisingly robust to the problems that might be expected when working with noisy audio. In our future work, we also plan to examine longitudinal data for individual subjects to see whether our techniques can detect subtle differences in recall and coherence between a recent retelling and a series of earlier baseline retellings. Since the metric commonly used to quantify the progression of dementia, the Clinical Dementia Rating, relies on observed changes in cognitive function over time, longitudinal analysis of performance on the Wechsler Logical Memory task may be the most promising application for our research.

References

Regina Barzilay and Kathleen R. McKeown. 2001. Extracting paraphrases from a parallel corpus. In *Proceeding of ACL*.

D.A. Bennett, J.A. Schneider, Z. Arvanitakis, J.F. Kelly, N.T. Aggarwal, R.C. Shah, and R.S. Wilson. 2006. Neuropathology of older persons without cognitive impairment from two community-based studies. *Neurology*, 66:1837–844.

Nicola Botting. 2002. Narrative as a tool for the assessment of linguistic and pragmatic impairments. *Child Language Teaching and Therapy*, 18(1).

Peter Brown, Vincent Della Pietra, Stephen Della Pietra, and Robert Mercer. 1993. The mathematics of statistical machine translation: Parameter estimation. *Computational Linguistics*, 19(2):263–311.

Chih-Chung Chang and Chih-Jen Lin. 2011. LIBSVM: A library for support vector machines. *ACM Transactions on Intelligent Systems and Technology*, 2(27):1–27.

Corinna Cortes, Mehryar Mohri, and Ashish Rastogi. 2007. An alternative ranking problem for search engines. In *Proceedings of WEA2007, LNCS 4525*, pages 1–21. Springer-Verlag.

Mona Diab and Philip Resnik. 2002. An unsupervised method for word sense tagging using parallel corpora. In *Proceedings of ACL*.

Kristy Dodwell and Edith L. Bavin. 2008. Children with specific language impair ment: an investigation of their narratives and memory. *International Journal of Language and Communication Disorders*, 43(2):201–218.

John C. Dunn, Osvaldo P. Almeida, Lee Barclay, Anna Waterreus, and Leon Flicker. 2002. Latent semantic analysis: A new method to measure prose recall. *Journal of Clinical and Experimental Neuropsychology*, 24(1):26–35.

James Egan. 1975. *Signal Detection Theory and ROC Analysis*. Academic Press.

Günes Erkan and Dragomir R. Radev. 2004. Lexrank: Graph-based lexical centrality as salience in text summarization. *J. Artif. Intell. Res. (JAIR)*, 22:457–479.

M. Folstein, S. Folstein, and P. McHugh. 1975. Minimental state - a practical method for grading the cognitive state of patients for the clinician. *Journal of Psychiatric Research*, 12:189–198.

Keyur Gabani, Melissa Sherman, Thamar Solorio, and Yang Liu. 2009. A corpus-based approach for the prediction of language impairment in monolingual English and Spanish-English bilingual children. In *Proceedings of NAACL-HLT*, pages 46–55.

Cheryl Glasgow and Judy Cowley. 1994. *Renfrew Bus Story test - North American Edition*. Centreville School.

H Goodglass, E Kaplan, and B Barresi. 2001. *Boston Diagnostic Aphasia Examination. 3rd ed.* Pro-Ed.

Dilek Hakkani-Tur, Dimitra Vergyri, and Gokhan Tur. 2010. Speech-based automated cognitive status assessment. In *Proceedings of Interspeech*, pages 258–261.

Mark Hall, Eibe Frank, Geoffrey Holmes, Bernhard Pfahringer, Peter Reutemann, and Ian H. Witten. 2009. The WEKA data mining software: An update. *SIGKDD Explorations*, 11(1).

David K. Johnson, Martha Storandt, and David A. Balota. 2003. Discourse analysis of logical memory recall in normal aging and in dementia of the alzheimer type. *Neuropsychology*, 17(1):82–92.

R.J. Kiernan, J. Mueller, J.W. Langston, and C. Van Dyke. 1987. The neurobehavioral cognitive status examination, a brief but differentiated approach to cognitive assessment. *Annals of Internal Medicine*, 107:481–485.

Marit Korkman, Ursula Kirk, and Sally Kemp. 1998. *NEPSY: A developmental neuropsychological assessment*. The Psychological Corporation.

Percy Liang, Ben Taskar, and Dan Klein. 2006. Alignment by agreement. In *Proceedings of HLT NAACL*.

Catherine Lord, Michael Rutter, Pamela DiLavore, and Susan Risi. 2002. *Autism Diagnostic Observation Schedule (ADOS)*. Western Psychological Services.

John Morris. 1993. The clinical dementia rating (CDR): Current version and scoring rules. *Neurology*, 43:2412–2414.

A Nordlund, S Rolstad, P Hellstrom, M Sjogren, S Hansen, and A Wallin. 2005. The goteborg mci study: mild cognitive impairment is a heterogeneous condition. *Journal of Neurology, Neurosurgery and Psychiatry*, 76:1485–1490.

Franz Josef Och and Hermann Ney. 2003. A systematic comparison of various statistical alignment models. *Computational Linguistics*, 29(1):19–51.

Franz Josef Och, Christoph Tillmann, , and Hermann Ney. 2000. Improved alignment models for statistical machine translation. In *Proceedings of ACL*, pages 440–447.

Jahna Otterbacher, Günes Erkan, and Dragomir R. Radev. 2009. Biased lexrank: Passage retrieval using random walks with question-based priors. *Inf. Process. Manage.*, 45(1):42–54.

Lawrence Page, Sergey Brin, Rajeev Motwani, and Terry Winograd. 1999. The pagerank citation ranking: Bringing order to the web. Technical Report 1999-66, Stanford InfoLab, November. Previous number = SIDL-WP-1999-0120.

Tapio Pahikkala, Antti Airola, Jorma Boberg, and Tapio Salakoski. 2008. Exact and efficient leave-pair-out cross-validation for ranking RLS. In *Proceedings of AKRR 2008*, pages 1–8.

Ronald Peterson, Glenn Smith, Stephen Waring, Robert Ivnik, Eric Tangalos, and Emre Kokmen. 1999. Mild cognitive impairment: Clinical characterizations and outcomes. *Archives of Neurology*, 56:303–308.

Emily T. Prud'hommeaux and Brian Roark. 2011a. Alignment of spoken narratives for automated neuropsychological assessment. In *Proceedings of ASRU*.

Emily T. Prud'hommeaux and Brian Roark. 2011b. Extraction of narrative recall patterns for neuropsychological assessment. In *Proceedings of Interspeech*.

Karen Ritchie and Jacques Touchon. 2000. Mild cognitive impairment: Conceptual basis and current nosological status. *Lancet*, 355:225–228.

Brian Roark, Margaret Mitchell, and Kristy Hollingshead. 2007. Syntactic complexity measures for detecting mild cognitive impairment. In *Proceedings of the ACL 2007 Workshop on Biomedical Natural Language Processing (BioNLP)*, pages 1–8.

Brian Roark, Margaret Mitchell, John-Paul Hosom, Kristina Hollingshead, and Jeffrey Kaye. 2011. Spoken language derived measures for detecting mild cognitive impairment. *IEEE Transactions on Audio, Speech and Language Processing*, 19(7):2081–2090.

Kenji Sagae, Alon Lavie, and Brian MacWhinney. 2005. Automatic measurement of syntactic development in child language. In *Proceedings of ACL*, pages 197–204.

Magnus Sahlgren and Jussi Karlgren. 2005. Automatic bilingual lexicon acquisition using random indexing of parallel corpora. *Natural Language Engineering*, 11(3).

F.A. Schmitt, D.G. Davis, D.R. Wekstein, C.D. Smith, J.W. Ashford, and W.R. Markesbery. 2000. Preclinical ad revisited: Neuropathology of cognitively normal older adults. *Neurology*, 55:370–376.

William R. Shankle, A. Kimball Romney, Junko Hara, Dennis Fortier, Malcolm B. Dick, James M. Chen, Timothy Chan, and Xijiang Sun. 2005. Methods to improve the detection of mild cognitive impairment. *Proceedings of the National Academy of Sciences*, 102(13):4919–4924.

Martha Storandt and Robert Hill. 1989. Very mild senile dementia of the alzheimers type: Ii psychometric test performance. *Archives of Neurology*, 46:383–386.

Helen Tager-Flusberg. 1995. Once upon a ribbit: Stories narrated by autistic children. *British journal of developmental psychology*, 13(1):45–59.

David Wechsler. 1997. *Wechsler Memory Scale - Third Edition Manual*. The Psychological Corporation.

Bootstrapping Biomedical Ontologies for Scientific Text using NELL

Dana Movshovitz-Attias
Carnegie Mellon University
5000 Forbes Avenue
Pittsburgh, PA 15213 USA
dma@cs.cmu.edu

William W. Cohen
Carnegie Mellon University
5000 Forbes Avenue
Pittsburgh, PA 15213 USA
wcohen@cs.cmu.edu

Abstract

We describe an open information extraction system for biomedical text based on NELL (the Never-Ending Language Learner) (Carlson et al., 2010), a system designed for extraction from Web text. NELL uses a coupled semi-supervised bootstrapping approach to learn new facts from text, given an initial ontology and a small number of "seeds" for each ontology category. In contrast to previous applications of NELL, in our task the initial ontology and seeds are automatically derived from existing resources. We show that NELL's bootstrapping algorithm is susceptible to ambiguous seeds, which are frequent in the biomedical domain. Using NELL to extract facts from biomedical text quickly leads to semantic drift. To address this problem, we introduce a method for assessing seed quality, based on a larger corpus of data derived from the Web. In our method, seed quality is assessed at each iteration of the bootstrapping process. Experimental results show significant improvements over NELL's original bootstrapping algorithm on two types of tasks: learning terms from biomedical categories, and named-entity recognition for biomedical entities using a learned lexicon.

1 Introduction

NELL (the Never-Ending Language Learner) is a semi-supervised learning system, designed for extraction of information from the Web. The system uses a coupled semi-supervised bootstrapping approach to learn new facts from text, given an initial ontology and a small number of "seeds", *i.e.*, labeled examples for each ontology category. The new facts are stored in a growing structured knowledge base.

One of the concerns about gathering data from the Web is that it comes from various un-authoritative sources, and may not be reliable. This is especially true when gathering scientific information. In contrast to Web data, scientific text is potentially more reliable, as it is guided by the peer-review process. Open access scientific archives make this information available for all. In fact, the production rate of publicly available scientific data far exceeds the ability of researchers to "manually" process it, and there is a growing need for the automation of this process.

The biomedical field presents a great potential for text mining applications. An integral part of life science research involves production and publication of large collections of data by curators, and as part of collaborative community effort. Prominent examples include: publication of genomic sequence data, *e.g.*, by the Human Genome Project; online collections of three-dimensional coordinates of protein structures; and databases holding data on genes. An important resource, initiated as a means of enforcing data standardization, are ontologies describing biological, chemical and medical terms. These are heavily used by the research community. With this wealth of available data the biomedical field holds many information extraction opportunities.

We describe an open information extraction system adapting NELL to the biomedical domain. We present an implementation of our approach, named *BioNELL*, which uses three main sources of information: (1) a public corpus of biomedical scientific text, (2) commonly used biomedical ontologies, and

High PMI Seeds			Random Seeds		
SoxN	achaete	cycA	cac	section 33	28
Pax-6	Drosomycin	Zfh-1	crybaby	hv	Bob
BX-C	Ultrabithorax	GATAe	ael	LRS	dip
D-Fos	sine oculis	FMRFa	chm	sht	3520
Abd-A	dCtBP	Antp	M-2	AGI	tou
PKAc	huckebein	abd-A	shanti	disp	zen
Hmgcr	Goosecoid	knirps	Buffy	Gap	Scm
fkh	decapentaplegic	Sxl	lac	Mercurio	REPO
abdA	naked cuticle	BR-C	subcosta	mef	Ferritin
zfh-1	Kruppel	hmgcr	Slam	dad	dTCF
tkv	gypsy insulator	Dichaete	Cbs	Helicase	mago
CrebA	alpha-Adaptin	Abd-B	Sufu	ora	Pten
D-raf	doublesex	gusA	pelo	vu	sb
MtnA	FasII	AbdA	sombre	domain II	TrpRS
Dcr-2	GAGA factor	dTCF	TAS	CCK	ripcord
fushi tarazu	kanamycin resistance	Ecdysone receptor	GABAA receptor	diazepam binding inhibitor	yolk protein
Tkv	dCBP		Debcl	arm	

Table 1: Two samples of fruit-fly genes, taken from the complete fly gene dictionary. *High PMI Seeds* are the top 50 terms selected using PMI ranking, and *Random Seeds* are a random draw of 50 terms from the dictionary. These are used as seeds for the *Fly Gene* category (Section 4.2). Notice that the random set contains many terms that are often not used as genes including *arm*, *28*, and *dad*. Using these as seeds can lead to semantic drift. In contrast, high PMI seeds exhibit much less ambiguity.

(3) a corpus of Web documents.

NELL's ontology, including categories and seeds, has been manually designed during the system development. Ontology design involves assembling a set of interesting categories, organized in a meaningful hierarchical structure, and providing representative seeds for each category. Redesigning a new ontology for a technical domain is difficult without non-trivial knowledge of the domain. We describe a process of merging source ontologies into one structure of categories with seed examples.

However, as we will show, using NELL's bootstrapping algorithm to extract facts from a biomedical corpus is susceptible to noisy and ambiguous terms. Such ambiguities are common in biomedical terminology (see examples in Table 1), and some ambiguous terms are heavily used in the literature. For example, in the sentence "We have cloned an induced *white* mutation and characterized the insertion sequence responsible for the mutant phenotype", *white* is an ambiguous term referring to the name of a gene. In NELL, ambiguity is limited using coupled semi-supervised learning (Carlson et al., 2009): if two categories in the ontology are declared mutually exclusive, instances of one category are used as negative examples for the other, and the two categories cannot share any instances. To resolve the ambiguity of *white* with mutual exclusion, we would have to include a *Color* category in the ontology, and declare it mutually exclusive with the *Gene* category. Then, instances of *Color* will not be able to refer to genes in the KB. It is hard to estimate what additional categories should be added, and building a "complete" ontology tree is practically infeasible.

NELL also includes a polysemy resolution component that acknowledges that one term, for example *white*, may refer to two distinct concepts, say a color and a gene, that map to different ontology categories, such as *Color* and *Fly Gene* (Krishnamurthy and Mitchell, 2011). By including a *Color* category, this component can identify that *white* is both a color and a gene. The polysemy resolver performs word sense induction and synonym resolution based on relations defined between categories in the ontology, and labeled synonym examples. However, at present, BioNELL's ontology does not contain relation definitions (it is based only on categories), so we cannot include this component in our experiments. Additionally, it is unclear how to avoid the use of polysemous terms as category seeds, and no method has been suggested for selecting seeds that are representative of a single specific category.

To address the problem of ambiguity, we introduce a method for assessing the desirability of noun phrases to be used as seeds for a specific target category. We propose ranking seeds using a Pointwise Mutual Information (PMI) -based collocation measure for a seed and a category name. Collocation is measured based on a large corpus of domain-independent data derived from the Web, accounting for uses of the seed in many different contexts.

NELL's bootstrapping algorithm uses the morphological and semantic features of seeds to propose new facts, which are added to the KB and used as seeds in the next bootstrapping iteration to learn more facts. This means that ambiguous terms may be added at any learning iteration. Since *white* really *is* a name of a gene, it is sometimes used in the same semantic context as other genes, and may be added to the KB despite not being used as an initial seed.

To resolve this problem, we propose measuring seed quality in a *Rank-and-Learn* bootstrapping methodology: after every iteration, we rank all the instances that have been added to the KB by their quality as potential category seeds. Only high-ranking instances are used as seeds in the next iteration. Low-ranking instances are stored in the KB and "remembered" as true facts, but are not used for learning new information. This is in contrast to NELL's approach (and most other bootstrapping systems), in which there is no distinction between acquired facts, and facts that are used for learning.

2 Related Work

Biomedical Information Extraction systems have traditionally targeted recognition of few distinct biological entities, focusing mainly on genes (*e.g.*, (Chang et al., 2004)). Few systems have been developed for fact-extraction of many biomedical predicates, and these are relatively small scale (Wattarujeekrit et al., 2004), or they account for limited subdomains (Dolbey et al., 2006). We suggest a more general approach, using bootstrapping to extend existing biomedical ontologies, including a wide range of sub-domains and many categories. The current implementation of BioNELL includes an ontology with over 100 categories. To the best of our knowledge, such large-scale biomedical bootstrapping has not been done before.

Bootstrap Learning and Semantic Drift. Carlson *et al.* (2010) use coupled semi-supervised bootstrap learning in NELL to learn a large set of category classifiers with high precision. One drawback of using iterative bootstrapping is the sensitivity of this method to the set of initial seeds (Pantel et al., 2009). An ambiguous set of seeds can lead to *semantic drift*, *i.e.*, accumulation of erroneous terms and contexts when learning a semantic class. Strict bootstrapping environments reduce this problem by adding boundaries or limiting the learning process, including learning mutual terms and contexts (Riloff and Jones, 1999) and using mutual exclusion and negative class examples (Curran et al., 2007).

McIntosh and Curran (2009) propose a metric for measuring the semantic drift introduced by a learned term, favoring terms different than the recent m learned terms and similar to the first n, (shown for n=20 and n=100), following the assumption that semantic drift develops in late bootstrapping iterations. As we will show, for biomedical categories, semantic drift in NELL occurs within a handful of iterations (< 5), however according to the authors, using low values for n produces inadequate results. In fact, selecting effective n and m parameters may not only be a function of the data being used, but also of the specific category, and it is unclear how to automatically tune them.

Seed Set Refinement. Vyas *et al.* (2009) suggest a method for reducing ambiguity in seeds provided by human experts, by selecting the tightest seed clusters based on context similarity. The method is described for an order of 10 seeds, however, in an ontology containing hundreds of seeds per class, it is unclear how to estimate the correct number of clusters to choose from. Another approach, suggested by Kozareva *et al.* (2010), is using only constrained contexts where both seed and class are present in a sentence. Extending this idea, we consider a more general collocation metric, looking at entire documents including both the seed and its category.

3 Implementation

3.1 NELL's Bootstrapping System

We have implemented BioNELL based on the system design of NELL. NELL's bootstrapping algorithm is initiated with an input ontology structure of categories and seeds. Three sub-components operate to introduce new facts based on the semantic and morphological attributes of known facts. At every iteration, each component proposes candidate facts, specifying the supporting evidence for each candidate, and the candidates with the most strongly supported evidence are added to the KB. The process and sub-components are described in detail by Carlson *et al.* (2010) and Wang and Cohen (2009).

3.2 Text Corpora

PubMed Corpus: We used a corpus of 200K full-text biomedical articles taken from the PubMed Central Open Access Subset (extracted in October 2010)[1], which were processed using the OpenNLP package[2]. This is the main BioNELL corpus and it

[1]http://www.ncbi.nlm.nih.gov/pmc/

[2]http://opennlp.sourceforge.net

is used to extract category instances in all the experiments presented in this paper.

Web Corpus: BioNELL's seed-quality collocation measure (Section 3.4) is based on a domain-independent Web corpus, the English portion of the ClueWeb09 data set (Callan and Hoy, 2009), which includes 500 million web documents.

3.3 Ontology

BioNELL's ontology is composed of six base ontologies, covering a wide range of biomedical subdomains: the Gene Ontology (GO) (Ashburner et al., 2000), describing gene attributes; the NCBI Taxonomy for model organisms (Sayers et al., 2009); Chemical Entities of Biological Interest (ChEBI) (Degtyarenko et al., 2008), a dictionary focused on small chemical compounds; the Sequence Ontology (Eilbeck et al., 2005), describing biological sequences; the Cell Type Ontology (Bard et al., 2005); and the Human Disease Ontology (Osborne et al., 2009). Each ontology provides a hierarchy of terms but does not distinguish concepts from instances.

We used an automatic process for merging base ontologies into one ontology tree. First, we group the ontologies under one hierarchical structure, producing a tree of over 1 million entities, including 856K terms and 154K synonyms. We then separate these into *potential categories* and *potential seeds*. *Categories* are nodes that are unambiguous (have a single parent in the ontology tree), with at least 100 descendants. These descendants are the category's *Potential seeds*. This results in 4188 category nodes. In the experiments of this paper we selected only the top (most general) 20 categories in the tree of each base ontology. We are left with 109 final categories, as some base ontologies had less than 20 categories under these restrictions. Leaf categories are given seeds from their descendants in the full tree of all terms and synonyms, giving a total of around 1 million potential seeds. Seed set refinement is described below. The seeds of leaf categories are later extended by the bootstrapping process.

3.4 BioNELL's Bootstrapping System

3.4.1 PMI Collocation with the Category Name

We define a seed quality metric based on a large corpus of Web data. Let s and c be a seed and a target category, respectively. For example, we can take s = "white", the name of a gene of the fruit-fly, and c = "fly gene". Now, let D be a document corpus (Section 3.2 describes the Web corpus used for ranking), and let D_c be a subset of the documents containing a mention of the category name. We measure the collocation of the seed and the category by the number of times s appears in D_c, $|Occur(s, D_c)|$. The overall occurrence of s in the corpus is given by $|Occur(s, D)|$. Following the formulation of Church and Hanks (1990), we compute the PMI-rank of s and c as

$$\text{PMI}(s, c) = \frac{|Occur(s, D_c)|}{|Occur(s, D)|} \tag{1}$$

Since this measure is used to compare seeds of the same category, we omit the log from the original formulation. In our example, as *white* is a highly ambiguous gene name, we find that it appears in many documents that do not discuss the fruit fly, resulting in a PMI rank close to 0.

The proposed ranking is sensitive to the descriptive name given to categories. For a more robust ranking, we use a combination of rankings of the seed with several of its ancestors in the ontology hierarchy. In (Movshovitz-Attias and Cohen, 2012) we describe this hierarchical ranking in more detail and additionally explore the use of the binomial log-likelihood ratio test (BLRT) as an alternative collocation measure for ranking.

We further note that some specialized biomedical terms follow strict nomenclature rules making them easily identifiable as category specific. These terms may not be frequent in general Web context, leading to a low PMI rank under the proposed method. Given such a set of high confidence seeds from a reliable source, one can enforce their inclusion in the learning process, and specialized seeds can additionally be identified by high-confidence patterns, if such exist. However, the scope of this work involves selecting seeds from an ambiguous source, biomedical ontologies, thus we do not include an analysis for these specialized cases.

3.4.2 Rank-and-Learn Bootstrapping

We incorporate PMI ranking into BioNELL using a *Rank-and-Learn* bootstrapping methodology. After every iteration, we rank all the instances that have been added to the KB. Only high-ranking instances

Learning System	Bootstrapping Algorithm	Initial Seeds	Corpus
BioNELL	Rank-and-Learn with PMI	PMI top 50	PubMed
NELL	NELL's algorithm	Random 50	PubMed
BioNELL+Random	Rank-and-Learn with PMI	Random 50	PubMed

Table 2: Learning systems used in our evaluation, all using the PubMed biomedical corpus and the biomedical ontology described in Sections 3.2 and 3.3.

are added to the collection of seeds that are used in the next learning iteration. Instances with low PMI rank are stored in the KB and are not used for learning new information. We consider a high-ranking instance to be one with PMI rank higher than 0.25.

4 Experimental Evaluation

4.1 Experimental Settings

4.1.1 Configurations of the Algorithm

In our experiments, we ran BioNELL and NELL with the following system configurations, all using the biomedical corpus and the ontology described in Sections 3.2 and 3.3, and all running 50 iterations, in order to evaluate the long term effects of ranking. Section 4.2 includes a discussion on the learning rate of the tested systems which motivates the reason for evaluating performance at the 50th iteration.

To expand a category we used the following systems, also summarized in Table 2: (1) the *BioNELL* system, using Rank-and-Learn bootstrapping (Section 3.4.2) initialized with the top 50 seeds using PMI ranking, (2) the *NELL* system, using NELL's original bootstrapping algorithm (Section 3.1) initialized with 50 random seeds from the category's potential seeds (NELL does not provide a seed selection method), and (3) in order to distinguish the contribution of Rank-and-Learn bootstrapping over ranking the initial seeds, we tested a third system, *BioNELL+Random*, using Rank-and-Learn bootstrapping initialized with 50 random seeds.

4.1.2 Evaluation Methodology

Using BioNELL we can learn *lexicons*, collections of category terms accumulated after running the system. One evaluation approach is to select a set of learned instances and assess their correctness (Carlson et al., 2010). This is relatively easy for data extracted for general categories like City or Sports Team. For example, it is easy to evaluate the statement "London is a City". This task becomes more difficult when assessing domain-specific facts such as "Beryllium is an S-block molecular entity" (in fact, it is). We cannot, for example, use the help of Mechanical Turk for this task. A possible alternative evaluation approach is asking an expert. On top of being a costly and slow approach, the range of topics covered by BioNELL is large and a single expert is not likely be able to assess all of them.

We evaluated lexicons learned by BioNELL by comparing them to available resources. Lexicons of gene names for certain species are available, and the Freebase database (Google, 2011), an open repository holding data for millions of entities, includes some biomedical concepts. For most biomedical categories, however, complete lexicons are scarce.

4.1.3 Data Sets

We compared learned lexicons to category *dictionaries*, lists of concept terms taken from the following sources, which we consider as a Gold Standard.

We used three lexicons of biomedical categories taken from Freebase: Disease (9420 terms), Chemical Compound (9225 terms), and Drug (3896 terms).

To evaluate gene names we used data from the BioCreative Challenge (Hirschman et al., 2005), an evaluation competition focused on annotations of genes and gene products. The data includes a dictionary of genes of the fruit-fly, *Drosophila Melanogaster*, which specifies a set of gene identifiers and possible alternative forms of the gene name, for a total of 7151 terms, which we consider to be the complete fly gene dictionary.

We used additional BioCreative data for a named-entity recognition task. This includes 108 scientific abstracts, manually annotated by BioCreative with gene IDs of fly genes discussed in the text. The abstracts contain either the gene ID or any gene name.

4.2 Extending Lexicons of Biomedical Categories

4.2.1 Recovering a Closed Category Lexicon

We used BioNELL to learn the lexicon of a closed category, representing genes of the fruit-fly,

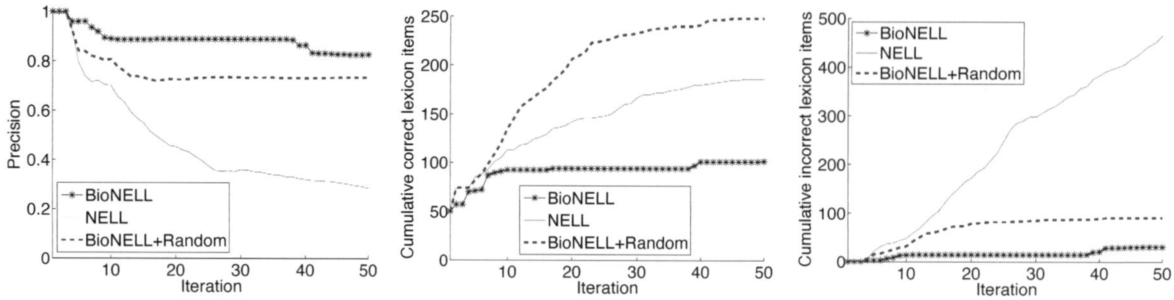

(a) Precision (b) Cumulative correct items (c) Cumulative incorrect items

Figure 1: Performance per learning iteration for gene lexicons learned using BioNELL and NELL.

Learning System	Precision	Correct	Total
BioNELL	**.83**	109	132
NELL	.29	186	**651**
BioNELL+Random	.73	**248**	338
NELL $_{\text{by size 132}}$	.72	93	130

Table 3: Precision, total number of instances (*Total*), and correct instances (*Correct*) of gene lexicons learned with BioNELL and NELL. BioNELL significantly improves the precision of the learned lexicon compared with NELL. When examining only the first 132 learned items, BioNELL has both higher precision and more correct instances than NELL (last row, NELL $_{\text{by size 132}}$).

D. Melanogaster, a model organism used to study genetics and developmental biology. Two samples of genes from the full fly gene dictionary are shown in Table 1: *High PMI Seeds* are the top 50 dictionary terms selected using PMI ranking, and *Random Seeds* are a random draw of 50 terms. Notice that the random set contains many seeds that are not distinct gene names including *arm*, *28*, and *dad*. In contrast, high PMI seeds exhibit much less ambiguity. We learned gene lexicons using the test systems described in Section 4.1.1 with the high-PMI and random seed sets shown in Table 1. We measured the precision, total number of instances, and correct instances of the learned lexicons against the full dictionary of genes. Table 3 summarizes the results.

BioNELL, initialized with PMI-ranked seeds, significantly improved the precision of the learned lexicon over NELL (29% for *NELL* to 83% for *BioNELL*). In fact, the two learning systems using Rank-and-Learn bootstrapping resulted in higher precision lexicons (83%, 73%), suggesting that constrained bootstrapping using iterative seed ranking successfully eliminates noisy and ambiguous seeds.

BioNELL's bootstrapping methodology is highly restrictive and it affects the size of the learned lexicon as well as its precision. Notice, however, that while *NELL*'s final lexicon is 5 times larger than *BioNELL*'s, the number of correctly learned items in it are less than twice that of *BioNELL*. Additionally, *BioNELL+Random* has learned a smaller dictionary than *NELL* (338 and 651 terms, respectively) with a greater number of correct instances (248 and 186).

We examined the performance of *NELL* after the 7th iteration, when it has learned a lexicon of 130 items, similar in size to *BioNELL*'s final lexicon (Table 3, last row). After learning 130 items, *BioNELL* achieved both higher precision (83% versus 72%) and higher recall (109 versus 93 correct lexicon instances) than *NELL*, indicating that *BioNELL*'s learning method is overall more accurate.

After running for 50 iterations, all systems recover only a small portion of the complete gene dictionary (109-248 instances out of 7151), suggesting either that, (1) more learning iterations are required, (2) the biomedical corpus we use is too small and does not contain (frequent) mentions of some gene names from the dictionary, or (3) some other limitations exist that prevent the learning algorithm from finding additional class examples.

Lexicons learned using BioNELL show persistently high precision throughout the 50 iterations, even when initiated with random seeds (Figure 1A). By the final iteration, all systems stop accumulating further significant amounts of correct gene instances (Figure 1B). Systems that use PMI-based Rank-and-Learn bootstrapping also stop learning incorrect

Learning System	Precision			Correct			Total		
	CC	Drug	Disease	CC	Drug	Disease	CC	Drug	Disease
BioNELL	**.66**	**.52**	**.43**	63	508	276	96	972	624
NELL	.15	.40	.37	**74**	**522**	**288**	**449**	**1300**	**782**
NELL $_{\text{by size}}$	.58	.47	.37	58	455	232	100	968	623

Table 4: Precision, total number of instances (*Total*), and correct instances (*Correct*) of learned lexicons of *Chemical Compound* (CC), *Drug*, and *Disease*. BioNELL's lexicons have higher precision on all categories compared with NELL, while learning a similar number of correct instances. When restricting NELL to a total lexicon size similar to BioNELL's, BioNELL has both higher precision and more correct instances (last row, NELL $_{\text{by size}}$).

instances (*BioNELL* and *BioNELL+Random*; Figure 1C). This is in contrast to *NELL* which continues learning incorrect examples.

Interestingly, the highest number of correct gene instances was learned using Rank-and-Learn bootstrapping with random initial seeds (248 items; *BioNELL+Random*). This may happen when the random set includes genes that are infrequent in the general Web corpus, despite being otherwise category-specific in the biomedical context. As such, these would result in low PMI rank (see note in Section 3.4.1). However, random seed selection does not offer any guarantees on the quality of the seeds used, and therefore will result in unstable performance. Note that *BioNELL+Random* was initiated with the same random seeds as *NELL*, but due to the more constrained Rank-and-Learn bootstrapping it achieves both higher recall (248 versus 186 correct instances) and precision (73% versus 29%).

4.2.2 Extending Lexicons of Open Categories

We evaluated learned lexicons for three open categories, *Chemical Compound* (*CC*), *Drug*, and *Disease*, using dictionaries from Freebase. Since these are open categories — new drugs are being developed every year, new diseases are discovered, and varied chemical compounds can be created — the Freebase dictionaries are not likely to contain complete information on these categories. For our evaluation, however, we considered them to be complete.

We used *BioNELL* and *NELL* to learn these categories, and for all of them *BioNELL*'s lexicons achieved higher precision than *NELL* (Table 4). The number of correct learned instances was similar in both systems (63 and 74 for *CC*, 508 and 522 for *Drug*, and 276 and 288 for *Disease*), however in *BioNELL*, the additional bootstrapping restrictions assist in rejecting incorrect instances, resulting in a smaller, more accurate lexicon.

We examined *NELL*'s lexicons when they reached a size similar to *BioNELL*'s final lexicons (at the 8th, 42nd and 39th iterations for *CC*, *Drug*, and *Disease*, respectively). *BioNELL*'s lexicons have both higher precision and higher recall (more correct learned instances) than the comparable *NELL* lexicons (Table 4, NELL $_{\text{by size}}$, last row).

4.3 Named-Entity Recognition using a Learned Lexicon

We examined the use of gene lexicons learned with BioNELL and NELL for the task of recognizing concepts in free text, using a simple strategy of matching words in the text with terms from the lexicon. We use data from the BioCreative challenge (Section 4.1.3), which includes text abstracts and the IDs of genes that appear in each abstract. We show that BioNELL's lexicon achieves both higher precision and recall in this task than NELL's.

We implemented an *annotator* for predicting what genes are discussed in text, which uses a gene lexicon as input. Given sample text, if any of the terms in the lexicon appear in the text, the corresponding gene is predicted to be discussed in the text. Following BioCreative's annotation format, the annotator emits as output the set of gene IDs of the genes predicted for the sample text.

We evaluated annotators that were given as input: the complete fly-genes dictionary, a filtered version of that dictionary, or lexicons learned using BioNELL and NELL. Using these annotators we predicted gene mentions for all text abstracts in the data. We report the average precision (over 108 text

Lexicon	Precision	Correct	Total
BioNELL	.90	18	20
NELL	.02	5	268
BioNELL+Random	.03	3	82
Complete Dictionary	.09	153	1616
Filtered Dictionary	.18	138	675

Table 5: Precision, total number of predicted genes (*Total*), and correct predictions (*Correct*), in a named-entity recognition task using a complete lexicon, a filtered lexicon, and lexicons learned with BioNELL and NELL. BioNELL's lexicon achieves the highest precision, and makes more correct predictions than NELL.

abstracts) and number of total and correct predictions of gene mentions, compared with the labeled annotations for each text (Table 5).

Many gene names are shared among multiple variants. For example, the name *Antennapedia* may refer to several gene variations, *e.g.*, *Dgua\Antp* or *Dmed\Antp*. Thus, in our precision measurements, we consider a prediction of a gene ID as "true" if it is labeled as such by BioCreative, or if it shares a synonym name with another true labeled gene ID.

First, we used the complete fly gene dictionary for the recognition task. Any dictionary gene that is mentioned in the text was recovered, resulting in high recall. However, the full dictionary contains ambiguous gene names that contribute many false predictions to the complete dictionary annotator, leading to a low precision of 9%.

Some ambiguous terms can be detected using simple rules, *e.g.*, short abbreviations and numbers. For example, *section 9* is a gene named after the functional unit to which it belongs, and abbreviated by the symbol *9*. Clearly, removing *9* from the full lexicon should improve precision without great cost to recall. We similarly filtered the full dictionary, removing one- and two-letter abbreviations and terms composed only of non-alphabetical characters, leaving 6253 terms. Using the filtered dictionary, precision has doubled (18%) with minor compromise to recall. Using complete or manually refined gene dictionaries for named-entity recognition has been shown before to produce similar high-recall and low-precision results (Bunescu et al., 2005).

We evaluated annotators on gene lexicons learned with BioNELL and NELL. *BioNELL*'s lexicon achieved significantly higher precision (90%) than other lexicons (2%-18%). It is evident that this lexicon contains few ambiguous terms as it leads to only 2 false predictions. Note also, that *BioNELL*'s lexicon has both higher precision and recall than *NELL*.

5 Conclusions

We have proposed a methodology for an open information extraction system for biomedical scientific text, using an automatically derived ontology of categories and seeds. Our implementation is based on constrained bootstrapping in which seeds are ranked at every iteration.

The benefits of iterative seed ranking have been demonstrated, showing that our method leads to significantly less ambiguous lexicons for all the evaluated biomedical concepts. BioNELL shows 51% increase over NELL in the precision of a learned lexicon of chemical compounds, and 45% increase for a category of gene names. Importantly, when BioNELL and NELL learn lexicons of similar size, BioNELL's lexicons have both higher precision and recall. We have demonstrated the use of BioNELL's learned gene lexicon as a high precision annotator in an entity recognition task (with 90% precision). The results are promising, though it is currently difficult to provide a similar quantitative evaluation for a wider range of concepts.

Many interesting improvements could be made in the current system, mainly discovery of relations between existing ontology categories. In addition, we believe that Rank-and-Learn bootstrapping and iterative seed ranking can be beneficial in general, domain-independent settings, and we would like to explore further use of this method.

Acknowledgments

This work was funded by grant 1R101GM081293 from NIH, IIS-0811562 from NSF and by a gift from Google. The opinions expressed in this paper are solely those of the authors.

References

M. Ashburner, C.A. Ball, J.A. Blake, D. Botstein, H. Butler, J.M. Cherry, A.P. Davis, K. Dolinski, S.S. Dwight, J.T. Eppig, et al. 2000. Gene ontology: tool for the unification of biology. *Nature genetics*, 25(1):25.

J. Bard, S.Y. Rhee, and M. Ashburner. 2005. An ontology for cell types. *Genome Biology*, 6(2):R21.

R. Bunescu, R. Ge, R.J. Kate, E.M. Marcotte, and R.J. Mooney. 2005. Comparative experiments on learning information extractors for proteins and their interactions. *Artificial Intelligence in Medicine*, 33(2).

J. Callan and M. Hoy. 2009. Clueweb09 data set. *http://boston.lti.cs.cmu.edu/Data/clueweb09/.*

A. Carlson, J. Betteridge, E.R. Hruschka Jr, T.M. Mitchell, and SP Sao Carlos. 2009. Coupling semi-supervised learning of categories and relations. *Semi-supervised Learning for Natural Language Processing*, page 1.

A. Carlson, J. Betteridge, B. Kisiel, B. Settles, E.R. Hruschka Jr, and T.M. Mitchell. 2010. Toward an architecture for never-ending language learning. In *Proceedings of the Twenty-Fourth Conference on Artificial Intelligence (AAAI 2010)*, volume 2, pages 3–3.

J.T. Chang, H. Schütze, and R.B. Altman. 2004. Gapscore: finding gene and protein names one word at a time. *Bioinformatics*, 20(2):216.

K.W. Church and P. Hanks. 1990. Word association norms, mutual information, and lexicography. *Computational linguistics*, 16(1):22–29.

J.R. Curran, T. Murphy, and B. Scholz. 2007. Minimising semantic drift with mutual exclusion bootstrapping. In *Proceedings of the 10th Conference of the Pacific Association for Computational Linguistics*.

K. Degtyarenko, P. De Matos, M. Ennis, J. Hastings, M. Zbinden, A. McNaught, R. Alcántara, M. Darsow, M. Guedj, and M. Ashburner. 2008. Chebi: a database and ontology for chemical entities of biological interest. *Nucleic acids research*, 36(suppl 1):D344.

A. Dolbey, M. Ellsworth, and J. Scheffczyk. 2006. Bioframenet: A domain-specific framenet extension with links to biomedical ontologies. In *Proceedings of KR-MED*, pages 87–94. Citeseer.

K. Eilbeck, S.E. Lewis, C.J. Mungall, M. Yandell, L. Stein, R. Durbin, and M. Ashburner. 2005. The sequence ontology: a tool for the unification of genome annotations. *Genome biology*, 6(5):R44.

Google. 2011. Freebase data dumps. http://download.freebase.com/datadumps/.

L. Hirschman, A. Yeh, C. Blaschke, and A. Valencia. 2005. Overview of biocreative: critical assessment of information extraction for biology. *BMC bioinformatics*, 6(Suppl 1):S1.

Z. Kozareva and E. Hovy. 2010. Not all seeds are equal: measuring the quality of text mining seeds. In *Human Language Technologies: The 2010 Annual Conference of the North American Chapter of the Association for Computational Linguistics*, pages 618–626. Association for Computational Linguistics.

J. Krishnamurthy and T.M. Mitchell. 2011. Which noun phrases denote which concepts? In *Proceedings of the 49th Annual Meeting of the Association for Computational Linguistics: Human Language Technologies*. Association for Computational Linguistics.

T. McIntosh and J.R. Curran. 2009. Reducing semantic drift with bagging and distributional similarity. In *Proceedings of the Joint Conference of the 47th Annual Meeting of the ACL and the 4th International Joint Conference on Natural Language Processing of the AFNLP: Volume 1-Volume 1*, pages 396–404. Association for Computational Linguistics.

D. Movshovitz-Attias and W.W. Cohen. 2012. Bootstrapping biomedical ontologies for scientific text using nell. Technical report, Carnegie Mellon University, CMU-ML-12-101.

J. Osborne, J. Flatow, M. Holko, S. Lin, W. Kibbe, L. Zhu, M. Danila, G. Feng, and R. Chisholm. 2009. Annotating the human genome with disease ontology. *BMC genomics*, 10(Suppl 1):S6.

P. Pantel, E. Crestan, A. Borkovsky, A.M. Popescu, and V. Vyas. 2009. Web-scale distributional similarity and entity set expansion. In *Proceedings of the 2009 Conference on Empirical Methods in Natural Language Processing: Volume 2-Volume 2*, pages 938–947. Association for Computational Linguistics.

E. Riloff and R. Jones. 1999. Learning dictionaries for information extraction by multi-level bootstrapping. In *Proceedings of the National Conference on Artificial Intelligence (AAAI-99)*, pages 474–479.

E. W. Sayers, T. Barrett, D. A. Benson, S. H. Bryant, K. Canese, V. Chetvernin, D. M. Church, M. DiCuccio, R. Edgar, S. Federhen, M. Feolo, L. Y. Geer, W. Helmberg, Y. Kapustin, D. Landsman, D. J. Lipman, T. L. Madden, D. R. Maglott, V. Miller, I. Mizrachi, J. Ostell, K. D. Pruitt, G. D. Schuler, E. Sequeira, S. T. Sherry, M. Shumway, K. Sirotkin, A. Souvorov, G. Starchenko, T. A. Tatusova, L. Wagner, E. Yaschenko, and J. Ye. 2009. Database resources of the National Center for Biotechnology Information. *Nucleic Acids Res.*, 37:5–15, Jan.

V. Vyas, P. Pantel, and E. Crestan. 2009. Helping editors choose better seed sets for entity set expansion. In *Proceeding of the 18th ACM conference on Information and knowledge management*. ACM.

R.C. Wang and W.W. Cohen. 2009. Character-level analysis of semi-structured documents for set expansion. In *Proceedings of the 2009 Conference on Empirical Methods in Natural Language Processing: Volume 3-Volume 3*, pages 1503–1512. Association for Computational Linguistics.

T. Wattarujeekrit, P. Shah, and N. Collier. 2004. Pasbio: predicate-argument structures for event extraction in molecular biology. *BMC bioinformatics*, 5(1):155.

Semantic distance and terminology structuring methods for the detection of semantically close terms

Marie Dupuch
CNRS UMR 8163 STL
Université Lille 1&3
59653 Villeneuve d'Ascq, France
dupuchm@hotmail.fr

Laëtitia Dupuch
Université Toulouse III Paul Sabatier
France
laetitialdupuch@hotmail.com

Thierry Hamon
LIM&BIO (EA3969) UFR SMBH
Université Paris 13, France
thierry.hamon@univ-paris13.fr

Natalia Grabar
CNRS UMR 8163 STL
Université Lille 1&3
59653 Villeneuve d'Ascq, France
natalia.grabar@univ-lille3.fr

Abstract

The identification of semantically similar linguistic expressions despite their formal difference is an important task within NLP applications (information retrieval and extraction, terminology structuring...) We propose to detect the semantic relatedness between biomedical terms from the pharmacovigilance area. Two approaches are exploited: semantic distance within structured resources and terminology structuring methods applied to a raw list of terms. We compare these methods and study their complementarity. The results are evaluated against the reference pharmacovigilance data and manually by an expert.

1 Introduction

When an automatic system is able to identify that different linguistic expressions convey the same or similar meanings, this is a positive point for several applications. For instance, when documents referring to *muscle pain* or *cephalgia* are searched, information retrieval system can also take advantage of the synonyms, like *muscle ache* or *headache*, to return more relevant documents and in this way to increase the recall. This is also a great advantage for systems designed for instance for text mining, terminology structuring and alignment, or for more specific tasks such as pharmacovigilance.

The pharmacovigilance area covers the identification of adverse drug reactions (ADRs) in order to improve the vigilance on the health products. Pharmacovigilance reports are traditionally encoded with normalised terms from the dedicated terminologies, such as MedDRA (Medical Dictionary for Drug Regulatory Activities) (Brown et al., 1999). MedDRA is a relatively fine-grained terminology with nearly 90,000 terms. This means that a given pharmacovigilance report can be coded with different terms which have close meaning (Fescharek et al., 2004), like *muscle pain* and *muscle ache* or *headache* and *cephalgia*: although formally different the terms from these pairs have the same meaning. The difficulty is then to detect their semantic closeness. Indeed, if this semantic information is available, reports from the phramacovigilance databanks and mentionning similar adverse events can be aggregated: the safety signal is intensified and the safety regulation process is improved.

In order to aggregate the pharmacovigilance reports, several types of semantic information from MedDRA are used: (1) different hierarchical levels of MedDRA between the five levels available; (2) the SMQs (Standardized MedDRA Queries) which group together terms associated to a given medical condition such as *Acute renal failure*, *Angioedema* or *Embolic and thrombotic events*; and (3) specific resources (Bousquet et al., 2005; Iavindrasana et al., 2006; Alecu et al., 2008; Jaulent and Alecu, 2009). The SMQs are defined by groups of experts through a long and meticulous work consisting of the manual study of the MedDRA structure and of the analysis of the scientific literature (CIOMS, 2004). 84 SMQs have been created so far. They become the gold standard data of the pharmacovigilance area. However, the SMQs currently suffer from the lack of exhaustivity (Pearson et al., 2009): the set of SMQs is not exhaustive because this is an ongoing work. We assume that automatic approaches can be exploited to systematize and accelerate the process of recruiting the semantically related MedDRA terms and to build the SMQs. We propose to exploit two approaches: methods dedicated to the terminology structuring and semantic distance approaches. We compare and combine the generated results. For the evaluation, we compare the results with the existing SMQs and also analyse them manually with an expert. Our work is different from previous work because we exploit the whole set of the available MedDRA terms, we apply several methods to cluster the terms and we perform several types of evaluation.

2 Material

We exploit two kinds of material: material issued from MedDRA and specific to the pharmacovigilance area (sections 2.1 and 2.3), and linguistic resources issued from general and biomedical languages (section 2.2). The MedDRA terms are structured into five hierarchical levels: *SOC* (*System Organ Class*) terms belong to the first and the highest level, while *LLT* (*Lowest Level Terms*) terms belong to the fifth and the lowest level. Terms from the fourth level *PT* (*Preferred Terms*) are usually exploited for the coding of the pharmacovigilance reports. They are also used for the creation of SMQs. A given PT term may belong to several SMQs.

2.1 Ontology ontoEIM

ontoEIM is an ontology of ADRs (Alecu et al., 2008) created through the projection of MedDRA to SNOMED CT (Stearns et al., 2001). This projection is performed thanks to the UMLS (NLM, 2011), where an important number of terminologies are already merged and aligned, among which MedDRA and SNOMED CT. The current rate of alignment of the *PT* MedDRA terms with SNOMED CT is weak (version 2011): 51.3% (7,629 terms). Projection of MedDRA to SNOMED CT allows to improve the representation of the MedDRA terms:

- the structure of the MedDRA terms is parallel to that of SNOMED CT, which makes it more fine-grained (Alecu et al., 2008). The number of hierarchical levels within the ontoEIM reaches 14, instead of five levels in MedDRA;
- the MedDRA terms receive formal definitions: semantic primitives which decompose the meaning. MedDRA terms can be described along up to four axes from SNOMED CT, exemplified here through the term *Arsenical keratosis*: (1) *Morphology* (type of abnormality): *Squamous cell neoplasm*; (2) *Topography* (anatomical localization): *Skin structure*; (3) *Causality* (agent or cause of the abnormality): *Arsenic AND OR arsenic compound*; and (4) *Expression* (manifestation of the abnormality): *Abnormal keratinization*. The formal definitions are not complete. For instance, only 12 terms receive formal definitions along these four axes and 435 along three axes. This is due to the incomplete alignment of the MedDRA terms and to the fact these four elements are not relevant for every term (their absence is not always problematic).

2.2 Linguistic resources

Linguistic resources provide three kinds of pairs of synonym words: (1) Medical synonyms extracted from the UMLS 2011AA (n=228,542) and then cleaned up (n=73,093); (2) Medical synonyms acquired from three biomedical terminologies thanks to the exploitation of their compositionality (Grabar and Hamon, 2010) (n=28,691); (3) Synonyms from the general language provided by WordNet (Fellbaum, 1998) (n=45,782). Among the pairs of words recorded in these resources, we can find {*accord, concordance*}, {*aceperone, acetabutone*}, {*adenazole, tocladesine*}, {*adrenaline, epinephrine*} or {*bleeding, hemorrhage*}. The last two pairs are provided by medical and general resources. However, the pair {*accord, concordance*} is provided only by medical resources.

2.3 Standardized MedDRA Queries

We exploit 84 SMQs as reference data. Among these SMQs, we distinguish 20 SMQs which are structured hierarchically. We also exploit 92 sub-SMQs, which compose these 20 hierarchical SMQs.

3 Methods

Our method consists into four main steps (figure 1): (1) computing of the semantic distance and similarity between the MedDRA terms and their clustering (section 3.1), (2) the application of the terminology structuring methods to acquire semantic re-

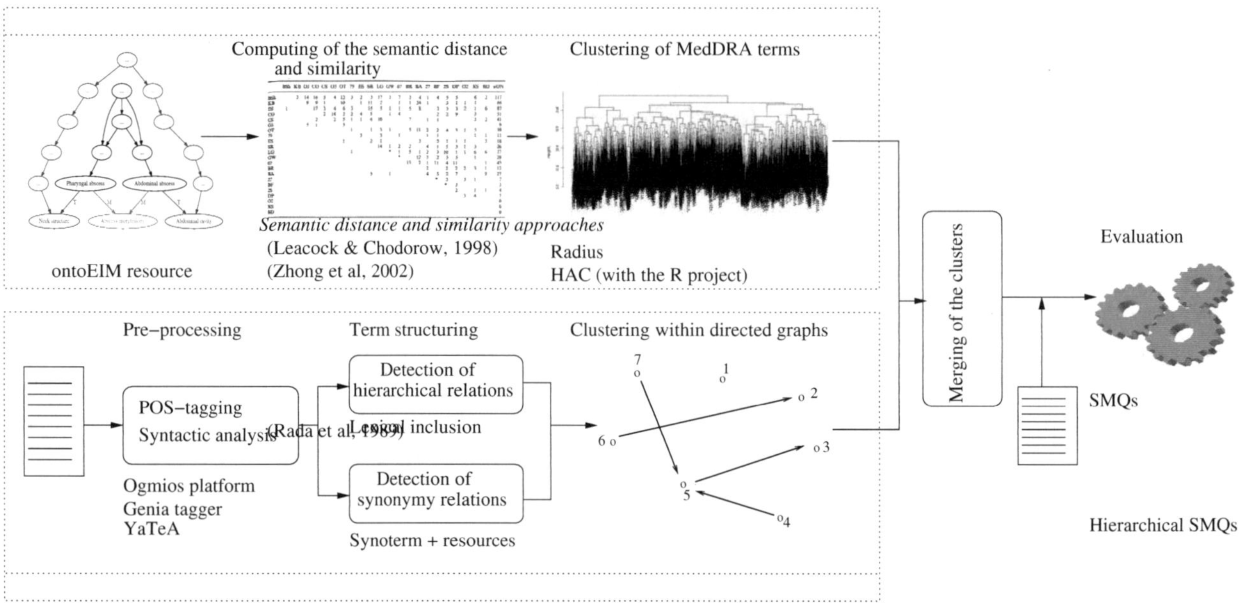

Figure 1: General schema of the experiment composed of four steps: (1) semantic distance approaches, (2) terminology structuring approaches, (3) their combination and (4) their evaluation

lations between MedDRA terms and their clustering (section 3.2), (3) the merging of these two sets of clusters (section 3.3), (4) the evaluation of the merged clusters (section 3.4). We exploit Perl language, R[1] project and several NLP tools.

3.1 Semantic distance approach

The semantic distance and similarity approach is applied to the 7,629 *PT* MedDRA terms and their formal definitions from ontoEIM. The two main steps are: computing the distance or similarity (section 3.1.1) and clustering of terms (section 3.1.2).

3.1.1 Computing the semantic distance

Because we work with a tree-structured resource, we exploit edge-based algorithms to compute the distance or similarity between two terms *t1* and *t2*: two semantic distances (*Rada* (Rada et al., 1989) and *Zhong* (Zhong et al., 2002)) and one semantic similarity (Leacock and Chodorow, 1998). In the following, we call them semantic distance algorithms. For each algorithm, three paths may be exploited: between the MedDRA terms but also between the elements of their formal definitions on two axes (morphology M and topography T often involved in diagnostics (Spackman and Campbell, 1998)). For the illustration, let's consider two MedDRA terms, *Abdominal abscess* and *Pharyngeal abscess* defined as follows:

- *Abdominal abscess*: M = *Abscess morphology*, T = *Abdominal cavity structure*
- *Pharyngeal abscess*: M = *Abscess morphology*, T = *Neck structure*

The shortest paths sp are computed between these two MedDRA terms and between their formal definitions, whose hierarchical structure is also inherited from SNOMED CT. The weight of edges is set to 1 because all the relations are of the same kind (hierarchical), and the value of each shortest path corresponds to the sum of the weights of all its edges. The semantic distance sd are then exploited to compute the unique distance between the ADR terms from MedDRA: $\frac{\sum_{i\in\{ADR,M,T\}} W_i * sd_i(t1,t2)}{\sum_{i\in\{ADR,M,T\}} W_i}$, where the three axes $\{ADR, M, T\}$ respectively correspond to terms meaning the ADR, axis Morphology M and axis Topography T; $t1$ and $t2$ are two ADR terms; W_i is the coefficient associated with each of the three axes; and sd_i is the semantic distance computed on a given axis. We carry out several ex-

[1] http://www.r-project.org

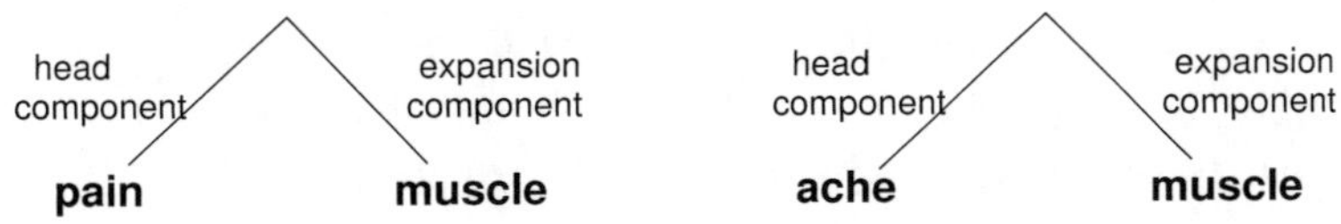

Figure 2: Syntactically analyzed terms (*muscle pain* and *muscle ache*) into their head and expansion components

periments. Semi-matrices 7629*7629 with semantic distance between the terms are built.

3.1.2 Clustering of terms

An unsupervised creation of clusters is applied to the semi-matrices. We exploit two approaches:

- R radius approach: every MedDRA term is considered a possible center of a cluster and its closest terms are clustered with it. The thresholds tested correspond to the following intervals: 2 and 3 for *Rada*, [0; 5.059] for *LCH* and [0; 0.49] for *Zhong*. The intersection of these clusters is not empty.
- *HAC* hierarchical ascendant classification is performed through the *R Project* tools (*hclust* function). Iteratively, this function chooses the best centers for terms and builds the hierarchy of terms by progressively clustering those which are closest to these centers. Then the unique cluster with all the terms is split up. Several splitting values between 100 and 7,000 are tested. These clusters are exclusive.

Clusters created with the radius approach are merged in order to eliminate smaller clusters included in bigger clusters and in order to aggregate clusters which have an important intersection between them. For the intersection, we test several intersection values within the interval [10; 90], which means that two compared clusters may have between 10% and 90% of common terms.

3.2 Terminology structuring approach

The terminology structuring methods are applied to a raw list of 18,209 MedDRA PTs. They allow the detection of semantic relations between these terms. The POS-tagging is done with Genia tagger (Tsuruoka et al., 2005) and the syntactic analysis with the YaTeA parser (Aubin and Hamon, 2006). Three kinds of methods are applied for the acquisition of synonymy and hierarchical relations: lexical inclusions (section 3.2.1), morpho-syntactic variants (section 3.2.2) and compositionality (section 3.2.3). The terms are then clustered (section 3.2.4).

3.2.1 Lexical inclusion and hierarchy

The lexical inclusion hypothesis (Kleiber and Tamba, 1990), which states that when a given term is lexically included at the head syntactic position in another term there is a semantic subsumption between them, allows to identify hierarchical relations between terms. For instance, on figure 2, the short term *pain* is the hierarchical parent and the long term *muscle pain* is its hierarchical child because *pain* is the syntactic head of *muscle pain*. The lexical inclusions are computed on POS-tagged and syntactically analyzed terms. We compute two kinds of lexical inclusions:

- syntactic dependencies on minimal syntactic heads: the parent term corresponds to the shortest lexical form of the syntactic head. For instance, within the term *kaolin cephalin clotting time*, the minimal head is *time*;
- syntactic dependencies on maximal syntactic heads: the parent term is the most complete lexical form of the syntactic head. Within the same term *kaolin cephalin clotting time*, the maximal head is *cephalin clotting time*.

Parent and child terms have to be MedDRA terms.

3.2.2 Morpho-syntactic variants

We exploit Faster (Jacquemin, 1996) for the indentification of morpho-syntactic variants between the PT terms. This tool applies several transformation rules, such as insertion (*cardiac disease/cardiac valve disease*), morphological derivation (*artery restenosis/arterial restenosis*) or permutation (*aorta coarctation/coarctation of the aorta*). Each transformation rule is associated with hierarchical or synonymy relations: the insertion introduces a hierarchical relation (*cardiac valve disease* is more specific than *cardiac disease*), while the permutation introduces a synonymy relation. When several transformations are involved, the detected relations may be ambiguous: *gland abscess* and *abscess of salivary gland* combines permutation (synonymy) and insertion (hierarchy) rules. In such cases the hierarchical relation prevails.

3.2.3 Compositionality and synonymy

The synonymy relations are acquired in two ways. First, the synonymy relation is established between two simple MedDRA terms if this relation is provided by the linguitistic resources. Second, the identification of synonym relations between complex terms relies on the semantic compositionality (Partee, 1984). Hence, two complex terms are considered synonyms if at least one of their components at the same syntactic position (head or expansion) are synonyms. For instance, on figure 2, given the synonymy relation between the two words *pain* and *ache*, the terms *muscle pain* and *muscle ache* are also identified as synonyms (Hamon and Nazarenko, 2001). Three transformation rules are applied: on the head component (figure 2), on the expansion component and on both of them. We perform several experiments: each medical synonymy resource is first used individually and then in combination with WordNet.

3.2.4 Clustering of terms

The sets of terms related through the lexical inclusions are considered as directed graphs: the terms are the nodes of the graph while the hierarchical relations are the directed edges. We partition these directed graphs and identify clusters of terms which could correspond to or be part of the SMQs. Among connected components and strongly connected components, we choose to generate the strongly connected components: they allow an intersection between clusters which means that a given term may belong to several clusters (this is also the case with the SMQs). Thus, within the directed graphs G we have to identify the maximal sub-graphs H of G where for each pair $\{x, y\}$ of the nodes from H, there exists a directed edge from x to y (or from y to x). To improve the coverage of the obtained clusters, we also add the synonyms: if a term has a synonymy relation with the term from a cluster then this term is also included in this cluster. From a graph theory point of view, the initial graph is augmented with two edges going from and to the synonyms.

Methods and relationships	#relations
Hierarchical relations	
Maximal syntactic head	3,366
Minimal syntactic head	3,816
Morpho-syntactic variants	743
Medical synonyms	
3 biomedical terminologies	1,879
UMLS/Filtered UMLS	190
Morpho-syntactic variants	100
Medical synonyms and WordNet	
3 biomedical terminologies	1,939
UMLS/Filtered UMLS	227

Table 1: Hierarchical and synonymy relations generated by terminology structuring methods

3.3 Merging of clusters from two approaches

We merge the clusters generated by the two approaches. The merging is performed on the intersection between the clusters. As previously, we test intersection values within the interval [10; 90].

3.4 Evaluation

We give judgments on: (1) the correctness of the generated relations, (2) their relevance according to the reference data, (3) their relevance according to the manual evaluation by an expert. The evaluation is performed with three measures: precision P (percentage of the relevant terms clustered divided by the total number of the clustered terms), recall R (percentage of the relevant terms clustered divided by the number of terms in the corresponding SMQ) and F-measure F_1. The association between the SMQs and the clusters relies on the best F_1.

4 Results

Semantic relations acquired with terminology structuring are indicated in table 1. There is a small difference between relations acquired through maximal and minimal syntactic heads, although the influence of medical resources for the acquisition of synonymy varies according to the resources. WordNet slightly increases the number of synonyms. Faster generates a large set of hierarchical and synonymy relations. MedDRA terms have also been processed with semantic distance and clustered. The best thresholds with the radius clustering are 2 for *Rada*,

Approach	Hierarchical SMQs			SMQs and sub-SMQs		
	#clusters	interval	mean	#clusters	interval	mean
Semantic distance	2,667	[2; 1,206]	73	2,931	[2; 546]	17
Structuring (hierarchical)	690	[1; 134]	3.69	748	[1; 117]	3.43
Structuring (hierarchical+synonymy)	690	[1; 136]	4.11	748	[1; 119]	3.82
Merging (hierarchical)	2,732	[1; 1,220]	72.40	2,998	[1; 563]	24.44
Merging (hierarchical+synonymy)	2,732	[1; 1,269]	75.94	2,998	[1; 594]	26.03

Table 2: Number of clusters and their size (the interval and the mean number of terms per cluster) for individual approaches and for their merging computed for hierarchical SMQs and also for SMQs and sub-SMQs

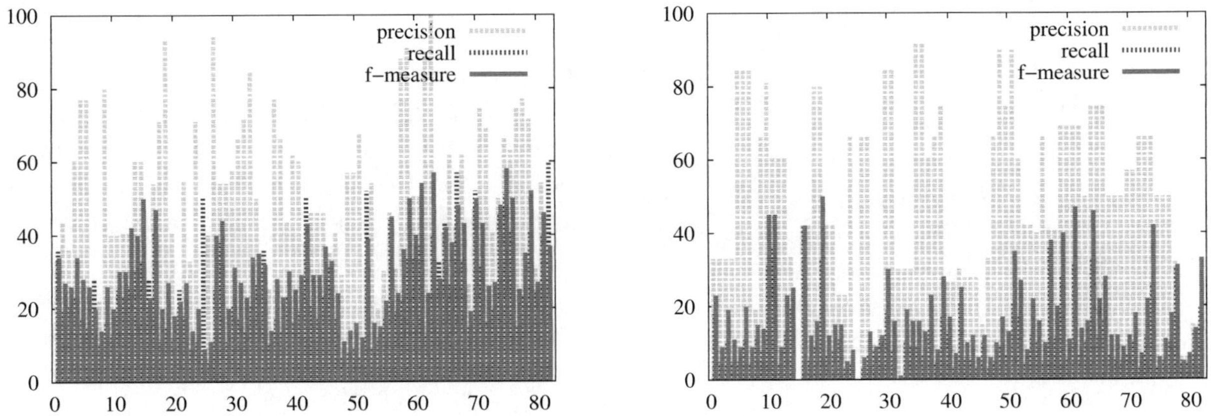

(a) Semantic distance

(b) Terminology structuring

Figure 3: Results (precision, recall and F-measure) for semantic distance and terminology structuring approaches

4.10 for *LCH* and 0 for *Zhong*. With the HAC, the best results are obtained with 300 classes (number of terms per class is within the interval [1; 98], mean number of terms per class is 25.34). Our results show that the best parameters for the semantic distance are the Rada distance, radius approach and no formal definitions, while the best parameters for the terminology structuring are maximal syntactic head with hierarchical relations by Faster augmented with synonyms. For the merging of the clusters we apply 50% intersection for hierarchical SMQs and 80% intersection for SMQs and sub-SMQs. We exploit and discuss these results. The percentage of the MedDRA terms involved by the terminology structuring is the 32% with hierarchical relations, it reaches 40% when the synonymy is also considered. With semantic distance, all the terms from ontoEIM (51% of the MedDRA) are used.

Table 2 provides information on clusters: number of clusters, number of terms per cluster (their interval and the mean number of terms per cluster). In table 2, we first indicate the results for the individual approaches, and then when the merging of the approaches is performed. We observe that the merging has a positive effect on the number and the size of clusters: data generated by the individual approaches (and by synonymy) are complementary.

4.1 Correctness of the semantic relations

A manual analysis of the generated hierarchical relations indicates that these relations are always correct: the constraint involved through the syntactic analysis guarantees correct propositions. Nevertheless, we observed a small number of syntactic ambiguities. They appear within 144 pairs (5%) with maximal syntactic heads and correspond to pairs like: {*anticonvulsant drug level*, *drug level*}, {*blood smear test*, *smear test*}, {*eye movement disorder*, *movement disorder*}. Thus, within the first

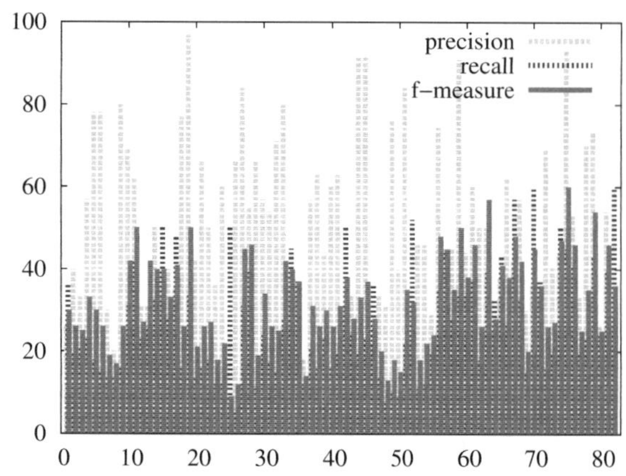

Figure 4: Results (precision, recall and F-measure) obtained when the two approaches are merged

pair, there is an ambiguity on *drug* as two dependencies seem possible: {*anticonvulsant drug level, drug level*} as proposed by the system and {*anticonvulsant drug level, level*}. But whatever the syntactic analysis performed, the semantic relations are correct.

4.2 Relevance of the generated clusters

Figures 3 and 4 provide quantitative evaluation of the clusters: semantic distance (figure 3(a)), terminology structuring (figure 3(b)), merging of these two sets (figure 4). On figure 3, we can observe that there is a great variability among the SMQs and the two approaches. The positive result is that these approaches are indeed complementary: their merging slightly increases performance. An analysis of the clusters generated with terminology structuring shows that: (1) hierarchical relations form the basis of the clusters: they correspond to 96% of the involved terms and show 69% precision. Only three clusters do not contain hierarchical relations; (2) Faster relations are involved in 50% of clusters and show precision between 75 and 85%; (3) one third of the clusters contains synonymy relations, which precision varies between 55 and 69%; (4) relations acquired with the UMLS resources are involved in 14% of clusters while their precision is only 38%.

We also performed a detailed qualitative analysis of several SMQs and clusters with an expert. Table 3 presents the analysis for three SMQs: *Angioedema, Embolic and thrombotic events, arterial* and *Haemodynamic oedema, effusions and fluid overload*. It indicates the number of terms in the SMQ and in the corresponding clusters *clu*, as well as the number of common terms between them *com* and the performance (precision P, recall R and F-measure F) when computed against the reference data *Reference* and also after the analysis performed by the expert *After expertise*. The results obtained with the two approaches are indicated: semantic distance *sd* and terminology structuring *struc*, as well as their merging *merg*. In the colums *Reference*, we can observe that the best F-measure values are obtained with the terminology structuring method for the SMQ *Haemodynamic oedema, effusions and fluid overload* (F=45) and with the semantic distance for the SMQ *Embolic and thrombotic events, arterial* (F=32). The merging of the two methods systematically improves the results: in the given examples, for all three SMQs.

A detailed analysis of the generated noise indicates that across the SMQs we have similar situations: we generate false positives (terms non relevant for the medical conditions, such as *Pulmonary oedema, Gestational oedema, Spinal cord oedema* for the SMQ *Angioedema*), but also the SMQs may contain non relevant terms or may miss relevant terms (thus, *Testicular oedema, Injection site urticaria, Bronchial eodema* are missing in the SMQ *Angioedema*). The expert evaluation (columns *After expertise* in table 3) attempts to analyse also the quality of the SMQs. The corrected performance of the clusters is improved in several points, which indicates that automatic approaches may provide a useful basis for the creation of SMQs.

5 Discussion

Despite the incompleteness of the ontoEIM resource, the semantic distance approach is quite efficient and provides the core terms for the building of the SMQs. Among the several algorithms tested, the most simple algorithm (Rada et al., 1989), which exploits the shortest path, leads to the best results, while the additional information on the hierarchical depth exploited by other algorithms appears non useful. The clustering method which allows the generation of non-disjoint clusters is the most efficient as MedDRA terms may belong to several SMQs.

SMQs	Number of terms			Reference			After expertise		
	SMQ	clu	com	P	R	F	P	R	F
$Angioedema_{sd}$	52	32	13	40	25	30	43	26	33
$Angioedema_{struc}$	52	31	19	61	36	45	61	36	45
$Angioedema_{merg}$	52	33	21	63	42	50	71	48	57
Embolic and thrombotic events...$_{sd}$	132	159	48	30	36	32	32	39	35.2
Embolic and thrombotic events...$_{struc}$	132	13	12	92	9	16	92	9	16
Embolic and thrombotic events...$_{merg}$	132	130	49	38	37	37.5	47	46	46.5
Haemodynamic oedema, effusions...$_{sd}$	36	22	7	32	20	24	54	33	41
Haemodynamic oedema, effusions...$_{struc}$	36	31	13	42	36	39	84	72	78
Haemodynamic oedema, effusions...$_{merg}$	36	35	16	46	44	45	86	83	84.5

Table 3: Comparison between the two approaches (semantic distance *sd* and terminology structuring *struc*) and the merging of the two approaches *merg* for three SMQs: *Angioedema*, *Embolic and thrombotic events, arterial* and *Haemodynamic oedema, effusions and fluid overload*

Traditionnal classification methods, which produce disjoint clusters, are less efficient for this task.

It has been surprising to observe that the contribution of the generated hierarchical relations is so important (table 1) and that these relations appear to be so often correct for the creation of SMQs. Indeed, because PT terms belong to the same hierarchical level of MedDRA, they should be hierarchically equivalent between them. In reality, within a cluster, we can find several hierarchical levels of the PT terms. This means that the hierarchical structure of MedDRA could be more fine-grained and that intermediate hierarchical levels could be created. As for the generated synonymy relations, their number is low and they contribute in a lesser way to the building of the clusters: this means that the PTs are semantically differentiated between them.

Finally, the merging of these two approaches is beneficial for the generation of clusters: the performance is improved, although slightly. The two approaches provide indeed complementary results. The low recall and F-measure are due to the material and methods exploited: ontoEIM contains only 51% of the MedDRA terms to be processed while the exploited terminology structuring methods are not able to detect more common features between the terms.

The difference between the results obtained against the reference data and after the expert evaluation (table 3) show that the reference data are not very precise. In previous work, it has already been observed that some important PT terms can be missing in the SMQs (Pearson et al., 2009). With the proposed automatic methods we could find some of these terms. It has been also demonstrated that the SMQs are over-inclusive (Mozzicato, 2007; Pearson et al., 2009). In the proposed analysis of the SMQs, we have also found terms which have too large meaning and which should not be included in the SMQs.

6 Conclusion and Perspectives

We have applied two different approaches to the clustering of pharmacovigilance terms with similar or close meaning. We performed a comparison of the results obtained with these two approaches and analysed their complementarity. Several experiments have been carried out in order to test different parameters which may influence the performance of the methods. Although the automatic creation of the SMQs is a difficult task, our results seem to indicate that the automatic methods may be used as a basis for the creation of new SMQs. The precision of the clusters is often satisfactory, while their merging leads to the improvement of their completeness. These approaches generate complementary data and their combination provides more performant results.

Future studies will lead to the identification of other parameters which influence the quality of clusters and also other factors which may be exploited for the merging of clusters. More robust distances and clustering methods will also be used in future work, as well as approaches for a better acquisition and evaluation of the hierarchical structure of SMQs. We plan also to design corpora-based methods which may also to increase the recall of the results. We will perform an exhaustive analysis of the nature of semantic relations which can be observed within the SMQs and propose other methods to further improve the coverage of the clusters. Different filters will be tested to remove the true false positive relations between terms. The results will also be evaluation by several experts, which will allow to assess the inter-expert variation and its influence on the results. Besides, the obtained clusters will also be evaluated through their impact on the pharmacovigilance tasks and through the exploring of the pharmacovigilance databases.

References

I Alecu, C Bousquet, and MC Jaulent. 2008. A case report: using snomed ct for grouping adverse drug reactions terms. *BMC Med Inform Decis Mak*, 8(1):4–4.

S Aubin and T Hamon. 2006. Improving term extraction with terminological resources. In *FinTAL 2006*, number 4139 in LNAI, pages 380–387. Springer.

C Bousquet, C Henegar, A Lillo-Le Louët, P Degoulet, and MC Jaulent. 2005. Implementation of automated signal generation in pharmacovigilance using a knowledge-based approach. *Int J Med Inform*, 74(7-8):563–71.

EG Brown, L Wood, and S Wood. 1999. The medical dictionary for regulatory activities (MedDRA). *Drug Saf.*, 20(2):109–17.

CIOMS. 2004. Development and rational use of standardised MedDRA queries (SMQs): Retrieving adverse drug reactions with MedDRA. Technical report, CIOMS.

C Fellbaum. 1998. A semantic network of english: the mother of all WordNets. *Computers and Humanities. EuroWordNet: a multilingual database with lexical semantic network*, 32(2-3):209–220.

R Fescharek, J Kübler, U Elsasser, M Frank, and P Güthlein. 2004. Medical dictionary for regulatory activities (MedDRA): Data retrieval and presentation. *Int J Pharm Med*, 18(5):259–269.

N Grabar and T Hamon. 2010. Exploitation of linguistic indicators for automatic weighting of synonyms induced within three biomedical terminologies. In *MEDINFO 2010*, pages 1015–9.

T Hamon and A Nazarenko. 2001. Detection of synonymy links between terms: experiment and results. In *Recent Advances in Computational Terminology*, pages 185–208. John Benjamins.

J Iavindrasana, C Bousquet, P Degoulet, and MC Jaulent. 2006. Clustering WHO-ART terms using semantic distance and machine algorithms. In *AMIA Annu Symp Proc*, pages 369–73.

Christian Jacquemin. 1996. A symbolic and surgical acquisition of terms through variation. In S. Wermter, E. Riloff, and G. Scheler, editors, *Connectionist, Statistical and Symbolic Approaches to Learning for Natural Language Processing*, pages 425–438, Springer.

MC Jaulent and I Alecu. 2009. Evaluation of an ontological resource for pharmacovigilance. In *Stud Health Technol Inform*, pages 522–6.

G Kleiber and I Tamba. 1990. L'hyperonymie revisitée : inclusion et hiérarchie. *Langages*, 98:7–32, juin.

C Leacock and M Chodorow, 1998. *Combining local context and WordNet similarity for word sense identification*, chapter 4, pages 305–332.

P Mozzicato. 2007. Standardised MedDRA queries: their role in signal detection. *Drug Saf*, 30(7):617–9.

NLM, 2011. *UMLS Knowledge Sources Manual*. National Library of Medicine, Bethesda, Maryland. www.nlm.nih.gov/research/umls/.

Barbara H. Partee. 1984. Compositionality. In F. Landman and F. Veltman, editors, *Varieties of formal semantics*. Foris, Dordrecht.

RK Pearson, M Hauben, DI Goldsmith, AL Gould, D Madigan, DJ O'Hara, SJ Reisinger, and AM Hochberg. 2009. Influence of the MedDRA hierarchy on pharmacovigilance data mining results. *Int J Med Inform*, 78(12):97–103.

R Rada, H Mili, E Bicknell, and M Blettner. 1989. Development and application of a metric on semantic nets. *IEEE Transactions on systems, man and cybernetics*, 19(1):17–30.

K Spackman and K Campbell. 1998. Compositional concept representation using SNOMED: Towards further convergence of clinical terminologies. In *Journal of American Medical Informatics Association (JAMIA)*, pages 740–744.

MQ Stearns, C Price, KA Spackman, and AY Wang. 2001. SNOMED clinical terms: overview of the development process and project status. In *AMIA*, pages 662–666.

Y Tsuruoka, Y Tateishi, JD Kim, T Ohta, J McNaught, S Ananiadou, and J Tsujii. 2005. Developing a robust part-of-speech tagger for biomedical text. *LNCS*, 3746:382–392.

J Zhong, H Zhu, J Li, and Y Yu. 2002. Conceptual graph matching for semantic search. In *10th International Conference on Conceptual Structures, ICCS2002, LNCS 2393, Springer Verlag*, pages 92–106.

Temporal Classification of Medical Events

Preethi Raghavan*, **Eric Fosler-Lussier***, **and Albert M. Lai**†
*Department of Computer Science and Engineering
†Department of Biomedical Informatics
The Ohio State University, Columbus, Ohio, USA
{raghavap, fosler}@cse.ohio-state.edu, albert.lai@osumc.edu

Abstract

We investigate the task of assigning medical events in clinical narratives to discrete time-bins. The time-bins are defined to capture when a medical event occurs relative to the hospital admission date in each clinical narrative. We model the problem as a sequence tagging task using Conditional Random Fields. We extract a combination of lexical, section-based and temporal features from medical events in each clinical narrative. The sequence tagging system outperforms a system that does not utilize any sequence information modeled using a Maximum Entropy classifier. We present results with both hand-tagged as well as automatically extracted features. We observe over 8% improvement in overall tagging accuracy with the inclusion of sequence information.

1 Introduction

There has been a lot of interest in building timelines of medical events from unstructured patient narratives (Jung et al., 2011; Zhou and Hripcsak, 2007). Creating a timeline from longitudinal clinical text requires learning temporal relations such as *before, simultaneous, includes, overlaps, begins, ends* and their inverses between medical events found within and across patient narratives (Allen, 1981). However, learning temporal relations for fine-grained temporal ordering of medical events in clinical text is challenging: the temporal cues typically found in clinical text may not always be sufficient for this task.

An important characteristic of a clinical narrative is that the medical events in the same narrative are more or less semantically related by narrative discourse structure. However, medical events in the narrative are not ordered chronologically. Thus, the clinical narrative structure is not always temporally coherent.

Moreover, extracting precise temporal features for highly accurate temporal ordering of medical events is difficult as the temporal relationship between medical events is varied and complicated. Zhou and Hripcsak (2007) identify six major categories of temporal expressions from a corpus of discharge summaries: "date and time," "relative date and time," "duration," "event-dependent temporal expression," "fuzzy time," and "recurring times." Their study of temporal expressions in clinical text indicates that relative time (e.g., ever since the *episode* 2 days ago) may be more prevalent than absolute time (e.g., 06/03/2007). Further, temporal expressions may be fuzzy where "history of *cocaine use*" may imply that *cocaine use* started 2 years ago or 10 years ago.

In this paper, we address a relatively simpler task of assigning medical events to coarsely defined time-bins. The time-bins, *way before admission, before admission, on admission, after admission, after discharge*, are defined based on the relative temporal distance of the medical event from the admission date, which is the only explicit date almost always found in each clinical narrative. We extract features based on narrative structure as well as temporal expressions to label a sequence of medical events from each clinical narrative with a highly probable

HISTORY PHYSICAL DATE: 06/03/2007
NAME: Smith Jack MR#: XXX-XX-XXXX
ATTENDING PHYSICIAN: Bill Payne MD DOB: 02/28/1960
CHIEF COMPLAINT
Chest pain and arm infection.
HISTORY OF PRESENT ILLNESS (1) (2) (3)
Patient is a 48-year-old male with history of cocaine use hypertension who presents with chest pain (4)
(5) which started 2 days ago . He did not having chest pain yesterday but ever since the episode 2 days ago
(6) he has felt a little weaker. He did have chest pain today and this is what prompted him to come to the
ER. He also notices that he has had some infections under his arms. He states that he had to have an (8)
(7) abscess I and D 3 or 4 months ago under his arm and 2 to 3 weeks ago he noticed some more spots and
these spots have now grown and now are under both arms. Currently he is chest pain free. His blood
pressure upon presentation was 189/106. (9)
REVIEW OF SYSTEMS
On exam initial blood pressure was 189/106 current blood pressure 148/83 with heart rate of 74 respirations 16. Heart regular rhythm. No murmurs. Arms: He does have tender areas right greater than left under the arm. Difficult to tell if there is any erythema but obvious cellulitis sludge abscess under the right arm which is tender.
ASSESSMENT/PLAN
1. Chest pain history of cocaine with T-wave inversions in the inferior leads. Currently he is chest pain free. We will check a 2-D echocardiogram. Consult Cardiology for a stress test.
2. Axillary abscesses. Consult Surgery for I and D. We will place on IV vancomycin pain control.
3. Cocaine abuse. Encouraged to quit.

Figure 1: Excerpt from a de-identified clinical narrative (cn1) written for a patient in 2007. Medical events are underlined. Enumerated events (in circles) are used as an example later in Table 1.

sequence of time-bins using Conditional Random Fields (CRFs). The learned time-bins can be used as an informative temporal feature for tasks such as fine-grained temporal ordering of medical events and medical event coreference resolution.

2 Motivation

Clinical narratives are medical reports that contain unstructured text documenting the medical history of the patient. Medical events are temporally-related concepts in clinical narratives that describe medical conditions affecting the patient's health, or tests and procedures performed on a patient. Sample excerpts from two different clinical notes (cn1 and cn2) of the same patient, generated over time, are shown in Figures 1 and 2. We can see from the examples that narrative structure moves back and forth in time and is not temporally coherent. We use cn1 and cn2 as running examples throughout the paper.

The medical events assigned to time-bins in each clinical narrative allow us to derive a coarse temporal order between medical events within and across the longitudinal medical history of the patient. Since we learn time-bins centered around admission in each narrative and we also know the admission date and perhaps the discharge dates in cn1 and cn2, we can derive a coarse partial order across the medi-

HISTORY PHYSICAL DATE: 06/17/2007
NAME: Black Jack MR#: XXX-XX-XXXX
ATTENDING PHYSICIAN: Jack Payne MD DOB: 02/28/1960
He is a 48-year-old African American gentleman with a history of cocaine use and hypertension. He has hidradenitis of both axilla resected. The patient is MRSA positive on IV antibiotics at the present time. The patient's physical condition is excellent but he had MRSA in the axilla for hidradenitis that was devastating. The wounds now are very large but he is wound vac and being changed to alginate. Both axilla show major wounds of 20-25 cm in diameter and 4-5 cm deep in overall size and he has excoriations on his chest from the tape. The plan is to change him from vac to alginate and see him in a week.

Figure 2: Excerpt from another de-identified clinical narrative (cn2) for the same patient written in later in 2007. Medical events are underlined.

cal events in cn1 and cn2. This is shown in Figure 3. Even if the discharge dates are not known, we still know that the admission date (A1) of cn1 is 6/03/2007 and A2 of cn2 is 06/17/2007. Thus, $A2 > A1$, and all the time-bins in cn2 that are on or after admission would have happened after A2. The partially ordered time-bins can now be used for tasks such as medical concept coreference resolution.

In cross narrative coreference resolution tasks, we can prune the space of candidate pairs of medical events by ruling out portions of clinical narratives that will not have any coreferring medical events. For example, in the timeline shown in Figure 3, the medical events in time-bins *admission*, *after admission* and *discharge* of cn2 will not corefer with any medical event in cn1. Further, when mentions of the same medical events occur in different time-bins, it could mean that they are the same instance of the medical event and they corefer. For instance, *cocaine abuse* and *cocaine use*. Similarly, *MRSA positive* is assigned to time-bin *on admission* whereas *MRSA* is assigned to *before admission* and both mentions of *MRSA* corefer.

3 Related Work

The Timebank (Pustejovsky et al., 2003) corpus of annotated newswire text is widely used for temporal relation learning. The TempEval challenges have often focused on extracting different types of temporal relations from Timebank (Verhagen et al., 2009). In Timebank, events are typically verbs that denote change in state. Since the notion of an event in Timebank is different from medical events in clinical text, it is not possible to directly train models on Timebank and apply them to clinical text. The THYME work (Savova et al., 2009) extends TimeML to the

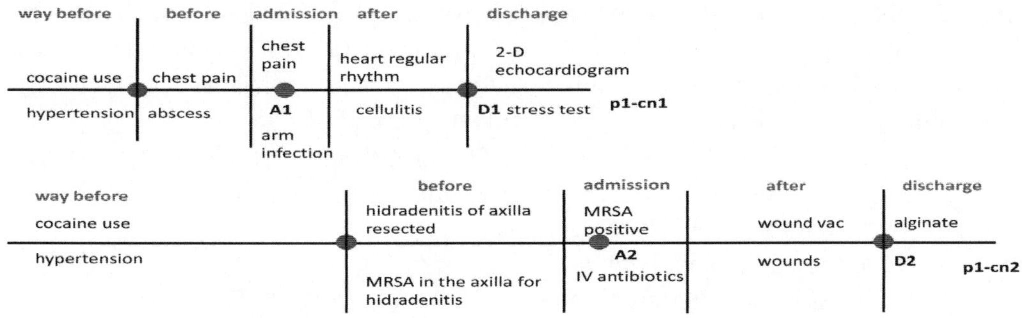

Figure 3: Medical events in clinical narratives cn1 and cn2 for patient p1 assigned to time-bins. A1 is the admission date in cn1 and D1 is the discharge date. Similarly A2 is the admission date in cn2 and D2 is the discharge date. Thus, we have, $A1 < D1, D1 < A2, A2 < D2$

medical domain to create layered annotation to be used for event linking. Boland et al. (2012) identify the temporal knowledge representation requirements of clinical eligibility criteria and develop a frame-based representation designed to support semantic annotation for temporal expressions in eligibility criteria. However, the nature of data found in eligibility criteria is different from clinical narratives.

Previous attempts at learning temporal relations between medical events in clinical text include Jung et al. (2011) and Zhou et al. (2006). Gaizauskas et al. (2006) learn the temporal relations *before, after, is_included* between events from a corpus of clinical text much like the event-event relation TLINK learning in Timebank (Pustejovsky et al., 2003). However, the corpora used in these studies are not freely available. A comprehensive survey of temporal reasoning in medical data is provided by Zhou and Hripcsak (2007).

The task addressed in this paper is at a higher level than the temporal relation learning or temporal ordering task. Without getting into fine-grained temporal ordering, we define coarse time-bins and classify medical events into one of the time-bins.

We work with a similar motivation of being able to answer clinical trial eligibility criteria with temporal constraints. However, while they model the temporal information in eligibility criteria, we process the temporal information and medical events in the clinical narrative to assign events to time-bins. The learned time-bins are a step towards fine-grained temporal ordering of medical events in clinical text. More importantly, we also demonstrate how automatic feature extraction for this task gives us promising results, though not as good as using hand-tagged features.

4 Problem Description

A patient could have multiple clinical narratives, generated over a period of time, representing the patient's longitudinal medical history. Returning to the examples in Figures 1 and 2, in this section we describe how such clinical narratives are translated into a temporal-bin assignment problem.

4.1 Medical event representation

Medical events in clinical narratives often have a time duration with a corresponding start and stop time, for example, *history of hypertension* (Zhou et al., 2006). In this example, *hypertension* started at some point before admission and is present to the current date. Time duration based representation is essential to learning the exact fine-grained temporal order of medical events within and across clinical narratives. In order to keep the task of classifying medical events into coarse time-bins relatively simple and easy to learn, we use a time-point notation for representing medical events. Each mention of a medical event is assigned to a time-bin without taking into consideration whether it denotes the beginning or end of that event. We also do not differentiate between coreferences of the same medical event. Thus, if *chest pain* is mentioned in the past medical history and the same *chest pain* continues to persist in the after admission time-bin, the two different mentions of chest pain get anchored to different time-bins. Similarly, *cocaine use* started in the history of the patient and *cocaine abuse* still persists. We assign the two different mentions of this medical event into different time-bins.

4.2 Time-bins

As mentioned earlier, we learn to classify medical events into one of the following time-bins: *way before admission, before admission, on admission, after admission, after discharge*. The intuition behind each time-bin label is as follows. The time-bin *way before admission* is intended to capture all medical events that happened in the past medical history of the patient but are not mentioned as being directly related to the present illness. *Before admission* captures events that occurred before admission and are related to the present illness. *On admission* captures medical events that occur on the day of admission. *After admission* captures medical events that occur between admission and discharge (during the hospital stay or clinic visit). Finally, medical events that are supposed to occur in the future after the patient is discharged belong to the class *after discharge*.

Further, the time duration of each time-bin varies based on the patient. For instance, the hospital stay of a patient could be 4 days or 1 month or a year. This makes it very difficult to define exact time-bins based on the intuitions described above. In order to make the problem more precise and consistent across different patients, we restrict *way before admission* to events that happened more than a year ago and *before admission* to events that occurred in the same year before admission. If it is unclear as to when in the past the medical event occurred, we assume it happened *way before admission*.

5 Learning time-bin assignments

We model the problem of classifying medical events to time-bins as a sequence tagging task using CRFs (Lafferty et al., 2001). CRFs are a joint model of label sequence conditioned on the observation.

For the task proposed in this paper, an observation sequence is composed of medical events in the order in which they appear in a clinical narrative, and the state sequence is the corresponding label sequence of time-bins. Each label in the label sequence could be any one of the time-bins *way before admission (wa), before admission (ba), on admission (a), after admission (aa), after discharge (ad)*. Thus, given a sequence of medical events in narrative order we learn a corresponding label sequence of time-bins {wb, b, a, aa, ad}.

The probability of time-bin (label) sequence y, given a medical event (input) sequence x, is given by,

$$P(Y|X) = \exp \sum_i (S(x, y, i) + T(x, y, i)) \quad (1)$$

where i is the medical event index and S and T are the state and transition features respectively. State features S consider the label of a single medical event and are defined as,

$$S(x, y, i) = \sum_j \lambda_j s_j(y, x, i) \quad (2)$$

Transition features consider the mutual dependence of labels y_{i-1} and y_i (dependence between the time-bins of the current and previous medical event in the sequence) and are given by,

$$T(x, y, i) = \sum_k \mu_k t_k(y_{i-1}, y_i, x, i) \quad (3)$$

where s_j and t_k are the state and transition feature functions. Above, s_j is a state feature function, and λ_j is its associated weight and t_j is a transition function, and μ_j is its associated weight. In contrast to the state function, the transition function takes as input the current label as well as the previous label, in addition to the data. The mutual dependence between the time-bins of the current and previous medical events is observed frequently in sections of the text describing the history of the patient. Around 40% of the medical events in gold standard corpus demonstrate such dependencies.

The Maximum Entropy (MaxEnt) model (Berger et al., 1996) estimates the probability of a time-bin given the observed medical event. In this case, we are interested in finding the time-bin with the maximum estimated probability.

6 Feature Space

We extract features from medical event sequences found in each clinical narrative. The extracted feature-set captures narrative structure in terms of the narrative type, sections, section transitions, and position in document. The medical event and the context in which it is mentioned is captured with the help of lexical features. The temporal features resolve temporal references and associate medical events with temporal expressions wherever possible.

6.1 Section-based features

Determining the document-level structure of a clinical narrative is useful in mapping medical events to time-bins. This can be achieved by identifying different sections in different types of clinical narratives and relating them to different time-bins. The section in which the medical event is mentioned tells us something about when it occurred. Li et al. (2010) train a hidden Markov model (HMM) to map a sequence of sections to 15 possible known section types in free-text narratives with high accuracy.

Commonly found sections in discharge summaries and history and physical reports include: "past medical history," "history of present illness," "findings on admission," "physical examination," "review of systems," "impression," and "assessment/plan." On the other hand, radiology notes tend to have sections describing "indication," "'comparison," "findings" and "impression". Similarly, pathology notes may have sections including "clinical history," "specimen received," "laboratory data" and "interpretation." While some sections talk about patient history, some other sections describe the patient's condition after admission, or plans after discharge. However, some clinical notes like cn2 in Figure 2 may not have any section information.

The combined feature representing the type of clinical narrative along with the section can be informative. Section transitions may also indicate a temporal pattern for medical events mentioned across those sections. For instance, "past medical history" (*way before admission*), followed by "history of present illness" (*way before admission*), followed by "findings on admission" (on admission), followed by "physical examination" (after admission), followed by "assessment/plan" (*discharge*). Medical events in different types of sections may also exhibit different temporal patterns. A "history of present illness" section may start with diseases and diagnoses 30 years ago and then proceed to talk about them in the context of a medical condition that happened few years ago and finally describe the patient's condition on admission.

In addition to the section information, we also use other features extracted from the clinical narrative structure such as the position of the medical concept in the section and in the narrative.

6.2 Lexical features

Bigrams are pairs of words that occur in close proximity to each other, and in a particular order. The bigrams preceding the medical event in the narrative can be useful in determining when it occurred. For instance, "**history of** *cocaine use and hypertension*," "**presents with** *chest pain*," "**have** *chest pain*," "**since the** *episode*," etc. If the preceding bigram contains a verb, we also extract the tense of the verb as a feature. However, tense is not always helpful in learning the time of occurrence of a medical event. Consider the following line from cn2 in Figure 2, "He has *hidradenitis of both axilla resected*." Though "has" is in present tense, the medical event has actually occurred in the history and is only being observed and noted now. Additionally, we also explicitly include the preceding bigrams and the tense of verb for the previous and next medical event as a feature for the current medical event.

Every medical event that occurs above a certain frequency threshold in all the clinical narratives of a particular patient is also represented as a binary feature. More frequent medical events tend to occur in the history of the patient, for example, *cocaine use*. We use a threshold of 3 in our experiments. The medical event frequency in also calculated in combination with other features such as the type of clinical narrative and section type.

6.3 Dictionary features

The UMLS[1] includes a large Metathesaurus of concepts and terms from many biomedical vocabularies and a lexicon that contains syntactic, morphological, and orthographic information for biomedical and common words in the English language. We map each medical event to the closest concept in the UMLS Metathesaurus and extract its semantic category. The semantic categories in UMLS include Finding, Disease or Syndrome, Therapeutic or Preventative procedure, Congenital abnormality,

[1] https://uts.nlm.nih.gov/home.html

and Pathologic Function. The intuition behind this is that medical events associated with certain semantic categories may be more likely to occur within certain time-bins. For instance, a medical event classified as "Congenital abnormality" may be more likely to occur *way before admission*.

6.4 Temporal features

Temporal features are derived from any explicit dates that fall in the same sentence as the medical concept. The gold-standard corpus contains annotations for temporal anchors for events. Although there are no explicit dates in cn1 and cn2, there may be narratives where there are mentions of dates such as *fever* on June 7th, 2007. In some cases, there may also be indirect references to dates, which tell us when the medical event occurred. The reference date with respect to which the indirect temporal reference is made depends on the type of note. In case of history and physical notes, the reference date is usually the admission date. For instance, *chest pain which started 2 days ago*, this would mean *chest pain* which started 2 days before admission. Since the admission date is 06/03/2007, *chest pain* would have started on 06/01/2007. Similarly, 3 to 4 months ago resolves to February 2007 or March 2007 and 2 to 3 weeks ago resolves to first or second week of May 2007. Whenever, the exact date is fuzzy, we assume the date that is farthest from the reference date as accurate. So in case of these examples, February 2007 and first week of May 2007 are assumed to be correct. We also calculate the difference between admission date and these dates associated with medical events. Another fuzzy temporal expression is "history of," where history could mean any time frame before admission. We assume that any medical event mentioned along with "history of" has occurred *way before admission*.

Other implicit temporal expressions can be found in phrases such as *upon presentation yesterday, today, at the present time,* and *now*. *Upon presentation, at the present time, today*, and *now* resolve to the admission date 06/03/2007 and *yesterday* resolves to the day before admission 06/02/2007. There are some other implicit temporal expressions expressed relative to medical events, for example, *ever since the episode 2 days ago he has felt a little weaker.* Here, *episode* refers to *chest pain* and since *chest pain* happened 2 days ago, ever since then up to the present time would resolve to the time period between 06/01/2007 and 06/03/2007. This time period is associated with *weaker*.

7 Corpus

We use annotators that are students or recently graduated students from diverse clinical backgrounds with varying levels of clinical experience to annotate a corpus of clinical narratives from the medical center. The corpus consists of narratives specifically from MRSA cases and consists of admission notes, radiology and pathology reports, history and physical reports and discharge summaries. The features marked by the annotators include medical events; corresponding time-bin; corresponding UMLS concept identifier; the UMLS semantic category; temporal expressions; the link between temporal expressions and medical events, if any; and the section under which the medical event is mentioned, if any. The annotators marked 1854 medical events across 5 patients and 51 clinical narratives. The annotation agreement across our team of annotators is high; all annotators agreed on 89.5% of the events and our overall inter-annotator Cohen's kappa statistic (Conger, 1980) for medical events was 0.865.

While we found the inter-annotator agreement for medical event UMLS concept identifiers to be lower than for medical events and temporal expressions, agreement was still very high. We discovered that in many cases there was either a discrepancy in the granularity to which the medical events were coded or whether or not clinical judgment was used in selecting the concept identifier. For example, all of our annotators marked "B-Cell CLL" as an event. Three of them coded this term as "C0023434: Chronic Lymphocytic Leukemia." Two others coded this event as "C0475774: B-cell chronic lymphocytic leukemia variant." While both could be considered correct annotations for "B-Cell CLL," C0475774 is the more specific term. In another example, all of the annotators marked the phrase "white blood cell count of 10,000." For this situation, one of them selected "C0750426: white blood cell count increased," while another selected "C0023508: White Blood Cell count procedure." In contrast, the other three selected different concept identifiers, applying clinical judgment to the medical events. One other annotator selected "C0860797: differential white blood cell count normal."

We use this gold-standard corpus for our experiments. We conduct two sets of experiments with the clinical narratives in this corpus: 1) Medical event, Time-bin experiments using hand-tagged features from the corpus and 2) Medical event, Time-bin experiments using automatically extracted features from the corpus.

8 Experiments

We first conducted experiments using the hand-tagged features in our corpus. Based on these features, we generated the section-based, lexical, dictionary and temporal features described in the previous sections. We used 10-fold cross validation in all our experiments. We use the Mallet[2] implementation of CRFs and MaxEnt. CRFs are trained by Limited-Memory Broyden-Fletcher-Goldfarb-Shanno (BFGS) for our experiments. The per-class accuracy values of both sequence tagging using CRFs and using a MaxEnt model are indicated in Table 3.

When modeled as a multi-class classification task using MaxEnt, we get an average precision of 81.2% and average recall of 71.4% whereas using CRFs we obtain an average precision of 89.4% and average recall of 79.2%. In order to determine the utility of temporal features, we do a feature ablation study with the temporal features removed. In this case the average precision of the CRF is 79.5% and average recall is 67.2%. Similarly, when we remove the section-based features, the average precision of the CRF is 82.7% and average recall is 72.3%. The section-based features seems to impact the precision of the *on admission* and *after admission* time-bins the most.

We compare our approach for classifying medical events to time-bins with the following baseline model. We assign medical events to time-bins based on the type of narrative, any explicit dates and section in which they occur. Each section is associated with a pre-defined time-bin. In the case of the sections in cn1, any medical event under "history of present illness" is *before admission*, "review of systems" is *after admission* and "assessment/plan" is *discharge*. If the narrative has a "past medical history" or a similar section, the events mentioned under it would be assigned to *way before admission*. Partial results of (medical event, time-bin) assignment in cn2 as per this baseline can be seen in Table 1. However, this baseline does not work for clinical narratives like cn2 that do not have any section information. This model gives us an average precision of 58.02% and average recall of 60.26% across the 5 time-bins. Per-class predictions for the baseline are shown in Table 2.

[2]http://mallet.cs.umass.edu/

Medical Event	Baseline	MaxEnt	CRF	Gold
①cocaine use	ba	wa	wa	wa
②hypertension	ba	wa	wa	wa
③chest pain	ba	ba	ba	ba
④episode	ba	ba	ba	ba
⑤chest pain	ba	ba	a	a
⑥infections	ba	wa	ba	ba
⑦abscess	ba	ba	ba	ba
⑧spots	ba	ba	ba	ba
⑨chest pain free	ba	wa	a	a

Table 1: Time-bin predictions by the section baseline method, MaxEnt model and CRF for a subset of medical events marked in cn1 in Figure 1.

Class(time-bin)	Section baseline	
	P	R
way before admission (wa)	56.3	61.4
before admission (ba)	60.2	57.5
on admission (a)	63.8	59.1
after admission (aa)	57.5	68.2
after discharge (ad)	52.3	55.1

Table 2: Per-class precision (P) and recall (R) for medical events, time-bins using hand-tagged extracted features.

The most common false positives for the *before admission* class are medical events belonging to *on admission*. This may be due to lack of temporal features to indicate that the event happened on the same day as admission. Frequently, medical events that belong to the *aa*, *ba* and *wa* time-bin get classified as *after discharge*. One of the reasons for this could be misleading section information in case of historical medical events mentioned in the assessment/plan section.

Next, we conduct experiments using automatically extracted features. This is done as follows. The medical events are extracted using MetaMap, which recognizes medical concepts and codes them using

Class(time-bin)	MaxEnt		CRF	
	P	R	P	R
way before admission (wa)	72.4	63.5	79.8	66.7
before admission (ba)	83.4	80.8	92.0	92.4
on admission (a)	76.6	72.1	87.5	75.2
after admission (aa)	88.6	82.1	93.6	99.1
after discharge (ad)	85.2	58.7	94.3	62.5

Table 3: Per-class precision (P) and recall (R) for medical events, time-bins using hand-tagged extracted features.

UMLS (Aronson, 2001). Based on this UMLS code, we can extract the semantic category associated with the code. Compared to the 1854 medical events marked by the annotators, MetaMap identifies 1257 medical events, which are a subset of the 1854. The UMLS coding by the annotators is more contextually relevant and precise. We use a rule-based algorithm to identify and extract document structure based features such as sections from clinical narratives. The rules are formulated based on commonly occurring sections in our corpus. We extract lines that are all upper-case, and longer than a word and use their stemmed representation to sort them by frequency of occurrence in the corpus. While parsing the text in each clinical narrative, on encountering a line that matches a section title from the frequent list, all subsequent lines are associated with that title until a new section title is encountered. In case of the lexical features, we extract bigrams and calculate the tense of the verb preceding the medical event using the Stanford NLP software.[3] The temporal features are extracted with the help of TimeText developed by Zhou and Hripcsak (2007) that automatically annotates temporal expressions in clinical text. However, it is not able to capture many of the implicit temporal references. Following this, a temporal expression is linked to a medical event if it occurs in the same sentence as the medical event.

The average precision and recall of the MaxEnt model using automatically extracted features is 74.3% and 66.5% respectively. Sequence tagging using CRFs gives us an average precision and recall of 79.6% and 69.7% respectively. Although the results are not as good as using hand-tagged features, they are certainly promising. One reason for the loss in accuracy could be because the automatically calculated temporal features are not as precise as the hand-tagged ones. These results are summarized in Table 4.

[3]`http://nlp.stanford.edu/software/`

Gold-standard Features		
	P	R
ME	81.2	71.4
CRF	89.4	79.2
CRF(no temp. feats)	79.5	67.2
CRF(no section feats)	82.7	72.3
Automatic Features		
	P	R
ME	74.3	66.5
CRF	79.6	69.7
Baseline (P;R)	58.02	60.26

Table 4: Overall Result Summary: Average precision (P) and recall (R) with manually annotated gold-standard features, automatically extracted features and the baseline.

9 Conclusion

We investigate the task of classifying medical events in clinical narratives to coarse time-bins. We describe document structure based, lexical and temporal features in clinical text and explain how these feature are useful in time-binning medical events. The extracted feature-set when used in a sequence tagging framework with CRFs gives us high accuracy when compared with a section-based baseline or a MaxEnt model. The learned time-bins can be used as an informative feature for tasks such as fine-grained ordering of medical events and medical event coreference resolution. We also experiment with hand-tagged vs. automatically extracted features for this task and observe that while automatically extracted features show promising results, they are not as good as using hand-tagged features for this task.

Acknowledgments

The project described was supported by the National Center for Research Resources, Grant UL1RR025755, KL2RR025754, and TL1RR025753, and is now at the National Center for Advancing Translational Sciences, Grant 8KL2TR000112-05, 8UL1TR000090-05, 8TL1TR000091-05. The content is solely the responsibility of the authors and does not necessarily represent the official views of the NIH.

References

James F. Allen. 1981. An interval-based representation of temporal knowledge. In *IJCAI*, pages 221–226.

Alan R. Aronson. 2001. Effective mapping of biomedical text to the UMLS Metathesaurus: the MetaMap program. *Proc of AMIA Symposium*, pages 17–21.

Adam Berger, Stephen Della Pietra, and Vincent Della Pietra. 1996. A maximum entropy approach to natural language processing. *Computational Linguistics*, 22:39–71.

Mary Regina Boland, Samson W. Tu, Simona Carini, Ida Sim, and Chunhua Weng. 2012. EliXR: An Approach to Eligibility Criteria Extraction and Representation. *Proc of AMIA Clinical Research Informatics Summit*.

Anthony J. Conger. 1980. Integration and generalization of kappas for multiple raters. In *Psychological Bulletin Vol 88(2)*, pages 322–328.

Rob Gaizauskas, Henk Harkema, Mark Hepple, and Andrea Setzer. 2006. Task-oriented extraction of temporal information: The case of clinical narratives. In *Proceedings of the Thirteenth International Symposium on Temporal Representation and Reasoning*, TIME '06, pages 188–195.

Hyuckchul Jung, James Allen, Nate Blaylock, Will de Beaumont, Lucian Galescu, and Mary Swift. 2011. Building timelines from narrative clinical records: initial results based-on deep natural language understanding. In *Proceedings of BioNLP 2011 Workshop*, BioNLP '11, pages 146–154.

John D. Lafferty, Andrew McCallum, and Fernando C. N. Pereira. 2001. Conditional random fields: Probabilistic models for segmenting and labeling sequence data. In *Proceedings of the Eighteenth International Conference on Machine Learning*, ICML '01, pages 282–289.

Ying Li, Sharon Lipsky Gorman, and Noemie Elhadad. 2010. Section classification in clinical notes using supervised hidden markov model. In *IHI*, pages 744–750.

James Pustejovsky, José M. Castaño, Robert Ingria, Roser Sauri, Robert J. Gaizauskas, Andrea Setzer, Graham Katz, and Dragomir R. Radev. 2003. TimeML: Robust specification of event and temporal expressions in text. In *New Directions in Question Answering '03*, pages 28–34.

Guergana Savova, Steven Bethard, Will Styler, James Martin, Martha Palmer, James Masanz, and Wayne Ward. 2009. Towards temporal relation discovery from the clinical narrative. *Proc of AMIA Symposium*, pages 568–572.

Marc Verhagen, Robert J. Gaizauskas, Frank Schilder, Mark Hepple, Jessica Moszkowicz, and James Pustejovsky. 2009. The TempEval challenge: identifying temporal relations in text. *Language Resources and Evaluation*, 43(2):161–179.

Li Zhou and George Hripcsak. 2007. Temporal reasoning with medical data - a review with emphasis on medical natural language processing. *Journal of Biomedical Informatics*, pages 183–202.

Li Zhou, Genevieve B. Melton, Simon Parsons, and George Hripcsak. 2006. A temporal constraint structure for extracting temporal information from clinical narrative. *Journal of Biomedical Informatics*, pages 424–439.

Analyzing Patient Records to Establish If and When a Patient Suffered from a Medical Condition

James Cogley, Nicola Stokes, Joe Carthy and John Dunnion
School of Computer Science and Informatics
University College Dublin
Dublin, Ireland
James.Cogley@ucdconnect.ie {Nicola.Stokes, Joe.Carthy, John.Dunnion}@ucd.ie

Abstract

The growth of digital clinical data has raised questions as to how best to leverage this data to aid the world of healthcare. Promising application areas include Information Retrieval and Question-Answering systems. Such systems require an in-depth understanding of the texts that are processed. One aspect of this understanding is knowing if a medical condition outlined in a patient record is recent, or if it occurred in the past. As well as this, patient records often discuss other individuals such as family members. This presents a second problem - determining if a medical condition is experienced by the patient described in the report or some other individual. In this paper, we investigate the suitability of a machine learning (ML) based system for resolving these tasks on a previously unexplored collection of *Patient History and Physical Examination reports*. Our results show that our novel Score-based feature approach outperforms the standard Linguistic and Contextual features described in the related literature. Specifically, near-perfect performance is achieved in resolving *if* a patient experienced a condition. While for the task of establishing *when* a patient experienced a condition, our ML system significantly outperforms the ConText system (87% versus 69% f-score, respectively).

1 Introduction

The growth of the digitization of clinical documents has fostered interest in how to best leverage this data in providing assistance in the world of healthcare, including novel information retrieval (Voorhees and Tong, 2010), question answering (Demner-Fushman and Lin, 2007; Patrick and Li, 2011) and clinical summarization systems (Feblowitz et al., 2011).

Given the richness of the language found in clinical reports, novel systems require a deeper understanding of this textual data. One aspect of this understanding is the *assertion status* of medical conditions (Demner-Fushman et al., 2011). The assertion status of a medical condition may refer to *Negation Resolution*, *Temporal Grounding* (deciding if a condition is currently or historical, and *Condition Attribution* (deciding if a condition is experienced by the patient described in the report or some other individual). The focus of this paper rests on the latter two tasks of *Temporal Grounding* and *Condition Attribution* as Negation has been satisfactorily addressed in Chapman et al. (2007).

Several approaches, ranging in complexity, have been proposed for resolving temporal information. Hripcsak et al. (2005) modeled the task as a constraint satisfaction problem. Another rule-based approach that achieved moderate results uses regular expressions matching occurrences of trigger terms (Chapman et al. 2007). A trigger term in this context refers to a term or phrase that provides strong evidence supporting the attribution (e.g., patient, family member) or temporality (e.g., current, past) of a condition. Given the limitations of solely using pre-composed trigger term lists, recent focus has been placed on the use of rule-based learning systems with different feature sets (Mowery et al., 2009). Section headers, tense and aspect are investigated as features, with promising results for the temporality task achieving an accuracy score of 89%. However, the authors' acknowledge that conclusions drawn must be tentative as a majority class classifier achieved an accuracy of 86.9% (only 13% of conditions in the dataset are historical).

This paper extends current work in the domain in the following ways. The dataset used in these experiments is generated from a collection of previously unannotated History & Physical (H&P) Examination reports. Prior work has focused on other report types such as discharge summaries and emergency department reports. In these cases the distribution of historical and recent conditions is often heavily skewed in favour of descriptions of recent conditions experienced by the patient. As H&P reports aim to provide a comprehensive picture of a patient's past and present state, a more uniform distribution of historical and recent conditions is present in this report type. This work extends previous work by exploring the use of machine learning (ML) as an alternative to hand-crafted rule based systems and rule-based ML approaches to resolving these tasks.

In this work, a comparative analysis of several ML algorithms from different paradigms are evaluated, in order to determine the best approach for our tasks. Building on this, the performance of four automatically extracted feature sets are evaluated, identifying their contributions and also their interactions. This work also extends existing work by automatically extracting features that were previously extracted manually as well as the proposal of a set of novel score-based features. The performance of the ML algorithms are compared to the rule-based system - ConText. Our results show that the ML approaches significantly outperform this rule-based system on the *Temporal Grounding* task.

2 Related Work

Natural Language Processing techniques have been shown to have many different uses in Clinical Text Analysis, such as in the representation (Sager et al., 1994) and understanding (Christensen, 2002) of clinical narratives, and frequently now in the context of more elaborate large-scale systems, such as a clinical decision support system (Demner-Fushman et al., 2009).

Given the sensitive nature of clinical narratives and the difficulty in obtaining data collections for experimental purposes, evaluation and comparison of NLP systems in this domain is difficult. However, recently anonymised data provided by the *Biomedical Language Understanding (BLU) Lab* at the University of Pittsburgh as well as datasets provided as part of the i2b2/VA 2010 challenge (Uzuner et al., 2011), has greatly aided NLP research into the processing of clinical narratives. The dataset provided by BLU Lab and used in this work consists of 101,711 reports of several different report types ranging from discharge summaries to surgical pathology reports. The report types differ in content, technical language and structure. For example, surgical pathology reports are very technical and explicit in the information that they convey, e.g. results of a biopsy, blood cell counts etc. In contrast, discharge summaries and consultation reports are more expressive, and aim to create a more complete patient profile, e.g. including work and personal circumstances. The BLU Lab have published a number of papers on the topic of resolving the assertion status of medical conditions (Chapman et al., 2007; Harkema et al., 2009; Mowery et al., 2009). Their `ConText` algorithm (Chapman et al., 2007) uses handcrafted regular expressions, along with trigger terms and termination terms to determine characteristics of a condition mention in a text. The condition characteristics investigated included negation, temporality (recent, historical, hypothetical) and experiencer (patient, other). Their approach worked very well on the negation and hypothetical temporality, achieving an f-score of 97% in determining negation and an f-score of 88% in resolving hypothetical conditions. However, the approach was less successful when determining historical conditions and their experiencer, with f-scores of 71% and 67%, respectively. These results were generated on emergency room reports only.

In more recent work, their algorithm was applied to 5 other types of clinical document, namely: surgical pathology, operative procedure, radiology, echocardiogram and discharge summaries (Harkema et al., 2009). Results achieved on these new datasets were largely the same, with f-scores for negation ranging between 75% and 95%, and for hypothetical conditions ranging between 76% and 96%. Again, a marked drop-off was seen for historical conditions, with few occurrences of conditions for other experiencers annotated in the datasets (i.e. relatives) making it difficult to draw definitive conclusions from this work.

Although this manual rule based approach has performed well and is advocated due to its ease of implementation (Meystre et al., 2008), Harkema et al. (2009) note its limitations in resolving historical conditions, and encourage the possibility of statistical classifiers in which information other than lexical items, are considered as features. Further work investigating the use of Machine Learning (Uzuner et al., 2009; Mowery et al., 2009) has seen positive breakthroughs in resolving the assertion status of medical conditions.

The 2010 i2b2 challenge (Uzuner et al., 2011) provided a rigid and standardized platform for evaluating systems in resolving the assertion status of medical conditions found in Discharge Summaries. The challenge consisted of three subtasks:*Concept Extraction*, *Assertion* and *Relation Identification*. The second subtask of Assertion involved the development of systems that resolved the assertion status of medical conditions. As part of the assertion task there were six possible assertion statuses: present, absent, uncertain, conditional, or not associated with the patient. Systems submitted to this challenge ranged from more simplistic pattern matching techniques to more complex supervised and semi-supervised approaches (de Bruijn et al., 2011; Clark et al., 2011). The datasets used in the 2010 i2b2 challenge were not available to non-participants at the time the experiments presented in this work were conducted. We plan to explore these datasets in future work. This research investigates patient vs. non-patient conditions as well as past vs. present conditions in order to fine tune feature-sets that may be generalized to further assertion statuses.

In the context of this paper, while the BLU Lab clinical report collection is available, the medical condition annotations are not. As already stated, our intention is to explore a machine learning approach to these tasks. For this purpose we annotated a portion of the consultation report section of the collection. There were two reasons for this - firstly, the BLU Lab have not reported results on this report type so there is no duplication of annotation effort and secondly, it turns out that the consultation reports are a much richer source of historical conditions and condition attribution than any of the report types annotated previously.

3 Method

3.1 Corpus

For this task, 120 H&P reports were randomly extracted from the BluLab's NLP repository. As already stated, this report type's fuller descriptions make it richer than previous datasets in instances of condition attribution and temporal grounding. A breakdown in the distributions of these annotations can be seen in Tables 1 and 2.

H&P reports may vary in the organization of content, but the content is mostly uniform, expressing the same information about patients (Sager et al., 1987). As well as this, many reports feature headings for different sections of the report (*past medical history*, *impression*), information which can be used as features in a classification task. Before annotating conditions found in the text, preprocessing was required in order to retain such information.

Table 1: Annotated Condition Attribution Occurrences

Class	Count
Patient	872
Other	93
Total	965

Table 2: Annotated Temporal Grounding Occurrences

Class	Count
Historical	448
Recent	424
Total	872

3.1.1 Preprocessing

Preprocessing of the data consisted of a simple Java program that extended Lingpipe[1] tools in order to correctly split sentences on this dataset, and extract the heading for the section in which the sentence is contained.

The preprocessing outputs the sentence number, followed by a separator, the sentence's contents and the heading under which the sentence features. Sentences were split for ease of annotation and also to allow parsing and part-of-speech tagging by the C&C[2] parsing tools. C&C was chosen for its scalabilty, speed and the accuracy of its biomedical language models. A cursory analysis of its output indicates that its performance is acceptable. As C&C does not provide a sentence splitter, Lingpipe's splitter was availed of for this task.

3.1.2 Annotation

Annotation of the dataset was performed by two annotators over a 60 hour period. The inter-annotator agreement was measured using the kappa statistic (Carletta, 1996). A kappa statistic of 0.78 was achieved. The annotators were presented with the collection, to annotate with an XML like tag "*CONDITION*". This tag must have two attributes, "*EXP*" representing condition attribution and "*HIST*"

[1] `http://alias-i.com/lingpipe/`

[2] `http://svn.ask.it.usyd.edu.au/trac/candc`

representing the temporal grounding of the condition.

- *HIST*: A value of 1 indicates the occurrence of a historical condition, where 0 describes a current or recent condition. e.g. *"The patient presented with <CONDITION NUM="1" HIST="0"> renal failure </CONDITION>"* would indicate the condition "renal failure" is current.
- *EXP*: A value of 1 implies the experiencer is the patient with 0 signifying "other". e.g. *"The patient has a family history of <CONDITION NUM="1" EXP="0">hypertension </CONDITION>"* signifies the condition "hypertension" is not experienced by the patient.

3.2 Machine Learning Algorithms

Early work in the resolution of assertion status primarily focused on the use of manually created rule-based systems, with more recent work focusing on statistical and ML methods. However, the domain of ML contains many sub-paradigms and varying approaches to classification. In this paper, three ML methods that have not been previously applied to this task are explored. These three classifiers, namely Naive Bayes, k-Nearest Neighbour and Random Forest represent the paradigms of probabilistic, lazy and ensemble learning, respectively.

Naive Bayes is a probabilistic classifier implementing Bayes Theorem. As a result, features implemented using this classifier are deemed to be independent. Despite this strong assumption it has been shown to be more than successful in text classification tasks such as spam filtering (Provost, 1999).

k-Nearest Neighbour (kNN) (Cover and Hart, 1967) is a simple pattern recognition algorithm that classifies an instance according to its distance to the *k* closest training instances. This algorithm has been chosen to represent the paradigm of lazy learning, i.e. there is no training phase as all computation is performed at the classification stage. Despite its simplicity, k-NN has often produce high accuracy results in comparison to other approaches (Caruana, 2006).

The final classifier chosen for this task represents the state-of-the-art in machine learning, namely the Random Forest algorithm (Breiman, 2001). A Random Forest consists of many different decision trees, combining bagging (Breiman, 1996), and random feature selection.

3.3 Features

In this section, a list of features contributing to this task are presented. All features are automatically extracted using a set of tools developed by the authors. Section 3.3.1 presents score-based features that are unique to this work. Section 3.3.2 describes the implementation of features outlined in Chapman et al (2007). Section 3.3.3 and Section 3.3.4 present features similar to those investigated in Mowery et al. (2009).

3.3.1 Score based features

Scored based features used in this system extend and reinforce Trigger List features by computing a normalized score for the number of occurrences of Trigger List terms[3]. This feature aims to add further granularity to the decision making of the ML algorithms, presenting a floating point number rather than a binary one.

The equation for computing these scores is defined as follows.

$$s = \frac{N_t}{(N_w - S_w)} \quad (1)$$

N_t represents the number of trigger terms found in the sentence that contains the condition, N_w is the total number of words in the sentence, with S_w being the number of stopwords[4]. These scores are calculated for each type of trigger term. For example, for trigger type *relative_mention*, a score is calculated using mentions of relatives in the sentence.

3.3.2 Trigger List Features

- `precededByHistTerm`: This feature performs a look-up for trigger terms from the historical word list, checking if it directly precedes the condition. An example historical trigger term would be "history of" as in "a history of diabetes". If a condition, such as diabetes, is modified by a historical trigger term, it will return 1, otherwise 0.
- `containsHistMention`: This is a weaker form of `precededByHistTerm`, checking simply if a trigger term from the historical list occurs in the same sentence as the condition. If one does, it will return 1 otherwise 0.
- `hasRelativeMention`: If the sentence which contains the condition also contains a trigger

[3]These trigger lists may be downloaded at `http://csserver.ucd.ie/~jcogley/downloads/wordlists.tar.gz`

[4]The list of stopwords may be downloaded at `http://csserver.ucd.ie/~jcogley/downloads/stopwords.txt`

term from the experiencer list such as 'mother', 'father' or 'uncle' it will return 1, otherwise 0.

- `hasPatientMention`: 1 if the sentence mentions the patient, otherwise 0.
- `containsDeath`: 1 if it contains the terms "deceased", "died" from the death trigger terms list otherwise 0. A sentence describing death is more likely to refer to a relative, rather than the patient.
- `mentionsCommunity`: 1 if one of "area", "community" from the geographical trigger list is mentioned, otherwise 0. If a sentence describes a community, as in "there has been a recent outbreak of flu in the area", it is not referring to the patient, therefore the condition should not be attributed to the patient.
- `precededByWith`: 1 if the condition is directly preceded by "with", otherwise 0. "with" was found to have higher frequency when describing patients rather than individuals other than the patient. e.g. "Patient presented with high blood pressure and fever."
- `containsPseudoTerms`: Pseudo-historical terms or phrases may mention a term that is found in the Historical list, but do not indicate that a condition mention in the same sentence is being used in a historical context. For example, "poor history" is a pseudo-historical trigger term. It uses a historical trigger term ("history"); however "poor history" refers to the incomplete nature of the patient's medical history, not the historical nature of their condition. This feature returns 1 on the occurrence of a pseudo trigger term, otherwise 0.

3.3.3 Contextual features

In resolving the textual context of conditions, it is important to look at what surrounds the condition beyond the lexical items. With these contextual features, we can capture that section in which a sentence occurs, and how many conditions occur in the sentence.

- `isInFamHist`: The importance of header information is motivated by the assumption that conditions that fall under explicit headings, are more than likely to have a greater affinity to the heading. This feature returns 1 if it is under *Family History*, 0 otherwise.
- `isInList`: A binary feature denoting whether a condition occurs as part of a list of conditions, with one condition per line. Returns 1 if it is a member of such a list, otherwise 0.
- `numOfConditions`: This feature represents the number of conditions present in a given sentence. A higher number of conditions indicates that the condition may be part of a list. Sentences that contain a list of conditions tend to list past conditions rather than recently suffered illnesses.

3.3.4 Linguistically motivated features

Three features were designed to monitor the effect of the verb tense on a condition. This feature has already been shown to aid the classification process (Mowery et al., 2009). For this task, linguistic features were extracted from the output of the C&C parsing tool, using both part-of-speech tags along with dependency information.

- `hasPastTense`: A binary feature with 1 indicating the sentence contains a past tense verb, 0 otherwise. e.g. "The patient previously suffered renal failure" would return 1. If a condition is modified by a past tense verb, it has occurred in the past.
- `hasPresentTense`: A binary feature with 1 indicating the sentence contains a present tense verb, 0 otherwise. If a condition is modified by a present tense verb, the condition is current. e.g. "the patient presents coughing".
- `containsModalVerb`: A binary feature with 1 indicating the sentence contains a modal verb, 0 otherwise. e.g. "palpitations may have been caused by anxiety". The presence of the modal "may" following the condition indicates the condition is currently being examined and is therefore recent.
- `tenseInClause`: Analyzes the tense found in the same syntactic clause as the condition being examined. For example, in "abdominal pain has ceased, but patient now complains of lower extremity pain", "abdominal pain" has a past tense within its clausal boundary, where the clause which contains "lower extremity pain" has a present tense verb.
- `tenseChange`: Determines whether the verb tense used in the clause that contains the condition differs with the verb in another clause in the sentence. e.g. "Migraines persist yet palpitations resolved". This feature allows finer granularity in resolving the tense surrounding conditions, such as the description of current conditions in the context of the patient's history.

4 Experiment Setup & Evaluation

There are two aims of the experiments reported in this section: firstly, to evaluate the performance of ML algorithms in resolving the assertion status of medical conditions. Secondly, to assess the performance of individual feature sets in order to discover the most contributory and informatory features or sets of features. To evaluate the latter, classifications using all possible combinations of feature sets (listed in Table 3) were performed.

Table 3: Feature-set Combinations

ID	Feature-Sets
TrigLingConScore	trigger, linguistic, score-based, contextual
TrigLingScore	trigger, linguistic, score-based
TrigLingCon	trigger, linguistic, contextual
TrigConScore	trigger, score-based, contextual
LingConScore	linguistic, score-based, contextual
TrigLing	trigger, linguistic
TrigScore	trigger, score-based
TrigCon	trigger, contextual
LingScore	linguistic, score-based
LingCon	linguistic, contextual
ConScore	score-based, contextual
Trigger	trigger
Ling	linguistic
Score	score-based
Con	contextual

4.1 Evaluation

The systems are evaluated by the metrics precision, recall and f-score:

$$precision = \frac{TP}{TP + FP}$$

$$recall = \frac{TP}{TP + FN}$$

$$f = \frac{2 \times Precision \times Recall}{Precision + Recall}$$

where TP is the number of true positives, FP is the number of false positives, FN is the number of false negatives.

Systems are evaluated using both cross-validation and hold-out methods. In the hold-out method there are two data sets, one used for training the classifier and a second for testing it on a blind sub-set of test material. 10-fold cross-validation is performed on the training sets and final results are reported in this paper on the held-out blind test set. Three hold-out classification splits were experimented with (i.e., train/test splits: 30%/70%; 50%/50%; 70%/30%). We found that results for each of the data splits and cross-validation experiments were largely uniform. To avoid repetition of results, Section 5 focuses on experiments using a held-out method training/test split of 70%/30%. All hold-out experiments were implemented using Weka's (Hall et al., 2009) Experimenter interface. Cross-Validation experiments were performed using a script developed by the authors in conjunction with Weka's API. This allowed the ML approaches and the ConText algorithm to be evaluated against the same test-folds.

4.1.1 Comparison with a rule-based system

ConText (Chapman et al., 2007) is a simple yet effective rule-based system designed to resolve the assertion status of medical conditions. Comparative analysis is performed between an implementation of ConText[5] and the ML approaches described in 3.2. The ML systems were trained on 70% of the dataset (610 conditions). The remaining 30% (262 conditions) was used as a test set for both ConText and the Machine Learning systems. For cross-validation experiments, ConText was evaluated against each test set fold. For the Condition Attribution experiments training was performed on 675 conditions with testing performed on 290 conditions. In evaluating Temporal Grounding the training set comprised of 610 conditions with the test-set containing 262 conditions.

5 Experimental Results

This section reports results of the experiments outlined in Section 4.

5.1 Condition Attribution

In a system that resolves the assertion status of medical conditions, it is of benefit to know who is experiencing the medical condition before resolving more complex information such as temporality. In this section, preliminary results on Condition Attribution are presented. The dataset used in evaluating the effectiveness of Condition Attribution was highly skewed, as shown in Table 1. This is a natural skew caused simply by the fact that reports discuss the patient more than other individuals (e.g., blood relatives). As a result the baseline score using a Majority Class classifier achieved an f-score of 95% (Table 4). Given these results, the contextual feature set contributes most, as shown by the removal of the contextual feature set in TrigLingScore coinciding with a drop in performance. However, the skewed dataset resulted in no statistical significance

[5] `http://code.google.com/p/negex/downloads/detail?name=GeneralConText.Java.v.1.0_10272010.zip`

between classifiers and feature-sets.

Table 4: Selected feature-sets (f-score) using Cross-Validation for the Condition Attribution task

ID	RFor	kNN	NB	Maj.
TrigLingConScore	100%	100%	100%	95%
TrigLingScore	96%	96%	96%	95%
TrigConScore	100%	100%	100%	95%
Con	100%	100%	100%	95%

In this task, ConText achieved an f-score of 99%. As there is little difference in scores achieved between ConText and the approaches in Table 4 - a manual rule-based system can be considered adequate for this task.

Taking a closer look at the performance of the feature sets, we see that the contextual feature set contributes most to the task with the removal of contextual features coinciding with a drop in performance e.g., TrigLingScore in Table 4. The strength of this feature set lies with the feature `isInFamHist`. This feature simply checks if the condition occurs under the heading "Family History". Its highly influential performance would indicate that its quite rare for the mention of another individual anywhere else in a clinical report. The Con run, which is solely composed of contextual features achieves near perfect performance, an indication that the contribution of other features to the task of Condition Attribution is minimal. Although this work used only H&P reports, this finding may be generalized to other report types such as Discharge Summaries which also explicitly mark sections pertaining to other individuals.

5.2 Temporal Grounding

The distribution of past and recent medical conditions is not skewed (as in the Condition Attribution task), and hence it presents a more challenging classification task. Despite the varying performance of individual classifiers and feature sets the results obtained are again largely similar across cross-validation and hold-out methods, with the performance of each training set fitting the distribution in Figure 1. Initial experiments investigated the use of another state-of-the-art classifier, the Support Vector Machine using a polykernel, however due to its relatively poor performance (70% f-score, with its precision soundly beaten by other approaches) it will not be discussed in further detail.

Random Forest proved to be the most effective classifier across almost all feature sets. However, kNN was a very near second place - Random Forest only significantly[6] outperformed kNN on two occasions (TrigLingConScore, LingConScore). In constrast, Naive Bayes performed poorly - despite having outperformed all other systems on the precision metric, it failed to outperform the baseline majority classifier on the recall.

Although the precision of ConText matches that of the Random Forest and kNN (Table 5), it is also let down by its recall performance. As a result, there is a statistical significant difference between its f-score and that of the Random Forest and kNN.

Table 5: Temporal Grounding ConText Comparison

System	Precision	Recall	F-score
`kNN`	80%	80%	80%
`RandomForest`	82%	84%	83%
`NaiveBayes`	91%	61%	72%
`ConText`	80%	61%	69%
`Majority`	55%	100%	71%

6 Discussion

The distribution of recent and historical conditions for the task of Temporal Grounding is more evenly distributed than that used in Condition Attribution, allowing for a more interesting comparison of the approaches and features employed.

Figure 1 shows the performance of each ML for each feature-set combination. Random Forest was expectedly the best performing algorithm. However, more surprising was the comparative performance of the often overlooked kNN algorithm. Both approaches statistically significantly outperformed the rule-based system ConText. Though the rule based system matched the high performing ML systems in terms of precision, it was significantly outperformed with respect to recall.

The most contributory feature set in the ML runs was the novel score-based feature set. This feature creates a normalized score for the occurrence of trigger terms in the same sentence as the medical condition in question. It was designed to reinforce the importance of trigger terms, by providing a numeric score to support the binary Trigger List feature. The addition of score-based features to any of the feature combinations coincided with a statistical significant boost in performance, with Score (composed solely of score-based features) outperforming half of all other feature combinations as seen in Figure 1,.

On the contrary, the effect of contextual features on the performance of the algorithms for Temporal

[6] Significance calculated by Paired T-Test with 95% confidence.

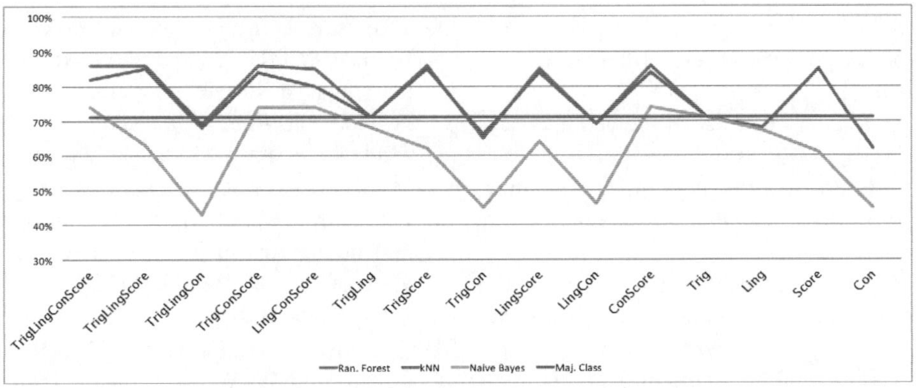

Figure 1: Temporal Grounding f-score performance with 70% Training Data

Grounding is minimal, or even detrimental to the task. For example, in Figure 1, the performance of the kNN algorithm increases from TrigLingConScore to TrigLingScore with the removal of contextual features. The performance of the Random Forest is unaffected by such detrimental features as it performs its own feature selection prior to classification. Though there are several feature set combinations reaching a high level of performance, the most effective approach combines trigger list terms, linguistic and score based features with the Random Forest algorithm.

These experiments extend previous work by providing a systematic, automated approach to feature extraction for the purpose of ML approaches to Temporal Grounding. They also indicate the high performance and contribution of our novel score-based features. These features are not designed to solely classify instances found in H&P reports and can be applied to other clinical reports such as Discharge Summaries and Emergency Department reports. Previous work has involved the use of the latter mentioned report types. Unfortunately, given the terms-of-use of these datasets they could not be made available to authors to facilitate comparative experiments.

7 Conclusion

In this paper, we proposed the use of machine learning (ML) in resolving if and when a patient experienced a medical condition. The implemented ML algorithms made use of features comprising of trigger terms, linguistic and contextual information, and novel score-based features. In an evaluation of these feature sets it was found that score-based features contributed to the task of resolving when a patient experienced a medical condition.

The ML approaches were also evaluated against the rule-based system ConText on a new annotated dataset of History & Physical (H&P) Examination Reports. In this evaluation it was discovered that the task of resolving *if a condition was experienced by the patient* was adequately solved by the ConText system, achieving an f-score of 99%. Although, the ML approaches proposed achieved perfect performance, there is no statistical significance between the result sets. However, the more challenging task of *deciding when a patient experienced a medical condition* is deemed to be best suited to a ML approach, with the top performing classifier *Random Forest* achieving an f-score of 87%, significantly outperforming ConText which achieved 69% on the same dataset .

The results achieved in these tasks have paved the way for several avenues of future work. We believe that the performance of these tasks is now sufficiently accurate to justify their inclusion in an Information Retrieval (IR) application. It is our intention to use our medical condition analysis techniques to annotate clinical documents and build an advanced IR system capable of taking advantage of this mark up in the context of the TREC Medical Records Track 2012[7]. With the availability of datasets such as that of the i2b2 Shared Task 2010 data, further work will include experimentation on these datasets as well as an investigation into further assertion statuses.

8 Acknowledgments

We are grateful to Dr Martina Naughton for her advice on many aspects of this paper. We also wish to acknowledge the support of Science Foundation Ireland, who fund this research under grant number 10/RFP/CMS2836.

[7] `http://groups.google.com/group/trec-med`

References

L. Breiman. 1996. Bagging predictors. *Machine Learning*, 24:123–140.

L. Breiman. 2001. Random forests. *Machine Learning*, 45:5–32.

J. Carletta. 1996. Assessing agreement on classification tasks: the kappa statistic. *Computational Linguistics*, 22(2):249 – 254.

R. Caruana. 2006. An empirical comparison of supervised learning algorithms. In *Proceedings of 23rd International Conference on Machine Learning*.

W. W. Chapman, D. Chu, and J. N. Dowling. 2007. Context: An algorithm for identifying contextual features from clinical text. In *BioNLP 2007: Biological, translational, and clinical language processing*, pages 81–88, June.

L. M. Christensen. 2002. Mplus: A probabilistic medical language understanding system. In *Proceedings of Workshop on Natural Language Processing in the Biomedical Domain*, pages 29–36.

C. Clark, J. Aberdeen, M. Coarr, D. Tresner-Kirsch, B. Wellner, A. Yeh, and L. Hirschman. 2011. Mitre system for clinical assertion status classification. *Journal of the American Medical Informatics Association*.

T. Cover and P. Hart. 1967. Nearest neighbor pattern classification. *Transactions on Information Theory*.

B. de Bruijn, C. Cherry, S. Kiritchenko, J. Martin, and X. Zhu. 2011. Machine-learned solutions for three stages of clinical information extraction: the state of the art at i2b2 2010. *Journal of the American Medical Informatics Association*.

D. Demner-Fushman and J. Lin. 2007. Answering clinical questions with knowledge-based and statistical techniques. In *Computational Linguistics*, pages 63–103.

D Demner-Fushman, W W. Chapman, and C J. McDonald. 2009. What can natural language processing do for clinical decision support? *Journal of Biomedical Informatics*, 42:760–772.

D. Demner-Fushman, S. Abhyankar, A. Jimeno-Yepes, R. Loane, B. Rance, F. Lang, N. Ide, E. Apostolova, and A. R. Aronson. 2011. A knowledge-based approach to medical records retrieval. In *TREC 2011 Working Notes*.

J. Feblowitz, A. Wright, H. Singh, L. Samal, and D. Sittig. 2011. Summarization of clinical information: A conceptual model. *Biomedical Informatics*.

M. Hall, F. Eibe, G. Holmes, B. Pfahringer, P. Reutemann, and I. H. Witten. 2009. The weka data mining software: An update. *SIGKDD Explorations*.

H. Harkema, J. N. Dowling, T. Thornblade, and W. W. Chapman. 2009. Context: An algorithm for identifying contextual features from clinical text. *Journal of Biomedical Informatics*, 42(5):839–851.

G. Hripcsak, L. Zhou, S. Parsons, A. K. Das, and S. B. Johnson. 2005. Modeling electronic discharge summaries as a simple temporal constraint satisfaction problem. *Journal of the American Medical Informatics Association*, 12(1):55–63, January.

S. M. Meystre, G. K. Savova, K. C. Kipper-Schuler, and J. F. Hurdle. 2008. Extracting information from textual documents in the electronic health record: a review of recent research. *IMIA Yearbook of Medical Informatics*, pages 128–144.

D. L. Mowery, H Harkema, J. N. Dowling, J. L. Lustgarten, and W. W. Chapman. 2009. Distinguishing historical from current problems in clinical reports-which textual features help? In *Proceedings of the Workshop on Current Trends in Biomedical Natural Language Processin*, pages 10–18. Association for Computational Linguistics.

J. Patrick and M. Li. 2011. An ontolotgy for clinical questions about the contents of patients notes. *Journal of Biomedical Informatics*.

J. Provost. 1999. Naive-bayes vs. rule-learning in classification of email. Technical report, The University of Texas at Austin.

N. Sager, C. Friedman, and M.S. Lyman. 1987. *Medical Language Processing: Computer Management of Narrative Data*. Addison-Wesley.

N. Sager, M. Lyman, C. Bucknall, N. Nhan, and L. J. Tick. 1994. Natural language processing and the representation of clinical data. *Journal of the American Medical Informatics Association*, 1:142–160.

O. Uzuner, X. Zhang, and T. Sibanda. 2009. Machine learning and rule-based approaches to assertion classification. *Journal of the American Medical Informatics Association*, 16(1):109–115.

Ö. Uzuner, BR. South, S. Shen, and SL. DuVall. 2011. 2010 i2b2/va challenge on concepts, assertions, and relations in clinical text. *Journal of the American Medical Informatics Association*.

E. Voorhees and R. Tong. 2010. Overview of the trec 2011 medical records track. preprint.

Alignment-HMM-based Extraction of Abbreviations from Biomedical Text

Dana Movshovitz-Attias
Carnegie Mellon University
5000 Forbes Avenue
Pittsburgh, PA 15213 USA
dma@cs.cmu.edu

William W. Cohen
Carnegie Mellon University
5000 Forbes Avenue
Pittsburgh, PA 15213 USA
wcohen@cs.cmu.edu

Abstract

We present an algorithm for extracting abbreviation definitions from biomedical text. Our approach is based on an alignment HMM, matching abbreviations and their definitions. We report 98% precision and 93% recall on a standard data set, and 95% precision and 91% recall on an additional test set. Our results show an improvement over previously reported methods and our model has several advantages. Our model: (1) is simpler and faster than a comparable alignment-based abbreviation extractor; (2) is naturally generalizable to specific types of abbreviations, *e.g.*, abbreviations of chemical formulas; (3) is trained on a set of unlabeled examples; and (4) associates a probability with each predicted definition. Using the abbreviation alignment model we were able to extract over 1.4 million abbreviations from a corpus of 200K full-text PubMed papers, including 455,844 unique definitions.

1 Introduction

Abbreviations and acronyms are commonly used in the biomedical literature for names of genes, diseases and more (Ambrus, 1987). Abbreviation definitions are a source of ambiguity since they may change depending on the context. The ability to recognize and extract abbreviations and map them to a full definition can be useful for Information Extraction tasks (Yu et al., 2007) and for the complete understanding of scientific biomedical text.

Yu et al. (2002) distinguish the two following uses of abbreviations: (1) *Common* abbreviations are those that have become widely accepted as synonyms, such as ⟨DNA, deoxyribonucleic acid⟩ or ⟨AIDS, acquired immunodeficiency syndrome⟩. These represent common fundamental and important terms and are often used, although not explicitly defined within the text (Fred and Cheng, 2003). In contrast, (2) *Dynamic* abbreviations, are defined by the author and used within a particular article. Such definitions can often overlap, depending on the context. For example, the term PBS most commonly abbreviates *Phosphate Buffered Saline*, but in other contexts may refer to the following: Pain Behavior Scale, Painful Bladder Syndrome, Paired Domain-Binding Site, Particle Based Simulation, Partitioned Bremer Support, Pharmaceutical Benefits Scheme, and more. Some abbreviations fall between these two definitions in the sense that they are normally defined in the text, however, they have become widely used, and therefore they do not normally overlap with other abbreviations. An example for this is the term ATP which, almost exclusively, abbreviates *adenosine triphosphate*, and is only rarely used in different contexts in biomedicine.

Gaudan et al. (2005) define two similar concepts, distinguishing *Global* and *Local* abbreviations. Global abbreviations are not defined within the document, similar to common abbreviation. Local abbreviations appear in the document alongside the long form, similar to dynamic abbreviations. The contextual ambiguity of dynamic, or local, abbreviations makes them an important target for abbreviation recognition tasks.

There is a great deal of variation in the way that different authors produce abbreviations. Our definition of *abbreviation* is quite flexible and can best be represented by the set of examples described in Table 1. These include simple acronyms, in which the first letter of every word from the long form is represented in the short form, as well as more complex cases such as: inner letter matches, missing short form characters, and specific substitutions (such as of a chemical element and its symbol). We generally assume that the abbreviated form contains some contraction of words or phrases from the full form. This definition is consistent with the one defined by many other extraction systems (see *e.g.*, (Schwartz and Hearst, 2002) and (Chang et al., 2002)).

We describe a method for extracting dynamic abbreviations, which are explicitly defined in biomedical abstracts. For each of the input texts, the task is to identify and extract ⟨*short form, long form*⟩ pairs of the abbreviations defined within the text. We also provide a mapping, formed as an alignment, between the characters of the two forms, and the probability of this alignment according to our model.

Our approach is based on dividing the abbreviation recognition task into the following stages: (1) Parsing the text and extracting *candidate* abbreviation pairs (long and short forms) based on textual cues, such as parentheses; (2) Recovering a valid alignment between the short and long form candidates (valid alignments are defined in Section 3.2). We perform a sequential alignment based on a pair-HMM; (3) Extracting a final short and long form from the alignment.

We will show that our approach is fast and accurate: we report 98% precision and 93% recall on a standard data set, and 95% precision and 91% recall on a validation set. The alignment model: (1) is simpler and faster than a comparable alignment-based abbreviation extractor; (2) is naturally generalizable to specific types of abbreviations; (3) is trained on a set of unlabeled examples; and (4) associates a probability with each predicted definition.

2 Related Work

A wide variety of methods have been introduced for recognizing abbreviations in biomedical context. Many utilize one of the following techniques: rule-based extraction, and extraction that relies on an alignment of the abbreviation and full definition. Abbreviation extraction methods have been used in two main contexts: to create online collections of abbreviations, normally extracted from PubMed abstracts (Zhou et al., 2006; Gaudan et al., 2005; Adar, 2004), and as part of larger learning frameworks, mainly for feature generation (Chowdhury et al., 2010; Huang et al., 2011).

Rule based extraction systems use a set of manually crafted pattern-matching rules to recognize and extract the pair of abbreviation and definition: Acrophile (Larkey et al., 2000) is an acronym recognition system that exploits morphological rules based on the case of the characters in the definitions. Unlike many of the other available systems, it recognized acronyms that are defined without parentheses; The Alice system (Ao and Takagi, 2005) is based on three extraction phases, each employing an elaborate set of over 15 rules, patterns and stop word lists. Liu and Friedman (2003) use a set of statistical rules to resolve cases in which an abbreviation is defined more than once with several different definitions. While these methods normally achieve high performance results, their main drawback is that they are difficult to implement and to extend. Rule development is normally based on a thorough investigation of the range of targeted abbreviations and the resulting heuristic patterns contain subtleties that are hard to recreate or modify.

Several extraction methods have been developed based on some variant of the Longest Common Subsequence algorithm (LCS) (Schwartz and Hearst, 2002; Chang et al., 2002; Taghva and Gilbreth, 1999; Bowden et al., 1997). These systems search for at least one possible alignment of an abbreviation and a full form definition.

The most widely used abbreviation extraction system is that presented by Schwartz and Hearst (2002). Their method scans the input text and extract pairs of candidate abbreviations from text surrounding parentheses. The algorithm scans the candidate definition from right to left, and searches for an implicit alignment of the definition and abbreviation based on few ad-hoc rules. This algorithm presents several constraints on the type of recognized abbreviations, the most restrictive being that every letter of the abbreviation must be matched during the process of scanning the definition. Of the variety of available extraction systems, this remains a popular choice due to its simplicity and speed. However, as the au-

Short	Long	Type of Abbreviation
AMS	Associated Medical Services	Acronym using the first letter of each long-form word.
PS	postsynaptic	Inner letters are represented in the abbreviation.
NTx	cross-linked N-telopeptides	1. Phonetic substitution (*cross* → *x*). 2. The short form is out-of-order. 3. Words from the long form are missing in the short form (*linked*).
EDI-2	Eating Disorders Inventory	Characters from the short form are missing in the long form (*-2*).
NaB	sodium butyrate	Substitution of a chemical element by its symbol (*sodium* → *Na*).
MTIC	5-(3-N-methyltriazen-1-yl)-imidazole-4-carboxamide	Chemical formula.
EBNA-1	Epstein-Barr virus (EBV) nuclear antigen 1	Recursive definition, in which the long form contains another abbreviation definition.
3-D	three-dimensional	Substitution of a number name and symbol (*three* → *3*).
A&E	accident and emergency	Substitution of a word and symbol (*and* → &).
anti-Tac	antibody to the alpha subunit of the IL-2 receptor	Synonym: the short form commonly represents the long form, although it is not a direct abbreviation of it.
R.E.A.L.	'Revised European-American Classification of Lymphoid Neoplasms'	The long- and/or short-forms contain characters that are not directly related to the abbreviation (*e.g.*, punctuation symbols).

Table 1: Examples of biomedical abbreviations.

thors report, this algorithm is less specific than other approaches and consequently results in lower recall. We will show that by performing an explicit alignment of the abbreviation using an alignment-HMM, our model results in more accurate predictions, and that the edit operations used in the alignment allow for natural extensions of the abbreviations domain.

Another frequently used alignment based approach is that of Chang *et al.* (2002), and it is closest to our approach. After calculating an abbreviation alignment, they convert the set of aligned terms into a feature vector which is scored using a binary logistic regression classifier. Using a correct threshold on the alignment scores produces a high performance abbreviation extractor. However this approach has several drawbacks. The run-time of this algorithm is fairly long (see Section 4.3), in part due to the steps following the alignment recovery, *i.e.*, calculating a feature vector, and generating an alignment score. Additionally, choosing a score threshold may depend on the genre of text, and different thresholds lead to a variety of quality in the results. We will show that presenting limitations on the range of available alignments can produce correct alignments more efficiently and quickly, maintaining high quality results, without the need for threshold selection. Our alignment method distinguishes and penalizes inner and leading gaps in the alignment, and it applies a set of constraints on the range of legal alignments. We will also show that relying solely on constrained alignments still allows for flexibility in the definition of the range of desired abbreviations.

Ristad and Yianilos (1998) proposed a single state alignment-HMM for learning string-edit distance based on matched strings. In later work, Bilenko and Mooney (2003) extend this model to include affine gaps, by including in their model separate states for Matches, Deletions and Insertions. McCallum et al. (2005) describe a discriminative string edit CRF, following a similar approach to that of Bilenko and Mooney. The CRF model includes two disjoint sets of states, each representing either "matching" or "mismatching" string pairs. Each of the sets is similar to the model described by Bilenko and Mooney. All of these models require labeled training examples, and the CRF approach also requires negative training examples, which train the "mismatching" states of the model. We describe an alignment HMM that is suited for aligning abbreviation long and short forms, and does not require any labeling of the input text or training examples.

3 Method

In the following sections we describe a method for extracting *candidate* abbreviation definitions from text, and an alignment model with affine gaps for

Description	Result
i. Input sentence:	"anti-sperm antibodies were studied by indirect mixed anti-globulin reaction test (MAR)"
ii. Candidate:	⟨MAR, by indirect mixed anti-globulin reaction test⟩
iii. Alignment:	

HMM States	`LG`	`LG`	`LG`	`LG`	`M`	`M`	`M`	`M`	`IG`	`M`	`M`	`M`	`IG`
Short Form					`M`		`A`				`R`		
Long Form	`by`		`indirect`		`mixed`		`anti`	`-`	`globulin`		`reaction`		`test`

iv. Abbreviation:	⟨MAR, mixed anti-globulin reaction test⟩

Table 2: Example of the processing steps of a sample sentence. (i) Input sentence containing a single abbreviation. (ii) Candidate ⟨short form, long form⟩ pair extracted from the sentence (after truncating the long-form). (iii) The most likely (Viterbi) alignment of the candidate pair, using our alignment model. Each state corresponds to a single edit-operation, which consumed the corresponding short-form and long-form characters in the alignment. (iv) Final abbreviation, extracted from the alignment by removing leading gaps.

matching the two forms of a candidate definition. Finally we describe how to extract the final abbreviation prediction out of the alignment.

3.1 Extracting candidate abbreviations

The process described below scans the text for textual cues and extracts a list of candidate abbreviation pairs, for every input document, in the form: ⟨*short form, long form*⟩. The following text also describes the restrictions and conditions of what we consider to be valid candidate pairs. The assumptions made in this work are generally less restrictive that those introduced by previous extraction systems and they lead to a larger pool of candidate definitions. We will later show that false candidates normally produce invalid alignment of their short and long forms, according to our alignment model, and so they are removed and do not affect the final results.

The parsing process includes a search for both single abbreviations, and abbreviation patterns. An example of a sentence with a single abbreviation can be seen in Table 2(i). We consider the following two cases of a single abbreviation definition: (1) "long form (short form)", and (2) "short form (long form)". Note that in some cases, the term within the parenthesis is parsed, *e.g.*, in the following text, *ELISA* is extracted from the parenthesis, by removing the text beyond the ';' symbol: "...human commercial enzyme-linked immunosorbent assay (ELISA; BioGen, Germany) ...".

We also consider *abbreviation patterns* which define multiple abbreviations simultaneously, as demonstrated by these examples:

- "anti-sperm (ASA), anti-phospholipid (APA), and antizonal (AZA) antibodies" – The main noun (*antibodies*) follows the pattern.
- "Epithelial-mesenchymal transition (EMT) and interaction (EMI)" – The main noun (*Epithelial-mesenchymal*) is at the head of the pattern.

Using textual cues (patterns and parentheses) we extract candidate short and long forms. Whenever possible, we consider the term within the parenthesis as the short form, and the text to the left of the parenthesis (until the beginning of the sentence) as the candidate long form. We consider valid short forms to be no longer than 3 words, having between 1 and 15 characters, and containing at least one letter. In the case that the candidate short form was found to be invalid by these definitions, we switch the assignment of long and short forms. The long-form string is truncated, following Park and Byrd (2001), to a length of $\min(|A| + 5, |A| * 2)$, where $|A|$ is the length of the short form.

The length of the candidate long form is estimated using the Park and Byrd formula, and it is therefore normally the case that the resulting candidate long form contains some leading characters that are not part of the abbreviation definition. Next, we define an alignment between short and long form strings

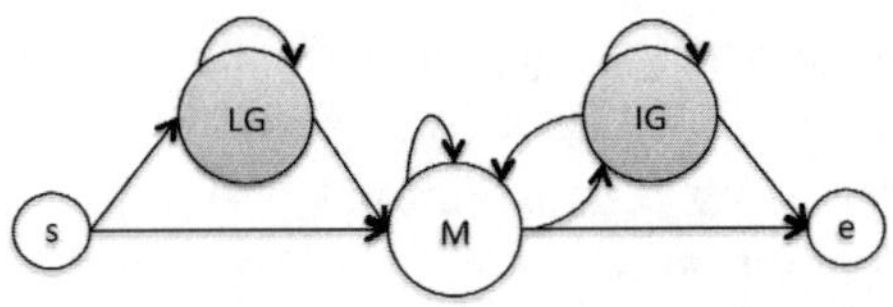

Figure 1: Abbreviation alignment HMM model with states: start (s), leading gaps (LG), match (M), inner gap (IG) and end (e).

Edit Operation	SF Match	LF Match	Valid States
LF deletion	ϵ	alpha-numeric char	LG, IG
LF deletion	ϵ	punct. symbol	LG, M
LF deletion	ϵ	word	LG, IG
SF deletion	digit or punct.	ϵ	IG
Match	char	(partial) word	M
Match	char	char	M
Substitution	'&'	'and'	M
Substitution	'1'-'9'	'one'-'nine'	M
Substitution	chem. symbol	chemical name	M

Table 3: Edit operations used in the alignment HMM model including, long form (LF) and short form (SF) deletions, matches and substitutions. We note the SF and LF characters consumed by each edit operation, and the HMM states in which it may be used.

which detects possible segments that are missing in the alignment in either string (gaps).

3.2 Aligning candidate long and short forms

For each of the candidate pairs produced in the previous step, we find the best alignment (if any) between the short and the long form strings. We describe an alignment HMM that is suited for abbreviation alignments. The model is shown in Figure 1, and Table 2 shows the parsing process of a sample sentence, including an alignment created for this sample using the model.

3.3 Abbreviation Alignment with Affine Leading and Inner Gaps

An alignment between a long and a short form of an abbreviation can be modeled as a series of edit operations between the two strings, in which characters from the short form may match a single or a series of characters from the long form. In previous work, Bilenko and Mooney (2003) describe a generative model for string edit distance with affine gaps, and an Expectation Maximization algorithm for learning the model parameters using a labeled set of matching strings. We propose a similar model for aligning the short and long form of an abbreviation, using an affine cost model for gaps

$$cost(g) = s + e \cdot l \tag{1}$$

where s is the cost of starting a gap, e is the cost of extending a gap and l is the length of the gap. In our method, we use extracted candidate pairs (candidate short and long forms) as training examples.

As described above, candidate long forms are formed by extracting text preceding parentheses and truncating it to some length. This process may lead to candidate long forms that contain leading characters that do not belong to the abbreviation, which will result in leading gaps in the final alignment. For example, the candidate long form presented in Table 2(ii) contains the leading text "by indirect ". While extra leading text is expected as an artifact of our candidates extraction method, inner alignment gaps are not expected to commonly appear in abbreviation alignments, and are usually an indication of a bad alignment. The example presented in Table 2 is of an abbreviation that *does* contain inner gaps (*e.g.*, *globulin*) despite being a valid definition.

We distinguish leading and inner alignment gaps using a model with five states: Leading Gap (LG), Match (M), Inner Gap (IG), and two "dummy" states for the beginning and end of an alignment (Figure 1). Since leading and inner gaps are represented by different states, their penalization is not coupled, *i.e.*, they are associated with different s, e and l costs. We use the EM algorithm to learn the model parameters, based on a set of unlabeled candidate pairs, following the assumption that many false-candidates will not produce a valid alignment, and will not affect training. This is in contrast to previous string edit distance models, which require labeled training examples.

The main effort in developing a successful abbreviation alignment model involves generating a meaningful set of edit operations. The edit operations used in our model, $E = E_d \cup E_m \cup E_s$, is shown in Table 3 and includes: E_d, deletions of characters or words from the long form, or of single characters from the short form; E_m, matches of a full of partial word from the long form to a character in the short form; and E_s, word substitutions in which a word from the long form is replaced by a symbol in the short form. Note that: (1) while all types of deletions from the long form are valid, deletions from the short form are limited to digits and punctuation symbols, and (2) deletion of non-alpha-numeric characters from the long form is not considered as opening a gap but as a match, as it is common for non-alpha-numeric characters to be missing in an abbreviation (*i.e.*, be "matched" with the empty string, ϵ).

Let $x = x_1 \ldots x_T$ be the short form candidate, $y = y_1 \ldots y_V$ be the long form candidate, and $a = \langle a_p \rangle_{p=1}^{n}$, $a_p = (e_p, q_p, ix_p, jy_p)$, be a possible alignment of the strings x and y. a represents as a sequence of HMM transitions, a_p, where $e_p \in E$ is an edit operation that consumes characters from x (deletion from the long form), y (deletion from the short form), or both (match or substitution), up to position ix_p in x and jy_p in y, and is associated with a transition in the model to state $q_p \in \{\text{LG}, \text{M}, \text{IG}, \text{e}\}$. Let $\pi(q, q')$ be the transition probability between states q and q', and let $\tau(q, e)$ be the emission probability of the edit operation e at state q. Given a candidate abbreviation pair $\langle x, y \rangle$, and the model parameters π and τ, the probability of an alignment is given by

$$p(a|x, y, \pi, \tau) = \prod_{p=1}^{|a|} \pi(q_{p-1}, q_p) \cdot \tau(q_p, e_p) \quad (2)$$

where q_0 is the start state. This probability can be calculated efficiently using dynamic programming with the forward-backward algorithm, and the most likely alignment corresponds to the Viterbi distance between x and y.

In our method, the model parameters, π and τ, are estimated using the EM algorithm on an unlabeled training set of candidate pairs that have been extracted from the text, without any further processing. At each EM iteration, we train on pairs that have valid alignments (see below) with non-zero probability under the model parameters at that iteration.

3.3.1 Valid Alignments

Given the edit operations defined above, the only valid way of matching a letter from the short form to the long form is by matching that letter to the beginning of a full or partial word, or by matching that letter using a substitution operation. There is no edit operation for deleting letters from the short form (only digits and punctuation symbols can be deleted). This means that for some candidate pairs there are no valid alignments under this model, in which case, no abbreviation will be predicted.

3.3.2 Extracting the Final Abbreviation

Given a valid alignment a between the candidate pair, x and y, we create a truncated alignment, a', by removing from a initial transitions in which $q_p = \text{LG}$. We consider a' valid if the number of matches in $a' = \langle a'_p \rangle_{p=1}^{n'}$ is greater than the number of deletions,

$$\sum_{p=1}^{n'} I(q'_p = \text{M}) > \sum_{p=1}^{n'} I(q'_p = \text{IG}) \quad (3)$$

where I is an indicator function.

The final abbreviation prediction is given by the portions of the x and y strings that are associated with a', named x' and y', respectively. These may be truncated compared to x and y, as leading alignment gaps are removed. The final alignment probability is given by $p(a'|x', y', \pi, \tau)$.

3.4 Substitution Edit Operations

In contrast to rule-based extraction algorithms, in our model, it is easy to introduce new types of edit operations, and adjust the model to recognize a variety of abbreviation types. As an example, we have added a number of substitution operations (see Table 3), including an operation for the commonly used convention of replacing a chemical element name (*e.g.*, *Sodium*) with its symbol (*Na*). These types of operations are not available using simpler models, such as that presented by Schwartz and Hearst (2002), making it impossible to recognize some important biomedical entities, such as chemical compounds (*e.g.*, $\langle NaB, SodiumButyrate \rangle$). In contrast, such additions are natural in our model.

4 Evaluation

4.1 Abbreviation Extraction Analysis

We evaluated the alignment abbreviation model over two data sets (Table 4). The method was tuned using

Data Set	Name	Abstracts	Abbreviations	Testing Method
Development (D)	Medstract	400	483	10-fold cross validation.
Validation (V)	PubMed Sample	50	76	Training on set D and testing on set V.

Table 4: Evaluation Data Sets.

Model	D (average %)			V (%)		
	P	R	F1	P	R	F1
Alignment HMM	98	**93**	**96**	95	91	**93**
SH	96	88	91	**97**	83	89
Chang $_{0.88}$	**99**	46	62	**97**	47	64
Chang $_{0.14}$	94	89	91	95	91	**93**
Chang $_{0.03}$	92	91	91	88	**93**	90
Chang $_{0}$	49	92	64	53	**93**	67

Table 5: Results on validation (V) and development (D) sets. Average results are shown for D set, which was tested using 10-fold cross-validation (results rounded to nearest percent, all standard deviations were < 0.1)

10 fold cross-validation over the publicly available Medstract corpus (Pustejovsky et al., 2002) which includes 400 Medline abstracts. The online version of the corpus was missing the Gold Standard annotations throughout the development of our algorithm, nor was it possible to get them through communication with the authors. We therefore hand-annotated the Medstract data, yielding 483 abbreviation definitions in the form of ⟨*short form*, *long form*⟩ pairs. In order to be consistent with previous evaluations over Medstract, our annotations include only definitions in which either the short or the long form appear in parenthesis, and it is assumed that there are no trailing gaps in the term preceding the parenthesis, although our model does detect such gaps.

We compare our results with two algorithms available for download: the Schwartz and Hearst (SH; (2002)) algorithm[1], and the Chang *et al.* (2002) algorithm[2] used at three score cutoffs reported in their paper (0.88, 0.14, 0.03). We also use a fourth score cutoff of 0 to account for any legal alignments produced by the Chang model.

In Table 5 we report precision (P), recall (R) and F1 scores for all methods, calculated by

$$P = \frac{\text{correct predicted abbreviations}}{\text{all predicted abbreviations}} \quad (4)$$

$$R = \frac{\text{correct predicted abbreviations}}{\text{all correct abbreviations}} \quad (5)$$

On the development set, our alignment model achieves 98% precision, 93% recall and 96% F1 (average values over cross-validation iterations, with standard deviations all under 0.03).

To test the final model we used a validation dataset consisting of 50 abstracts, randomly selected out of a corpus of 200K full-text biomedical articles taken from the PubMed Central Open Access Subset (extracted in October 2010)[3]. These were hand-annotated, yielding 76 abbreviation definitions.

On the validation set, we predicted 69 out of 76 abbreviations, with 4 false predictions, giving 95% precision, 91% recall and 93% F1. Our alignment model results in higher F1 score over all baselines in both datasets (with $\text{Chang}_{0.14}$ giving equal results on the validation set). Our results are most comparable with the Chang model at a score cutoff of 0.14, though our model does not require selecting a score cutoff, and as we will show, it is considerably faster. Interestingly, our model results in lower recall than precision on both data sets. This may be due to a limited scope of edit operations.

In order to evaluate the usability of our method, we used it to scan the 200K full-text documents of the PubMed Central Open Access Subset corpus. The process completed in under 3 hours, yielding over 1.4 million abbreviations, including 455,844 unique definitions. A random sample of the extracted abbreviations suggests a low rate of false positive predictions.

4.2 Error Analysis

Our model makes 4 incorrect predictions on the validation set, 3 of which are partial matches to the

[1] Taken from http://biotext.berkeley.edu/software.html

[2] Taken from http://abbreviation.stanford.edu

[3] http://www.ncbi.nlm.nih.gov/pmc/

Description	D	V
Letters in short form are missing (*e.g.*, ⟨GlyRalpha2, glycine alpha2⟩)	5	3
Abbreviation missed due to extraction rules.	6	1
Abbreviation is a synonym (*e.g.*, ⟨IRX-2, natural cytokine mixture⟩)	5	1
Abbreviation letters are out-of-order (*e.g.*, ⟨VSV-G, G glycoprotein of vesicular stomatitis virus⟩)	4	1
Correct alignment was found but it is invalid due to many inner gaps (see Section 3.3.1).	5	0
Abbreviations of chemical formulas or compounds.	4	0

Table 6: Abbreviations missed in development (D) and validation (V) sets.

correct definitions, *e.g.*, we predict the pair ⟨GlOx, glutamate oxidase⟩ instead of ⟨GlOx, L-glutamate oxidase⟩. On the development set, 3 out of 5 incorrect predictions are partial matches.

Our model did not extract 7 of the abbreviations from the validation set and 33 from the development set. Many of these abbreviations (6 from the validation set and 29 from the development set) had one of the properties described in Table 6. The remaining 5 definitions have been missed due to miscellaneous issues. Note that while we added several substitution operations for chemical formula recognition, the elaborate set of operations required for recovering the full range of chemical formulas was not included in this work, leading to 4 chemical formula abbreviations being missed.

4.3 Run-Time Analysis

We provide an estimated comparison of the run time of our method and the baseline algorithms. This analysis is especially interesting for cases in which an abbreviation extraction model is included within a larger learning framework (Chowdhury et al., 2010; Huang et al., 2011), and may be used in it in an online fashion. Run time was evaluated on an Apple iMac with 4GB 1333 MHz RAM, and a 3.06 GHz Core i3, double-core processor, by running all models on a random set of 400 abstracts. In order to evaluate the run time contribution of the substitution operations introduced in our model we ran it both with (88 $\frac{\text{docs}}{\text{sec}}$) and without (98 $\frac{\text{docs}}{\text{sec}}$) the use of substitution operations. We find that using substitutions did not have considerable effect on run time, adding under 1 ms for processing each document. We should note that the performance of the substitution-less model on this test data was similar to that of the original model, as substitutions were relevant to only a smaller portion of the abbreviations. As expected, the SH algorithm is considerably faster (6451 $\frac{\text{docs}}{\text{sec}}$) than our model, as it is based on only a number of simple rules. The Chang model, however, is slower (4 $\frac{\text{docs}}{\text{sec}}$) as it includes processing steps following the discovery of an abbreviation alignment, which means that our model provides comparable results to the Chang model and runs an order-of-magnitude faster.

5 Conclusions and Discussion

We presented a method for extracting abbreviation definitions with high precision and high recall (95% precision, 91% recall and 93% F1 on a validation set). Our model achieves higher F1 on both the development and validation data sets, when compared with two popular extraction methods.

Our approach is based on a sequential generative model, aligning the short and long form of an abbreviation. Using the proposed method we extracted 1.4 million abbreviations from a corpus of 200K PubMed articles. This data can be valuable for Information Extraction tasks and for the full understanding of biomedical scientific data.

The alignment abbreviation extractor can be easily extended by adding edit-operations over short and long forms. This was demonstrated by including substitutions of chemical elements and their symbols, which facilitates recognition of chemical formulas and compounds.

We have identified the main classes of abbreviation definitions missed by our approach. These include out-of-order matches, synonym-like abbreviations, and short forms with excess letters. It may be possible to address some of these issues by including "global" information on abbreviations, such as the occurrence of frequent definitions.

Acknowledgments

This work was funded by grant 1R101GM081293 from NIH, IIS-0811562 from NSF and by a gift from Google. The opinions expressed in this paper are solely those of the authors.

References

E. Adar. 2004. Sarad: A simple and robust abbreviation dictionary. *Bioinformatics*, 20(4):527–533.

JL Ambrus. 1987. Acronyms and abbreviations. *Journal of medicine*, 18(3-4):134.

H. Ao and T. Takagi. 2005. Alice: an algorithm to extract abbreviations from medline. *Journal of the American Medical Informatics Association*, 12(5):576–586.

M. Bilenko and R.J. Mooney. 2003. Adaptive duplicate detection using learnable string similarity measures. In *Proceedings of the ninth ACM SIGKDD international conference on Knowledge discovery and data mining*, pages 39–48. ACM.

P.R. Bowden, P. Halstead, and T.G. Rose. 1997. Dictionaryless english plural noun singularisation using a corpus-based list of irregular forms. *LANGUAGE AND COMPUTERS*, 20:339–352.

J.T. Chang, H. Schütze, and R.B. Altman. 2002. Creating an online dictionary of abbreviations from medline. *Journal of the American Medical Informatics Association*, 9(6):612–620.

M. Chowdhury, M. Faisal, et al. 2010. Disease mention recognition with specific features. In *Proceedings of the 2010 Workshop on Biomedical Natural Language Processing*, pages 83–90. Association for Computational Linguistics.

H.L. Fred and T.O. Cheng. 2003. Acronymesis: the exploding misuse of acronyms. *Texas Heart Institute Journal*, 30(4):255.

S. Gaudan, H. Kirsch, and D. Rebholz-Schuhmann. 2005. Resolving abbreviations to their senses in medline. *Bioinformatics*, 21(18):3658–3664.

M. Huang, J. Liu, and X. Zhu. 2011. Genetukit: a software for document-level gene normalization. *Bioinformatics*, 27(7):1032–1033.

L.S. Larkey, P. Ogilvie, M.A. Price, and B. Tamilio. 2000. Acrophile: an automated acronym extractor and server. In *Proceedings of the fifth ACM conference on Digital libraries*, pages 205–214. ACM.

H. Liu, C. Friedman, et al. 2003. Mining terminological knowledge in large biomedical corpora. In *Pac Symp Biocomput*, pages 415–426.

A. McCallum, K. Bellare, and F. Pereira. 2005. A conditional random field for discriminatively-trained finite-state string edit distance. In *Conference on Uncertainty in AI (UAI)*.

Y. Park and R.J. Byrd. 2001. Hybrid text mining for finding abbreviations and their definitions. In *Proceedings of the 2001 conference on empirical methods in natural language processing*, pages 126–133.

J. Pustejovsky, J. Castano, R. Sauri, A. Rumshinsky, J. Zhang, and W. Luo. 2002. Medstract: creating large-scale information servers for biomedical libraries. In *Proceedings of the ACL-02 workshop on Natural language processing in the biomedical domain-Volume 3*, pages 85–92. Association for Computational Linguistics.

E.S. Ristad and P.N. Yianilos. 1998. Learning string-edit distance. *Pattern Analysis and Machine Intelligence, IEEE Transactions on*, 20(5):522–532.

A.S. Schwartz and M.A. Hearst. 2002. A simple algorithm for identifying abbreviation definitions in biomedical text. In *Pacific Symposium on Biocomputing 2003: Kauai, Hawaii, 3-7 January 2003*, page 451. World Scientific Pub Co Inc.

K. Taghva and J. Gilbreth. 1999. Recognizing acronyms and their definitions. *International Journal on Document Analysis and Recognition*, 1(4):191–198.

H. Yu, G. Hripcsak, and C. Friedman. 2002. Mapping abbreviations to full forms in biomedical articles. *Journal of the American Medical Informatics Association*, 9(3):262–272.

H. Yu, W. Kim, V. Hatzivassiloglou, and W.J. Wilbur. 2007. Using medline as a knowledge source for disambiguating abbreviations and acronyms in full-text biomedical journal articles. *Journal of biomedical informatics*, 40(2):150–159.

W. Zhou, V.I. Torvik, and N.R. Smalheiser. 2006. Adam: another database of abbreviations in medline. *Bioinformatics*, 22(22):2813.

Medical diagnosis lost in translation – Analysis of uncertainty and negation expressions in English and Swedish clinical texts

Danielle L Mowery
University of Pittsburgh
200 Meyran Ave
Pittsburgh, PA 15260
dlm31@pitt.edu

Sumithra Velupillai
Stockholm University
164 40 Kista
Stockholm, Sweden
sumithra@dsv.su.se

Wendy W Chapman
University of California San Diego
10100 Hopkins Dr
La Jolla, CA 92093
wwchapman@ucsd.edu

Abstract

In the English clinical and biomedical text domains, negation and certainty usage are two well-studied phenomena. However, few studies have made an in-depth characterization of uncertainties expressed in a clinical setting, and compared this between different annotation efforts. This preliminary, qualitative study attempts to 1) create a clinical uncertainty and negation taxonomy, 2) develop a translation map to convert annotation labels from an English schema into a Swedish schema, and 3) characterize and compare two data sets using this taxonomy. We define a clinical uncertainty and negation taxonomy and a translation map for converting annotation labels between two schemas and report observed similarities and differences between the two data sets.

1 Introduction and Background

Medical natural language processing techniques are potentially useful for extracting information contained in clinical texts, such as emergency department reports (Meystre et al., 2008). One important aspect to take into account when developing accurate information extraction tools is the ability to distinguish negated, affirmed, and uncertain information (Chu et al., 2006). Several research studies have targeted this problem and created annotation schemas and manually annotated reference standards for uncertainty and negation occurrence in news documents (Saurí and Pustejovsky (2009), Wiebe et al. (2001), Rubin et al. (2006)), biomedical research articles (Wilbur et al. (2006), Vincze et al. (2008)), and clinical narratives (Uzuner et al. (2011) and Uzuner et al. (2009)). There are encoding tools developed for automatic identification of uncertainty and negation in English, such as ConText (Harkema et al., 2009), which relies on heuristics and keyword lists, and MITRE's CARAFE (Clark et al., 2011), which combines heuristic and statistical techniques.

However, most relevant annotation schemas, reference standards, and encoding tools are built for English documents. For smaller languages, such as Swedish, resources are scarce.

We present a pilot, qualitative study to compare two different annotation schemas and subsequent annotated corpora for uncertainty modeling of disorder mentions, e.g., signs, symptoms, and diseases, in clinical texts, for two different languages: English and Swedish. We compare these annotation schemas and their instantiation in the two languages in an attempt to gain a deeper understanding of how uncertainty and negation are expressed in different clinical texts with an emphasis on creating a portable uncertainty and negation application that generalizes among clinical texts of different languages.

This pilot study is motivated for at least two reasons. First, little attention has been given to mapping, characterizing, or comparing annotation schemas built for different languages or to characterizing different types of uncertainty expressions and the intention underlying those expressions. Such knowledge is needed for building information extraction tools that can accurately identify or track differential diagnoses over time, particularly when medical reasoning can be laden with uncertainty about a disorder's existence or change over time.

Second, building new resources for small languages is time consuming. Utilizing existing tools and techniques already developed for one language, such as English, could be an efficient way of developing new useful tools for other less exploited languages, such as Swedish.

Our overall goal is to move towards improving automatic information extraction from clinical texts by leveraging language differences and similarities. In order to address this issue, our aims in this study are to 1) create a taxonomy for deepened characterization of how uncertainty and negation is expressed in clinical texts, 2) compare two existing uncertainty and negation annotation schemas from this perspective, and 3) compare differences and similarities in expressions of uncertainty and negation between two languages: English and Swedish.

2 Methods

In this pilot, qualitative comparison study, we used grounded theory (Strauss and Corbin, 1990) to inductively identify themes that characterize clinical uncertainty and negation expressed in both English (University of Pittsburgh Medical Center) and Swedish (Karolinska University Hospital) research data sets derived from emergency department reports.

2.1 Uncertainty/negation annotation schemas

Two independently developed annotation schemas were used to annotate disorder mentions in the clinical texts: a schema developed for English reports (Mowery et al. (2012)) and one for Swedish (Velupillai et al. (2011)). Each disorder mention was pre-annotated and constituted the input to a separate set of annotators, who assigned values to a set of attributes defined in the schema. For instance, in the sentence "Patient with possible pneumonia.", annotators for the English data set assigned **values** to four attributes for the instance of pneumonia:

- Existence(**yes**, no): whether the disorder was ever present
- AspectualPhase(initiation, continuation, culmination, **unmarked**): the stage of the disorder in its progression
- Certainty(low, **moderate**, high, unmarked): amount of certainty expressed about whether the disorder exists
- MentalState(yes, **no**): whether an outward thought or feeling about the disorder's existence is mentioned

In the Swedish schema, annotators assigned **values** to two attributes:

- Polarity(**positive**, negative): whether a disorder mention is in the positive or negative polarity, i.e., affirmed (positive) or negated (negative)
- Certainty(**possibly**, probably, certainly): gradation of certainty for a disorder mention, to be assigned with a polarity value.

2.2 Data Sets

The English data set included 30 de-identified, full-length emergency department reports annotated with 283 disorders related to influenza-like illnesses by a board-certified infectious disease physician. Each disorder was annotated with four attributes – existence, aspectual phase, certainty and mental state – by two independent annotators (including DM) who came to consensus after reviewing disagreements.

The Swedish data set included 1,297 assessment sections from emergency department reports annotated with approx. 2,000 disorders, automatically marked from a manually created list of approximately 300 unique disorders by two physicians. The two physicians annotated each disorder mention with attributes of polarity and certainty. A random subset of approx. 200 annotated disorder mentions from the data set were used for this qualitative study.

2.3 Study Process

In order to better understand how physicians describe uncertainty of the presence or absence of a disorder, we evaluated the annotations from the two data sets as follows: 1) created a clinical uncertainty and negation taxonomy, 2) developed a translation map for mapping attributes and values from the English schema into the Swedish schema, and 3) characterized and compared both data sets and languages using the taxonomy.

To create the uncertainty and negation taxonomy, we conducted a literature review of recent annotation schemas (e.g. Vincze et al. (2008)), assignment applications (e.g. Uzuner et al. (2011), Harkema et al. (2009), Clark et al. (2011), Chapman et al. (2011)), and observational studies (Lingard et al., 2003) about uncertainty or negation in the clinical domain. From our review, we created a clinical taxonomy describing notable characteristics of uncertainty and negation, which were added to and refined using grounded theory, by inspecting the disorder annotations in our data sets and documenting emerging themes consistent with issues found from the literature review. For instance, one characteristic of negation annotations found in the literature and in our data sets is the existence of a lexical cue indicating that a disorder is negated, and the lexical cue can occur before, within, or after the disorder mention. The characteristics included in the taxonomy represent *features* describing the attributes of uncertainty and negation in the data sets (see Section 3.1).

To develop the translation map between certainty and negation values from each annotation schema, authors DM and SV jointly reviewed each annotated disorder mention from the English data set and assigned a Swedish polarity and certainty label, then devised a map from the English schema into the Swedish schema.

To characterize and compare manifestations of uncertainty and negation using annotations from the two data sets, DM and SV annotated each disorder mention in both data sets with the features in the clinical uncertainty and negation taxonomy. In the English data set, each disorder was annotated by DM and adjudicated by SV. In the Swedish data set, each disorder was annotated by SV then translated into English for adjudication by DM.

3 Results

3.1 Clinical Uncertainty and Negation Taxonomy

We developed a clinical uncertainty and negation taxonomy to characterize the linguistic manifestations of uncertainty and negation in clinical text (Figure 1). We found three high-level features in the literature and in our data sets: *position of lexical cue* (i.e., position of the lexical expression indicating uncertainty or negation with respect to the disorder), *opinion source* (i.e. person believing there is absence, presence, or uncertainty), and *evidence evaluation* (i.e., reason for the uncertainty or negation belief).

Position of lexical cue demonstrated itself in the data sets in three non-mutually exclusive ways:

- *pre-disorder* (lexical cue precedes the disorder) "Patient *denies* chest pain."
- *intra-disorder* (lexical cue occurs within the name of the disorder) "x-ray...*possibly be indicative of* pneumonia."
- *post-disorder* (lexical cue occurs after the disorder) "abdominal cramping..is *unlikely*."

Opinion source exhibited the following values:

- *dictating physician* (dictating physician alone expressed presence, absence, or uncertainty regarding the disorder) "*I* suspect bacterial pneumonia."
- *dictating physician with consultation* (dictating physician explicitly includes other clinical professional in statement) "*Discussing with Dr. **NAME***, pneumonia can not be excluded."
- *other clinical care providers* (other clinical team members explicitly stated as expressing presence, absence or uncertainty regarding the disorder) "*per patient's primary doctor*, pneumonia is suspected."
- *patient* (patient expressed presence, absence, or uncertainty regarding the disorder) "*Pt* doesn't think she has pneumonia."
- *unknown* (ambiguous who is expressing presence, absence, or uncertainty regarding the disorder) "there was a short episode of coughing."

Evidence evaluation includes a modified subset of values found in the model of uncertainty proposed by Lingard et al. (2003) to connote perceived reasons for the provider uncertainty (and negation) about the disorder mention as used in our data sets.

- *limits of evidence* (data limitations for hypothesis testing), *one diagnosis*
  - *evidence contradicts* (data contradicts expected hypothesis), "*Blood test normal, but we still think* Lyme disease."
  - *evidence needed* (evidence unavailable to test hypothesis) "*Waiting for x-ray results* to determine if it's a femur fracture."
  - *evidence not convincing, but diagnosis asserted* (data doesn't fully support proposed hypothesis), "*Slightly elevated levels of WBCs suggests* infection."
- *limits of evidence, more than one diagnosis*
  - *differential diagnoses enumerated* (competing diagnoses reasoned), "bacterial infection *vs.* viral infection."
- *limits in source of evidence* (untrusted evidence)
  - *non-clinical source* (from non-provider source), "*Pt* can't remember if she was diagnosed with COPD."
  - *clinical source* (from provider source), "*I do not agree with Dr. X's* diagnosis of meningitis."
  - *test source* (from test e.g., poor quality), "*We cannot determine from the x-ray* if the mass is fluid or a tumor."
- *limitless possibilities* (large number of likely diagnoses so diagnosis defaulted to most likely), "This is probably an infection *of some sort*."
- *other* (no evidence limitation)
  - *asserting a diagnosis or disorder as affirmed* (positive case), "*Confirms* nausea."
  - *asserting a diagnosis or disorder as negated* (negative case), "*No* vomiting."

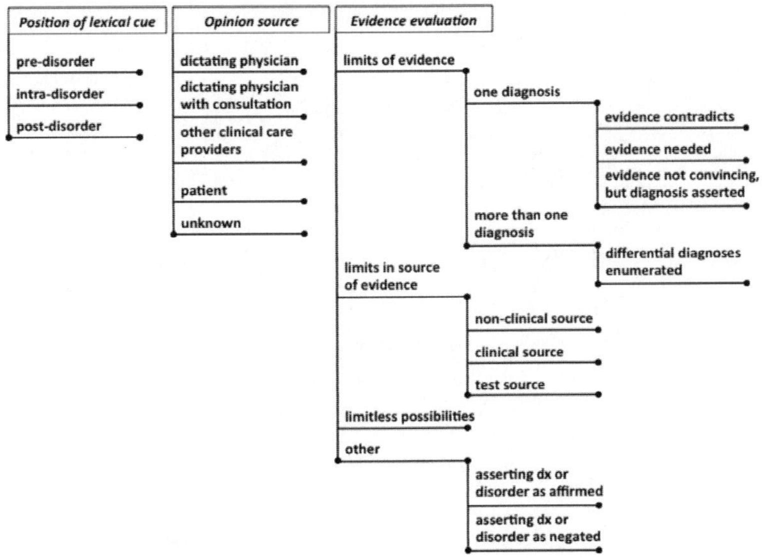

Figure 1: Uncertainty and negation taxonomy with features – *Position of lexical cue*, *Opinion source* and *Evidence evaluation* – with corresponding values (nested lines and sub-lines).

3.2 Translation Map

In order to compare annotations between the data sets, we developed a mapping procedure for converting the four annotated attribute values from the English schema into the two annotated attribute values from the Swedish schema. This mapping procedure uses two normalization steps, negation and certainty (see Figure 2).

Using Figure 2, we explain the mapping procedure to convert English annotations into Swedish annotations. Our steps and rules are applied with precedence, top down and left to right. For "I have no suspicion for bacterial infection for this patient", English annotations are Existence(no) AND AspectualPhase(null) AND Certainty(high) AND MentalState(yes), and Swedish annotations are Polarity(negative) AND Certainty(probably). The mapping procedure applies two normalization steps, negation and uncertainty, with the following rules.

The **first step** is negation normalization to *convert Existence and Aspectual Phase into Polarity annotations*. In this example, **Existence(no) → Polarity(negative)**.

The **second step** is certainty normalization with up to two sub steps. For Certainty mapping, in summary, *map English NOT Certainty(unmarked) to Swedish Certainty level*, e.g., **Certainty(high) → Certainty(probably)**. For MentalState mapping, *if English Certainty(unmarked) AND MentalState(yes), map to either Swedish Certainty(probably) OR Certainty(possibly) using your best judgment; otherwise, map to Certainty(certainly)*. For our example sentence, Certainty mapping was sufficient to map from the English to the Swedish Certainty levels.

We found that these two schemas were mappable. Despite the binary mapping splits from English Certainty(Moderate) → Swedish Certainty(possibly) OR Certainty(probably) and judgment calls necessary for MentalState mapping, few annotations were not easily mapped.

3.3 Characterization of English and Swedish Data sets

In this study, we characterized our data sets according to a clinical uncertainty and negation taxonomy comprised of three concepts – *position of lexical cue*, *opinion source*, and *evidence evaluation*.

3.3.1 Position of lexical cue

In Table 1, we show examples of phrases from each data set representing the Polarity and Certainty levels in the taxonomy. In our data set, we did not explicitly annotate markers for the highest certainty levels in the positive polarity, such as "definitely has". We did not encounter any of these cases in the

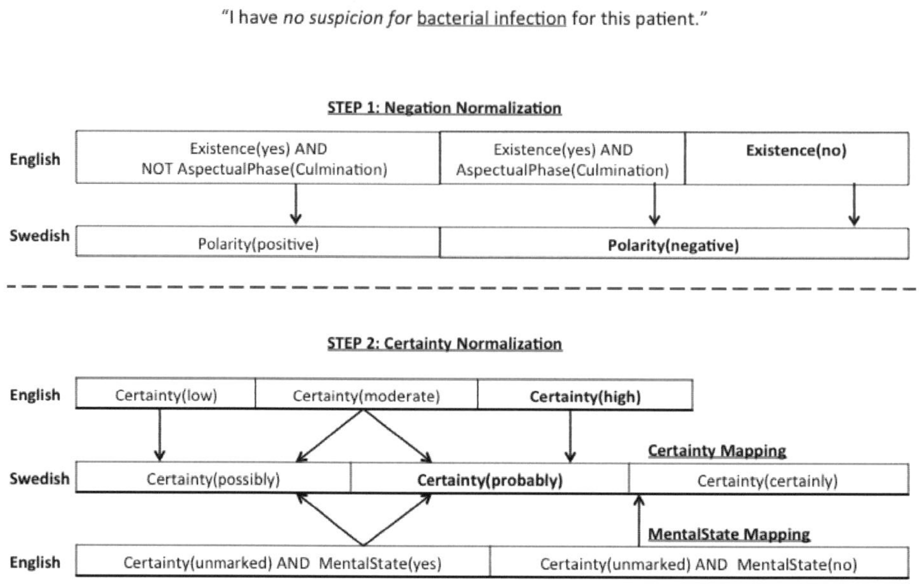

Figure 2: Map between values for attributes in Swedish and English schemas. Bolded rules indicate the rules used to assign values to the example sentence (English sentence on top).

data set. We observed that most uncertainty expressions precede a disorder mention. Few expressions both precede and follow the disorder mention, or within the disorder mention itself. We observed that most expressions of uncertainty are conveyed using positive polarity gradations such as "probably" and "possibly", for example "likely", "appears to have", "signs of". Lexical cues of low levels of certainty in the negative polarity were rare.

3.3.2 Opinion source

In Table 2, we report examples of the various individuals – *dictating physician, dictating physician with consultation, other clinical care providers, patient, unknown* – that are the source of the belief state for uncertainty about a disorder. We observed explicit judgments or mental postulations e.g., "I judge" or implied speculations in which the physician was not the subject and passive expressions were used e.g., "patient appears to have". In cases of dictating physician with consultation, the physician speculated about the disorder using references to other providers consulted to strengthen the assessment e.g., "Discussing with Dr...". In cases of other clinical care providers, there was no ownership on the part of the dictating physician, but of other members of the clinical care team e.g., "Consulting Attending (Infection) thinks...". In cases for patient, the patient is conveying statements of confusion with respect to self-diagnosing e.g., "Pat. reports that she finds it difficult to discern...". We observed no expressions of uncertainty owned by the patient in the English data set or by a relative in the Swedish data set. In the unknown case, it is unclear from the context of the report whether the speculation is on the part of the physician to believe the symptom reported or the relative unsure about reporting the symptoms e.g., "there was apparently".

3.3.3 Evidence evaluation

Below we list examples of the different reasons for uncertainties that were identified. Not all types were observed in both corpora (Not observed).

limits of evidence, one diagnosis

- evidence contradicts – **English**: "Likely upper GI bleed *with elevated bun, but normal h and h*."; **Swedish**: "Konsulterar infektionsjour som anser viros vara *osannolikt med tanke på normalt leverstatus*. (Consulting Attending (infection) who thinks that virosis is *improbable given normal liver status*.)"
- *evidence needed* – **English**: "*chest x-ray was ordered* to rule out TB."; **Swedish**: "Diskuterar med RAH-jour; *vi börjar utredning med CT-skalle med kontrast* på misstanke om metastaser och någon form av epileptiskt anfall (Discussion with Attendant [CLINIC]; *we start inves-*

	Certainly		Probably		Possibly	
	English	**Swedish**	**English**	**Swedish**	**English**	**Swedish**
Positive	<default>	<default>	• likely • suspect • thought • it was felt • appears to have • there was apparently • most likely	• förmodligen, troligen(probably) • troligtvis, troligen (probably/likely) • [mest] sannolikt ([most] probable) • tecken på (signs of) • oklar (unclear)	• **this is not unlikely** • rule out • differential includes possible • *could possible be indicative of* • possible • I think this is probably • probable	• möjlig[en|tvis], (possibly) • eventuell, ev, möjlig (possible) • misstanke [på] (suspicion [for]) • skulle kunna vara (could be) • **kan [ej|inte] uteslutas (cannot be ruled out)**
Negative	• denies; denies any recent • no; not • has never had; has not had • no further episodes of • no prior episode • **stopped** • resolution of • **has abated** • **resolved; resolves; had resolved** • **clear; free** • unremarkable for evidence of	• ingen • misstanke [om| för] (no suspicion for) • ing[en|a] (no) • inga hållpunkter för (no indication of) • utesluter (rule out)	• no suspicion for	• ingen stark [klinisk]misstanke [om] (no strong clinical suspicion for) • ej visar tecken på (does not show signs for)	• am not convinced‡	

Table 1: Common lexical cues and their relative position to the disorder mention: Pre-disorder: uncertainty marker before disorder, *Intra-disorder*: uncertainty marker inside disorder, **Post-disorder**: uncertainty marker after disorder, ‡= schema compatibility/neutral case.

Opinion source	English	Swedish
Dictating physician	• *it seems to me there is some active upper GI bleeding.*	• *...jag bedömer dock inte att besvären tyder på någon ungemboli... (...however I do not judge that the problems indicate any pulmonary embolism...)*
Dictating physician with consultation	• *..evaluation by myself and Dr. **NAME. It was felt that the patient was suffering from viral pharyngitis.*	• *Diskuterar med RAH-jour; vi börjar utredning med CT-skalle med kontrast på misstanke om metastaser och någon form av epileptiskt anfall (Discussion with oncologist on call; we start investigation with CT-brain with contrast on suspicion of metastases and some kind of epileptic seizure.)*
Other clinical care provider	• *..per his primary care doctor..they thought it was a viral syndrome of unknown type.*	• *Konsulterar infektionsjour som anser viros vara osannolikt med tanke på normalt leverstatus. (Consulting Attending (infection) who thinks that virosis is improbable given normal liver status.)*
Patient	–	• *... hon själv tror att det kan röra sig om panikångestattacker...(... she herself thinks that it can be panick attacks...)*
Unknown	• *there was apparently a short episode of apnea..*	--

Table 2: Opinion source of uncertainty or negation types with English and Swedish examples.

tigation with CT-brain with contrast on suspicion for metastasis and some form of epileptic seizure.)"

- *evidence not convincing, but diagnosis asserted* – **English**: Not observed; **Swedish**: "Förmodligen en viros *eftersom man kan se en viss lymfocytopeni i diff* (Probably a virosis since *there is some lymphocyte in blood cell count*.)"

limits of evidence, more than one diagnosis

- *differential diagnoses enumerated* – **English**: "*questionable right-sided increased density on the right side of the chest x-ray* that could possibly be indicative of a pneumonia versus increased pulmonary vasculature"; **Swedish**: "Förefaller neurologiskt, blödning? Infarkt? (Appears neurological, bleeding? Infarction?)"

limits in source of evidence

- *non-clinical source* – **English**: "*I am not convinced that he is perfectly clear on his situation*.."; **Swedish**: "*Pat uppger att hon har svårt att skilja på* panikångest och andra symtom. (*Pat. reports that she finds it difficult to discern* panick disorder from other symptoms...)"
- *clinical source* – **English**: "there was no definite diagnosis and *they thought* it was a viral syndrome of unknown type.."; **Swedish**: Not observed
- *test source* – **English**: "..confusion was possible related *a TIA without much facial droop appreciated on my physical exam*"; **Swedish**: "Ter sig mest sannolikt som reumatoid artrit *både klinisk och lab-mässigt* (Seems like it most probably is rheumatoid arthritis *both clinically and lab-wise*.)"

limitless possibilities – **English**: "*I think this is probably* a viral problem."; **Swedish**: "Pat *bedömes ha en förkylning*, troligen virusinfektion. (Patient *is evaluated as having a cold*, probably a virus infection.)"

other

- *asserting dx or disorder as affirmed* – **English**: *"I suspect that* colon cancer is both *the cause* of the patient's bleeding.."; **Swedish**: Not observed
- *asserting dx or disorder as negated* – **English**: "...her fever *has abated*."; **Swedish**: Not observed

In many cases, the local context was sufficient for understanding the evidential origins for uncertainty. When a single disorder was mentioned, uncertainty was due to data insufficient to make a definitive diagnosis because it contradicted a hypothesis, was unavailable, or was not convincing. For instance, data was to be ordered and the opportunity to interpret it had not presented itself, such as "..was ordered to rule out TB" or "..start investigation with CT-brain with contrast..". In few cases, more than one diagnosis was being enumerated due to a limitation in the evidence or data gathered e.g., "Appears neurological, bleeding? Infarction?". We observed cases in which the source of the evidence produced uncertainty including both non-clinical and clinical sources (care providers consulted and tests produced). In cases of limitless possibilities, the physician resorted to a common, default diagnosis e.g., "probably a virus infection". Limitations of evidence from a clinical source were not found in the Swedish data set and few were found in the English data set. We expect that more examples of this category would be found in e.g. radiology reports in which the quality of the image is a critical factor in making an interpretation.

4 Discussion and Conclusion

From the resulting clinical taxonomy and characterization, we observe some general differences and similarities between the two data sets and languages. The Swedish assessment entries are more verbose compared to the English medical records in terms of a more detailed account of the uncertainty and what is being done by whom to derive a diagnosis from a disorder mention. This might reflect cultural differences in how documentation is both produced and used. Differential diagnoses are often listed with question marks ("?") in the Swedish set, e.g., "Disorder 1? Disorder 2? Disorder 3?", whereas in the English data set enumerations are either listed or competing, e.g., "disorder 1 vs. disorder 2". Despite these differences, there are many similarities between the two data sets.

Mapping observations from the English schema into the Swedish schema was not complicated despite the difference in the modeled attributes. In most cases, we determined that designating attribute-value rules for negation and certainty normalization steps was sufficient to accurately map observations between the language schemas without changing an observation's semantics. This finding suggests that simple heuristics can be used to translate annotations made from English trained tools into the Swedish schema values.

The majority of the lexical markers are prepositioned in both languages, and the majority of these markers are similar across the two languages, e.g., "likely", "possible", "suspicion for". However, inflections and variants are more common in Swedish, and the language allows for free word order, this relation needs to be studied further. The default case, i.e. affirmed, or certainly positive, was rarely expressed through lexical markers.

When it comes to the opinion source of an uncertainty or negation, we observed a pattern in the use of passive voice, e.g. "it was felt", indicating avoidance to commitment in a statement. Accurate extraction of the opinion source of an expression has important implications for a system that, for instance, tracks the reasoning about a patient case over time by source. This has been recognized and incorporated in other annotation efforts, for example for news documents (Saurí and Pustejovsky, 2009). In the English data set, no cases of self-diagnosing are found, i.e. a patient owning the expressed uncertainty. In both data sets, an implicit dictating physician source is most common, i.e. there is no explicit use of pronouns indicating the opinion holder. In most cases it is clear that it is the writer's (i.e. the dictating physician's) opinion that is expressed, but in some cases, a larger context is needed for this knowledge to be resolved.

Reviewing the evidential origins or reason for expressed uncertainty, for both the Swedish and English data sets, the category "limits of evidence" is most common. This reflects a clinical reality, where many disorders require test results, radiology findings and other similar procedures before ascertaining a diagnosis. Although most cases of uncertainty are manifested and strengthened through a lexical marker, there are also instances where the uncertainty is evident without such explicit markers, e.g. the ordering of a test may in itself indicate uncertainty.

4.1 Limitations

There are several limitations of this study. The Swedish data set only contains parts of the medical record and the English data set is very small. In the creation of the taxonomy and characteristics, we have not focused on studying uncertainty *levels*, i.e. distinctions between "possibly" and "probably". The values of our taxonomy are preliminary and may change as we develop the size of our data set. Additionally, we only studied emergency department reports. We need to study other report types to evaluate the generalizability of the taxonomy.

The two compared languages both origin from the same language family (Germanic), which limits generalizability for other languages. Furthermore, the definitions of disorders in the two sets differ to some extent, i.e., English disorders are related to specific influenza-like illnesses and Swedish to more general disorders found in emergency departments.

4.2 Comparison to related work

Annotation schemas and reference standards for uncertainty and negation have been created from different perspectives, for different levels and purposes. The BioScope Corpus, for instance, contains sentence-level uncertainty annotations with token-level annotations for speculation and negation cues, along with their linguistic scope (Vincze et al., 2008). In Wilbur et al. (2006), five qualitative dimensions for characterizing biomedical articles are defined, including levels of certainty. In the 2010 i2b2/VA Challenge on concepts, assertions and relations in clinical texts, medical problem concepts were annotated. The assertion task included six annotation classes (present, absent, possible, hypothetical, conditional, not associated with the patient), to be assigned to each medical problem concept (Uzuner et al., 2011). Vincze et al. (2011) present a quantitative comparison of the intersection of two English corpora annotated for negation and speculation (BioScope and Genia Event) from two different perspectives (linguistic and event-oriented).

We extend these schemas by characterizing the underlying meaning and distinctions evident by the linguistic expressions used to indicate uncertainty and negation in the clinical domain and by exploring the relationship between uncertainty and negation, through an analysis and comparison of two different annotation schemas. However, this study is not a proposal for mapping to these schemas or others.

From an application perspective, uncertainty and negation handling have been included in rule-based systems such as NegEx and ConText, applied on disorder mentions. In Chapman et al. (2011), a generalized version of ConText is presented, with uncertainty values (probably, definitely) linked to either a positive or negative assertion, with an added indeterminate value. A previous study has shown promising results for adapting NegEx to Swedish (Skeppstedt, 2011), indicating that further extensions and adaptations between the two languages for e.g. uncertainty modeling should be viable. Machine-learning based approaches outperform rule-based for assertion classification according to results presented in Uzuner et al. (2009). A machine-learning approach was also used in the top performing system in the 2010 i2b2/VA Challenge assertion task (de Bruijn et al., 2011).

4.3 Implications and future work

With uncertainty lexicons for both Swedish and English, we hypothesize that we will be able to extend ConText to handle uncertainties in English as well as in Swedish. This enables both improvements over the existing system and the possibilities of further comparing system performances between languages. We will also experiment with machine-learning approaches to detect and annotate uncertainty and negation. We plan to extend both data sets, the English data set using semi-automatically translated disorders marked in the Swedish data set to encode new disorder mentions, and the Swedish data set by extracting the full medical records, thus creating a larger set for comparison. We will extend the taxonomy as needed e.g., syntactic and semantic patterns, and investigate how to integrate the clinical taxonomy to inform ConText by providing more granular descriptions of the motivation behind the uncertainty, thus bringing us closer to natural language understanding.

Acknowledgments

For the English and Swedish data sets, we obtained approval from the University of Pittsburgh IRB and the Regional Ethical Review Board in Stockholm (Etikprövningsnämnden i Stockholm). The study is part of the Interlock project, funded by the Stockholm University Academic Initiative and partially funded by NLM Fellowship 5T15LM007059. Lexicons and probabilities will be made available and updated on the iDASH NLP ecosystem under Resources: http://idash.ucsd.edu/nlp/natural-language-processing-nlp-ecosystem.

References

B. E. Chapman, S. Lee, H. Peter Kang, and W. W. Chapman. 2011. Document-level Classification of CT Pulmonary Angiography Reports Based on an Extension of the ConText Algorithm. *Journal of Biomedical Informatics*, 44:728–737.

D. Chu, J.N. Dowling, and WW Chapman. 2006. Evaluating the Effectiveness of Four Contextual Features in Classifying Annotated Clinical Conditions in Emergency Department Reports. In *AMIA Annu Symp Proc*, pages 141–145.

C. Clark, J. Aberdeen, M. Coarr, D. Tresner-Kirsh, B. Wellner, A. Yeh, and L. Hirschman. 2011. MITRE system for Clinical Assertion Status Classification. *J Am Med Inform Assoc*, 11(18):563–567.

B. de Bruijn, C. Cherry, S. Kiritchenko, J. Martin, and X. Zhu. 2011. Machine-learned Solutions for Three Stages of Clinical Information Extraction: The State of the Art at i2b2 2010. *Journal of the American Medical Informatics Association*, 18:557–562.

H. Harkema, J. N. Dowling, T. Thornblade, and W. W. Chapman. 2009. ConText: An Algorithm for Determining Negation, Experiencer, and Temporal Status from Clinical Reports. *Journal of Biomedical Informatics*, 42:839–851.

L. Lingard, K. Garwood, C. F. Schryer, and M. M. Spafford. 2003. A Certain Art of Uncertainty: Case Presentation and the Development of Professional Identity. *Social science medicine*, 56(3):603–616.

S. M. Meystre, G. K. Savova, K. C. Kipper-Schuler, and John E. Hurdle. 2008. Extracting Information from Textual Documents in the Electronic Health Record: A Review of Recent Research. *IMIA Yearbook of Medical Informatics 2008. 47 Suppl 1:138-154.*

D. Mowery, P. Jordan, J.M. Wiebe, H. Harkema, and W.W. Chapman. 2012. Semantic Annotation of Clinical Text: A Pilot Study. Unpublished.

V. L. Rubin, E. D. Liddy, and N. Kando. 2006. Certainty Identification in Texts: Categorization Model and Manual Tagging Results. In *Computing Affect and Attitutde in Text: Theory and Applications*. Springer.

R. Saurí and J. Pustejovsky. 2009. FactBank: A Corpus Annotated with Event Factuality. *Language Resources and Evaluation*, 43(3):227–268–268, September.

M. Skeppstedt. 2011. Negation Detection in Swedish Clinical Text: An Adaptation of NegEx to Swedish. *Journal of Biomedical Semantics*, 2(Suppl. 3):S3.

A. L. Strauss and J. Corbin. 1990. *Basics of Qualitative Research: Grounded Theory Procedures and Techniques*. Sage.

Ö. Uzuner, X. Zhang, and T. Sibanda. 2009. Machine Learning and Rule-based Approaches to Assertion Classification. *Journal of the American Medical Informatics Association*, 16(1):109–115.

Ö. Uzuner, B. R. South, S. Shen, and S. L. DuVall. 2011. 2010 i2b2/VA Challenge on Concepts, Assertions, and Relations in Clinical Text. *JAMIA*, 18(5):552–556.

S. Velupillai, H. Dalianis, and M. Kvist. 2011. Factuality Levels of Diagnoses in Swedish Clinical Text. In A. Moen, S. K. Andersen, J. Aarts, and P. Hurlen, editors, *Proc. XXIII International Conference of the European Federation for Medical Informatics (User Centred Networked Health Care)*, pages 559 – 563, Oslo, August. IOS Press.

V. Vincze, G. Szarvas, R. Farkas, G. Móra, and J. Csirik. 2008. The BioScope Corpus: Biomedical Texts Annotated for Uncertainty, Negation and Their Scopes. *BMC Bioinformatics*, 9(S-11).

V. Vincze, G. Szarvas, G. M'ora, T. Ohta, and R. Farkas. 2011. Linguistic Scope-based and Biological Event-based Speculation and Negation Annotations in the BioScope and Genia Event Corpora. *Journal of Biomedical Semantics*, 2(Suppl. 5):S8.

J. Wiebe, R. Bruce, M. Bell, M. Martin, and T. Wilson. 2001. A Corpus Study of Evaluative and Speculative Language. In *Proceedings of the Second SIGdial Workshop on Discourse and Dialogue - Volume 16*, SIGDIAL '01, pages 1–10, Stroudsburg, PA, USA. Association for Computational Linguistics.

J. W. Wilbur, A. Rzhetsky, and H. Shatkay. 2006. New Directions in Biomedical Text Annotation: Definitions, Guidelines and Corpus Construction. *BMC Bioinformatics*, 7:356+, July.

A Hybrid Stepwise Approach for De-identifying Person Names in Clinical Documents

Oscar Ferrández[1,2], Brett R. South[1,2], Shuying Shen[1,2], Stéphane M. Meystre[1,2]
[1] Department of Biomedical Informatics, University of Utah, Salt Lake City, Utah, USA
[2] IDEAS Center SLCVA Healthcare System, Salt Lake City, Utah, USA
oscar.ferrandez@utah.edu,
{brett.south,shuying.shen,stephane.meystre}@hsc.utah.edu

Abstract

As Electronic Health Records are growing exponentially along with large quantities of unstructured clinical information that could be used for research purposes, protecting patient privacy becomes a challenge that needs to be met. In this paper, we present a novel hybrid system designed to improve the current strategies used for person names de-identification. To overcome this task, our system comprises several components designed to accomplish two separate goals: 1) achieve the highest recall (no patient data can be exposed); and 2) create methods to filter out false positives. As a result, our system reached 92.6% F_2-measure when de-identifying person names in Veteran's Health Administration clinical notes, and considerably outperformed other existing "out-of-the-box" de-identification or named entity recognition systems.

1 Introduction

Electronic Healthcare Records are invaluable resources for clinical research, however they contain highly sensitive Protected Health Information (PHI) that must remain confidential. In the United States, patient confidentiality is regulated by the Health Insurance Portability and Accountability Act (HIPAA). To share and use clinical documents for research purposes without patient consent, HIPAA requires prior removal of PHI. More specifically, the HIPAA "Safe Harbor"[1] determines 18 PHI categories that have to be obscured in order to consider clinical data de-identified.

An ideal de-identification system should recognize PHI accurately, but also preserve relevant non-PHI clinical data, so that clinical records can later be used for various clinical research tasks.

Of the 18 categories of PHI listed by HIPAA, one of the most sensitive is patient names, and all person names in general. Failure to de-identify such PHI involves a high risk of re-identification, and jeopardizes patient privacy.

In this paper, we describe our effort to satisfactorily de-identify person names in Veteran's Health Administration (VHA) clinical documents. We propose improvements in person names de-identification with a pipeline of processes tailored to the idiosyncrasies of clinical documents. This effort was realized in the context of the development of a best-of-breed clinical text de-identification system (nicknamed "BoB"), which will be released as an open source software package, and it started with the implementation and evaluation of several existing de-identification and Named Entity Recognition (NER) systems recognizing person names. We then devised a novel methodology to better tackle this task and improve performance.

2 Background and related work

In many aspects de-identification resembles traditional NER tasks (Grishman and Sundheim, 1996). NER involves detecting entities such as person names, locations, and organizations. Consequently, given the similar entities targeted by both tasks, NER systems can be relevant to de-identify documents. However, most named entity recognizers were developed for newswire articles, and not for clinical narratives. Clinical records are character-

[1] GPO US: 45 C.F.R. § 164 Security and Privacy. http://www.access.gpo.gov/nara/cfr/waisidx_08/45cfr164_08.html Further details about the 18 HIPAA Safe Harbor PHI identifiers can be also found in (Meystre et al., 2010).

ized by fragmented and incomplete utterances, lack of punctuation marks and formatting, as well as domain specific language. These complications, in addition to the fact that some entities can appear both as PHI and non-PHI in the same document (e.g., “Mr. Epley” vs. “the Epley maneuver”), make clinical text de-identification a challenging task. Therefore, although person names de-identification is essentially NER, the unique characteristics of clinical texts make it more interesting and challenging than recognizing names in news articles, which also enhance the motivation for this study.

Several different approaches were proposed to deal with de-identification of clinical documents, and for named entity recognition of person names. These approaches are mainly focused on either pattern matching techniques, or statistical methods (Meystre et al., 2010), as exemplified below.

Beckwith et al. (2006) developed a de-identification system for pathology reports. This system implemented some patterns to detect dates, locations, and ID numbers, as well as a database of proper names and well-known markers such as ‘Mr.’ and ‘PhD’ to find person names.

Friedlin and McDonald (2008) described the Medical De-identification System (MeDS). It used a combination of methods including heuristics, pattern matching, and dictionary lookups to identify PHI. Pattern matching through regular expressions was used to detect numerical identifiers, dates, addresses, ages, etc.; while for names, MeDS used lists of proper names, common usage words and predictive markers, as well as a text string nearness algorithm to deal with typographical errors.

Neamatullah et al. (2008) proposed another rule-based de-identification approach focused on pattern matching via dictionary lookups, regular expressions and context checks heuristics denoting PHI. Dictionaries made up of ambiguous names and locations that could also be non-PHI, as well as dictionaries of common words were used by this system to disambiguate PHI terms.

Other de-identification systems such as (Aberdeen et al., 2010; Gardner and Xiong, 2009) use machine learning algorithms to train models and predict new annotations. The key aspect of these systems is the selection of the learning algorithm and features. Both (Aberdeen et al., 2010) and (Gardner and Xiong, 2009) use an implementation of Conditional Random Fields (CRF) and a set of learning features based on the morphology of the terms and their context. One disadvantage of these systems is the need for large amounts of annotated training examples.

As mentioned previously, for detecting person names, we could also use traditional newswire-trained NER systems. NER has long been studied by the research community and many different approaches have been developed (Tjong Kim Sang and De Meulder, 2003; Doddington et al., 2004). One successful and freely available named entity recognizer is the Stanford NER system (Finkel et al., 2005), which provides an implementation of linear chain CRF sequence models, coupled with well-engineered feature extractors for NER, and trained with newswire documents.

3 Methods

As already mentioned, we first selected and ran several existing de-identification and NER systems detecting person names in our clinical documents. Afterwards, we devised and present here a novel pipeline of processes designed to improve the PHI recognition task.

3.1 Existing de-identification and NER systems

Five available de-identification systems, as well as one newswire-trained named entity recognizer, were selected for an “out-of-the-box” evaluation. The aim of this evaluation was to compare the performance of the various methods and resources when de-identifying person names in our clinical documents.

We included three rule-based de-identification approaches:

- HMS Scrubber (Beckwith et al., 2006);
- MeDS (Friedlin and McDonald, 2008); and
- MIT deid system (Neamatullah et al., 2008).

And two systems based on machine learning classifiers:

- The MITRE Identification Scrubber Toolkit (MIST) (Aberdeen et al., 2010); and
- The Health Information DE-identification (HIDE) system (Gardner and Xiong, 2009).

Regarding NER systems, we chose the Stanford NER system (Finkel et al., 2005), which has reported successful results when detecting person names. These systems were described in Section 2, when we presented related work.

3.2 Our best-of-breed approach

Our names de-identification approach consists of a novel pipeline of processes designed to improve the current strategies for person names de-identification. This system is being developed as an Apache UIMA[2] pipeline, with two main goals:

1) Obtain the highest recall (i.e., sensitivity), regardless of the impact on precision; and
2) Improve overall precision by filtering out the false positives produced previously.

These goals correspond to the implementation of the main components of our system. When we tested existing systems (we will present results for these systems in Table 1), we observed that recall was better addressed by rule-based approaches, while precision was higher applying machine learning-based algorithms. We therefore used this knowledge for the design of our system: goal#1 is then accomplished mainly using rule-based techniques, and goal#2 implementing machine learning-based approaches.

Moreover, recall is of paramount importance in de-identification (patient PHI cannot be disclosed). And this was also a reason that motivated us to first focus on achieving high recall, and filtering out false positives afterwards as a separate procedure.

Unlike other de-identification and NER systems that tackle the classification problem from one perspective (i.e., rule-based or machine learning-based) or from a limited combined approach (e.g., learning features extracted using regular expressions), the design of our system allows us to take advantage of the strong points of both techniques separately. And more importantly, our classifiers for filtering out false-positives (goal#2) are trained using correct and incorrect annotations derived from previous modules implemented in goal#1. Thus, they do not predict if every token in the document is or belongs to a PHI identifier, they instead decide if an actual annotation is a false or true positive. This design makes our classifiers better with less learning examples, which is a restriction we have to deal with, and it also allows us to create methods that can be only focused on maximizing recall regardless of the amount of false-positives introduced (goal#1). To the best of our knowledge, this perspective has not been exploited before, and as we will show in the evaluation section, it empirically demonstrates more robustness than previous approaches.

The design of our system integrates different components described below. Figure 1 depicts an overview of our system's architecture and workflow.

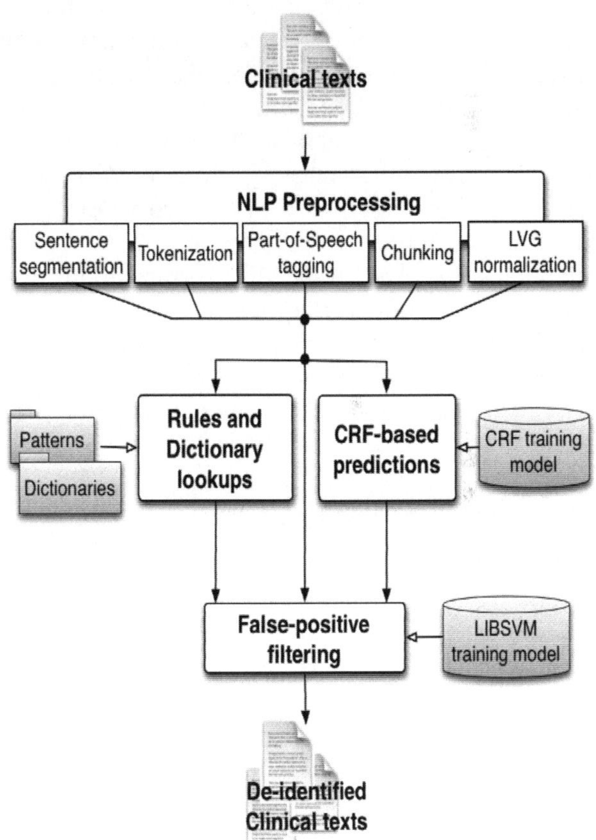

Figure 1. System's architecture.

3.2.1 NLP preprocessing steps

This NLP preprocessing prepares the input for the main components of our system. It includes sentence segmentation, tokenization, part-of-speech tagging, chunking, and word normalization based

[2] http://uima.apache.org/

on Lexical Variant Generation (LVG)[3]. The output of this preprocessing will be used by subsequent pattern matching techniques and features for machine learning algorithms. For these processes, we adapted several cTAKES (Savova et al., 2010) components.

3.2.2 Rules and dictionary lookups

We created a pattern matching component supported by contextual keyword searches (e.g., "Dr.", "Mr.", "M.D.", "R.N.", "L.C.S.W."), dictionaries of person names[4], and a simple disambiguation procedure based on a list of common words and the capitalization of the entity. We adapted some of the techniques implemented in (Beckwith et al., 2006; Friedlin and McDonald, 2008; Neamatullah et al., 2008) to our documents, and developed new patterns. For dictionary lookups, we used Lucene[5] indexing, experimenting with keyword and fuzzy dictionary searches. Each word token is compared with our indexed dictionary of names (last and first names from the 1990 US Census[4]), considering all matches as candidate name annotations. However, candidates that also match with an entry in our dictionary of common words[6] and do not contain an initial capital letter are discarded from this set of candidate name annotations.

With this component, we attempt to maximize recall, even if precision is altered.

3.2.3 CRF-based predictions

To further enhance recall, we created another component based on CRF models. We incorporated this component in our system considering that machine learning classifiers are more generalizable and can detect instances of names that are not supported by our rules or dictionaries. Therefore, although we knew the individual results of a CRF classifier at this level were not enough for de-identification, at this point our main concern is to obtain the highest recall. Thus, adding a machine learning classifier into this level we could help the system predicting the PHI formats and instances that could not be covered by our patterns and dictionaries.

To develop this component, we used the CRF classifier implementation provided by the Stanford NLP group[7]. We carried out a feature selection procedure using greedy forward selection. It provided us with the best learning feature set, which consisted of: the target word, 2-grams of letters, position in the document, part-of-speech tag, lemma, widely-used word-shape features (e.g., initial capitals, all capitals, digits inside, etc.), features from dictionaries of names and common words, a 2-word context window, and combinations of words, word-shapes and part-of-speech tags of the word and its local context.

The learning features considered before and after the selection procedure are shown in Table 1.

3.2.4 False-positive filtering

The two previous components' objective is maximal recall, producing numerous false positives. The last component of our pipeline was therefore designed to filter out these false positives and consequently increase overall precision. We built a machine learning classifier for this task, based on LIBSVM (Chang and Lin, 2001), a library for Support Vector Machines (SVM), with the RBF (Radial Basis Function) kernel. We then trained this classifier with reference standard text annotations, as well as the correct and incorrect annotations made by the previous components. We used our training document set (section 4.1) for this purpose.

Features for the LIBSVM machine learning model were: the LVG normalized form of the target annotation, three words before and after, part-of-speech tags of the words within the annotation and the local context, number of tokens within the annotation, position in the document, 40 orthographic features (denoting capitals, digits, special characters, etc.), features from dictionaries of names and common words, and the previous strategy used to make the annotation (i.e., rules, dictionary lookups or CRF-based predictions).

[3] http://lexsrv2.nlm.nih.gov/LexSysGroup/Projects/lvg/current/web/index.html

[4] Frequently Occurring Names from the 1990 Census. http://www.census.gov/genealogy/names.

[5] http://lucene.apache.org/java/docs/index.html

[6] We used the dictionary of common words from Neamatullah et al. (2008).

[7] http://nlp.stanford.edu/software/corenlp.shtml

Feature	Description	Selected*
target word	The word to classify as person name	Yes
2-grams of letters	Features from the 2-grams of letters from the word	Yes
3-grams of letters	Features from the 3-grams of letters from the word	No
4-grams of letters	Features from the 4-grams of letters from the word	No
lowercase n-grams	Features from the n-grams of letters from the word in lowercase (considering 2-, 3-, and 4-grams separately)	No
position	Position of the word within a sentence	Yes
PoS	Part-of-speech tag of the word	Yes
lemma	Lemma of the word	Yes
word shape	Initial capital	Yes
	All capitals	
	Mix of uppercase and lowercase letters	
	Digits inside	
	All digits	
	Has dash	
	End dash	
	Alpha-numeric	
	Numeric-alpha (starts with a number)	
	Contains punctuation mark	
dictionaries	Does the word match with an entry of the dictionary of names?	Yes
	Does the word match with an entry of the dictionary of common words?	Yes
2-word window	The two preceding and following words in the context	Yes
3-,4-,5-word window	The three, four and five preceding and following words in the context	No
word-pairs	Combinations of the word and the next and previous words in the context window, preserving direction but not position (considering separate features for the different combinations of the context and the target word)	No
titles	Match the word against a list of name titles (Mr, Mrs, etc.)	No
lemma_context	Lemma of the words inside the contextual window	No
PoS_context	Individual features from the part-of-speech tags of the contextual window	Yes
PoS_sequence	Sequence of the part-of-speech tags of the 2-word contextual window and the target word	Yes
word_shape_context	Word shape features of the contextual window	Yes
word-tag	Combination of the word and part-of-speech	No

Table 1. Set of learning features for the CRF-based prediction module. (* = selected in the best learning features set)

4 Evaluation and discussion

Our evaluation consists of: 1) "out-of-the-box" evaluation of the systems presented in Section 3.1; and 2) evaluation of the performance of our person names de-identification pipeline.

4.1 Data

We manually annotated all person names (including patients, relatives, health care providers, and other persons) in a corpus of various types of Veteran's Health Administration (VHA) clinical notes. These notes were selected using a stratified random sampling approach with documents longer than 500 words. Then, the 100 most frequent VHA note types were used as strata for sampling, and the same number of notes was randomly selected in each stratum. Two reviewers independently annotated each document, a third reviewer adjudicated their disagreements, and a fourth reviewer eventually examined ambiguous and difficult adjudicated cases.

The evaluation corpus presented here comprises a subset of 275 VHA clinical notes from the aforementioned corpus. For training, 225 notes were randomly selected (contained 748 person name annotations), and the remaining 50 notes (with 422 name annotations) were used for testing the systems.

4.2 Experiments and results

We present results in terms of precision, recall and F-measure (harmonic mean of recall and precision). We used a weight of 2 when calculating the F_2-measure giving recall more (twice) importance than precision (Jurafsky and Martin, 2009). This reflects our emphasis on recall for de-identification. To our understanding, due to legal and privacy issues, a good de-identification system should be tailored to prioritize recall, and consequently patient confidentiality. It is not the scope of this paper to judge or modify the development design adopted by other de-identification systems.

Moreover, we considered correct predictions at least overlapping with the entire PHI annotation in the reference standard (i.e., exact match with the reference annotation, or more than the exact match). We can therefore assure complete redaction of PHI.

Table 2 illustrates "out-of-the-box" evaluation results of the systems described in Section 3.1. For this evaluation, we trained MIST and HIDE with our 225 notes training corpus, while the Stanford NER was run using the trained models available with its distribution[8]. Testing was realized using our 50 notes testing corpus.

Table 3 shows the performance of our names de-identification approach. We provide results for different configurations of our pipeline:

- **Rules & Dictionaries**. Results of the rules and dictionary lookups component described in Section 3.2.2, in this case using a keyword-search strategy for dictionary lookups.
- **R&D with fuzzy searches**. Results from the rules and dictionary lookups component using Lucene's Fuzzy Query engine for dictionary searches. It implements a fuzzy search based on the Levenshtein (edit distance) algorithm[9] (Levenshtein, 1966), which has to surpass a similarity threshold in order to produce a match. We carried out a greedy search on the training corpus for the best similarity threshold. We found 0.74 to be the best threshold.
- **CRF-based w/FS**. The CRF-based predictions component results after selecting the best set of features (see Section 3.2.3). The CRF classifier was trained using our 225-document training corpus.
- **R&D + CRF w/FS**. The cumulative results from the rules and dictionary lookups (not implementing fuzzy dictionary searches) and the CRF-based predictions components.
- **R&D + CRF w/FS + FP-filtering**. Includes all components together, adding the false-positive filtering component (Section 3.2.4) at the end of the pipeline. The SVM model for this last component was created using our training corpus.

System	Prec.	Rec.	F_2
HMS Scrubber	0.150	0.675	0.397
MeDS	0.149	0.768	0.419
MIT deid	0.636	0.893	0.826
MIST	0.865	0.319	0.356
HIDE	0.975	0.376	0.429
Stanford NER	0.692	0.723	0.716

Table 2. "Out-of-the-box" evaluation of existing de-identification and NER systems (Prec.=precision; Rec.=recall; F_2= F_2-measure).

System	Prec.	Rec.	F_2
Rules & Dictionaries	0.360	0.962	0.721
R&D + fuzzy	0.171	0.969	0.502
CRF-based w/FS	0.979	0.874	0.893
R&D + CRF w/FS	0.360	0.988	0.732
R&D + CRF w/FS + FP-filtering	0.774	0.974	**0.926**

Table 3. Cumulative results of our pipeline of processes.

[8] Further details about these models can be found at http://nlp.stanford.edu/software/CRF-NER.shtml

[9] http://www.merriampark.com/ld.htm

4.3 Analysis

Our novel names de-identification pipeline significantly outperforms all other systems we evaluated "out-of-the-box" or trained with our VHA notes corpus.

Among the five existing systems we evaluated (Table 1), only one achieved noteworthy recall around 89%. However, none of them obtained any remarkable F_2-measure. Most errors produced by the pattern matching systems (i.e., HMS Scrubber, MeDS, and MIT deid system) were due to false positive annotations of medical eponyms (e.g., "Achilles", "Guyon", etc.), as well as acronyms denoting medical facilities (e.g., "ER" and "HCS"). The false negatives consisted of ambiguous person names (e.g., "Bill" and "Chase"), some formats not covered by the patterns (e.g., "[LastName], [FirstName] [Initial]"), and a few names not found in the dictionaries.

Among machine learning-based systems, the two de-identification applications (i.e., MIST and HIDE) obtained good precision, but quite low recall. The size of our training corpus was somewhat limited, and these results probably indicate a need for more sophisticated learning features, as well as feature selection procedures (rather than using the "out-of-the-box" feature specification that comes with these systems) for better performance. With improved learning features, we could mitigate the relative lack of training examples. Interestingly, the NER system, which was trained on newswire documents, performed even better than some de-identification systems, although a need for improvement is still present.

We acknowledge that the comparison with Stanford NER is not completely fair due to the different source of documents used for training. However, we considered it interesting information, and although clinical notes contain characteristics not present in newswire corpora, they also have similarities regarding person names (e.g., titles "Mr.", "Dr.", "PhD", part-of-speech, verb tenses). Therefore, we think that only for names recognition, a newswire trained NER can provide interesting results, and this was actually what we observed.

Table 2 points out that the combination of our components produces successful cumulative results. Using the training corpus to create a simple component made up of rules, dictionary lookups, and few heuristics for disambiguation allowed for recall values of 0.96. This demonstrates the need to adapt these techniques to the target documents, instead of employing systems "out-of-the-box".

Our experiments with fuzzy dictionary lookups did not allow for a significant increase in recall, but caused a decrease in precision (-19%). It suggests that there was no need for considering person name misspellings.

The component based on CRF predictions alone achieved good performance, especially in precision. It obtained the best F_2-measure (0.89), clearly higher than the other "out-of-the-box" systems based on CRF models. It proves that selecting suitable learning features mitigates to some extent the scarcity of training examples.

Our next experiment combined the rules and dictionaries and CRF components. It improved the overall recall to about 0.99, which means that CRF-based predictions recognized some person names that were missed by our pattern matching components, but didn't increase the precision. We reached here our first goal of high recall or sensitivity.

Finally, we added the false-positive filtering component to our system. This component was able to filter out 622 (84%) false positives from a total of 742, improving the precision to 0.77 (+41%); but also causing a slight decrease in recall (-1.4%). This application of our pipeline was successful, reaching an F_2-mesure of 0.93, and was an effective way of training the SVM model for false-positives filtering.

5 Conclusions

We designed and evaluated a novel person names de-identification system with VHA clinical documents. We also presented an "out-of-the-box" evaluation of several available de-identification and NER systems; all of them were surpassed by our approach.

With our proposal, we showed that it is possible to improve the recognition of person names in clinical records, even when the corpus for training machine learning classifiers is limited. Furthermore, the workflow of our pipeline allowed us to tackle the de-identification task from an intuitive but powerful perspective, i.e. facing the achievement of high recall and precision as two separate goals implementing specific techniques and components.

Packaging this two-step procedure as a bootstrapping learning or adding the rules to define learning features would not allow us to use the qualities of the R&D and CRF components (i.e., obtain the highest recall by any means). Moreover, considering the small size of our manually annotated examples, these approaches would not work much better than existing systems.

As future efforts, we plan to improve the precision of the rules and dictionary lookups component by adding more sophisticated person names disambiguation procedures. Such procedures should deal with the peculiar formatting of clinical records as well as integrate enriched knowledge from biomedical resources. We also plan to evaluate the portability of our approach by using other sets of clinical documents, such as the 2006 i2b2 de-identification challenge corpus (Uzuner et al., 2007).

Acknowledgments

Funding provided by the Department of Veterans Affairs Health Services Research & Development Services Consortium for Healthcare Informatics Research grant (HIR 08-374).

References

John Aberdeen, Samuel Bayer, Reyyan Yeniterzi, Ben Wellner, Cheryl Clark, David Hanauer, Bradley Malin, and Lynette Hirschman. 2010. *The MITRE Identification Scrubber Toolkit: design, training, and assessment.* International journal of medical informatics, 79 (12) (December): 849-59.

Bruce A. Beckwith, Rajeshwarri Mahaadevan, Ulysses J. Balis, and Frank Kuo. 2006. *Development and evaluation of an open source software tool for deidentification of pathology reports.* BMC medical informatics and decision making, 6 (1) (January): 12.

Chih-Chung Chang and Chih-Jen Lin. (2001). *LIBSVM: a library for support vector machines.* Computer, 1-30.

George Doddington, Alexis Mitchell, Mark Przybocki, Lance Ramshaw, Stephanie Strassel, and Ralph Weischedel. 2004. *The Automatic Content Extraction (ACE) Program - Tasks, Data, and Evaluation.* Proceedings of LREC 2004: 837-840.

Jenny Rose Finkel, Trond Grenager, and Christopher Manning. 2005. *Incorporating non-local information into information extraction systems by Gibbs sampling.* Proceedings of the 43rd Annual Meeting on Association for Computational Linguistics: 363-370.

F. Jeff Friedlin and Clement J. McDonald. 2008. *A software tool for removing patient identifying information from clinical documents.* Journal of the American Medical Informatics Association : JAMIA 15 (5) (January 1): 601-10.

James Gardner and Li Xiong. 2009. *An integrated framework for de-identifying unstructured medical data.* Data & Knowledge Engineering 68 (12) (December): 1441-1451.

Ralph Grishman and Beth Sundheim. 1996. *Message understanding conference-6: A brief history.* Proceedings of the 16th conference on Computational linguistics - Volume 1: 466-471. Association for Computational Linguistics, Copenhagen, Denmark.

Daniel Jurafsky and James H. Martin. 2009. *Speech and Language Processing: An Introduction to Natural Language Processing, Speech Recognition, and Computational Linguistics.* 2nd edition. Prentice-Hall. Upper Saddle River, NJ, USA.

V.I. Levenshtein. 1966. *Binary Codes Capable of Correcting Deletions, Insertions and Reversals.* Soviet Physics - Doklady 10: 707–710.

Stephane M. Meystre, F. Jeffrey Friedlin, Brett R. South, Shuying Shen, and Matthew H. Samore. 2010. *Automatic de-identification of textual documents in the electronic health record: a review of recent research.* BMC medical research methodology 10 (1) (January): 70.

Ishna Neamatullah, Margaret M. Douglass, Li-wei H. Lehman, Andrew Reisner, Mauricio Villarroel, William J. Long, Peter Szolovits, George B. Moody, Roger G. Mark, and Gari D. Clifford. 2008. *Automated de-identification of free-text medical records.* BMC medical informatics and decision making 8 (1) (January): 32.

Guergana K. Savova, James J. Masanz, Philip V. Ogren, Jiaping Zheng, Sunghwan Sohn, Karin C. Kipper-Schuler, and Christopher G. Chute. 2010. *Mayo clinical Text Analysis and Knowledge Extraction System (cTAKES): architecture, component evaluation and applications.* Journal of the American Medical Informatics Association : JAMIA 17 (5): 507-13.

Erik F. Tjong Kim Sang, and Fien De Meulder. 2003. *Introduction to the CoNLL-2003 Shared Task: Language-Independent Named Entity Recognition.* In Proceedings of the seventh conference on Natural language learning at HLT-NAACL 2003 - Volume 4: 142-147.

Özlem Uzuner, Yuan Luo, and Peter Szolovits. 2007. *Evaluating the State-of-the-Art in Automatic De-identification.* Journal of the American Medical Informatics Association : JAMIA 14(5):550-563.

Active Learning for Coreference Resolution

Timothy A. Miller and **Dmitriy Dligach** and **Guergana K. Savova**
Children's Hospital Boston
and Harvard Medical School
300 Longwood Ave.
Enders 141
Boston, MA 02115, USA
{Timothy.Miller,Dmitriy.Dligach,Guergana.Savova}@childrens.harvard.edu

Abstract

Active learning can lower the cost of annotation for some natural language processing tasks by using a classifier to select informative instances to send to human annotators. It has worked well in cases where the training instances are selected one at a time and require minimal context for annotation. However, coreference annotations often require some context and the traditional active learning approach may not be feasible. In this work we explore various active learning methods for coreference resolution that fit more realistically into coreference annotation workflows.

1 Introduction

Coreference resolution is the task of deciding which entity mentions in a text refer to the same entity. Solving this problem is an important part of the larger task of natural language understanding in general. The clinical domain offers specific tasks where it is easy to see that correctly resolving coreference is important. For example, one important task in the clinical domain is template filling for the Clinical Elements Model (CEM).[1] This task involves extracting various pieces of information about an entity and fitting the information into a standard data structure that can be reasoned about. An example CEM template is that for *Disease* with attributes for *Body Location*, *Associated Sign or Symptom*, *Subject*, *Negation*, *Uncertainty*, and *Severity*. Since a given entity may have many different attributes and relations, it may be mentioned multiple times in a text. Coreference resolution is important for this task because it must be known that all the attributes and relations apply to the same entity so that a single CEM template is filled in for an entity, rather than creating a new template for each mention of the entity.

2 Background

2.1 Coreference Resolution

Space does not permit a thorough review of coreference resolution, but recent publications covered the history and current state of the art for both the general domain and the clinical domain (Ng, 2010; Pradhan et al., 2011; Zheng et al., 2011).

The system used here (Zheng et al., 2012) is an end-to-end coreference resolution system, meaning that the algorithm receives no gold standard information about mentions, named entity types, or any linguistic information. The coreference resolution system is a module of the clinical Textual Analysis and Knowledge Extraction System (cTAKES) (Savova et al., 2010) that is trained on clinical data. It takes advantage of named entity recognition (NER) and categorization to detect entity mentions, and uses several cTAKES modules as feature generators, including the NER module, a constituency parser module, and a part of speech tagging module.

The system architecture is based on the pairwise discriminative classification approach to the coreference resolution problem. In that paradigm, pairs of mentions are classified as coreferent or not, and then some reconciliation must be done on all of the

[1] http://intermountainhealthcare.org/cem

links so that there are no conflicts in the clusters. The system uses support vector machines (SVMs) as the pairwise classifiers, and conflicts are avoided by only allowing an anaphor to link with one antecedent, specifically that antecedent the classifier links with the highest probability.

There are separate pairwise classifiers for named entity and pronominal anaphor types. In the domain of clinical narratives, person mentions and personal pronouns in particular are not especially challenging – the vast majority of person mentions are the patient. In addition, pronoun mentions, while important, are relatively rare. Thus we are primarily interested in named entity coreference classification, and we use that classifier as the basis of the work described here.

The feature set of this system is similar to that used by Ng and Cardie (2002). That system includes features based on surface form of the mentions, shallow syntactic information, and lexical semantics from WordNet. The system used here has a similar feature set but uses Unified Medical Language System (UMLS)[2] semantic features as it is intended for clinical text, and also incorporates several syntactic features extracted from constituency parses extracted from cTAKES.

To generate training data for active learning simulations, mention detection is run first (cTAKES contains a rule-based NER system) to find named entities and a constituency parser situates entities in a syntax tree). For each entity found, the system works backwards through all other mentions within a ten sentence window. For each candidate anaphor-antecedent pair, a feature vector is extracted using the features briefly described above.

2.2 Active Learning

Active Learning (AL) is a popular approach to selecting unlabeled data for annotation (Settles, 2010) that can potentially lead to drastic reductions in the amount of annotation that is necessary for training an accurate statistical classifier. Unlike passive learning, where the data is sampled for annotation randomly, AL delegates data selection to the classifier. AL is an iterative process that operates by first training a classifier on a small sample of the data known as the seed examples. The classifier is subsequently applied to a pool of unlabeled data with the purpose of selecting additional examples the classifier views as informative. The selected data is annotated and the cycle is repeated, allowing the learner to quickly refine the decision boundary between classes. One common approach to assessing the informativeness is *uncertainty sampling* (Lewis and Gale, 1994; Schein and Ungar, 2007), in which the learner requests a label for the instance it is most uncertain how to label. In this work, we base our instance selection on the distance to the SVM decision boundary (Tong and Koller, 2002), assuming that informative instances tend to concentrate near the boundary.

Most AL work focuses on *instance selection* where the unit of selection is one instance represented as a feature vector. In this paper we also attempt *document selection*, where the unit of selection is a document, typically containing multiple coreference pairs each represented as a feature vector. The most obvious way to extend a single instance informativeness metric to the document scenario is to aggregate the informativeness scores. Several uncertainty metrics have been proposed that follow that route to adapt single instance selection to multiple instance scenarios (Settles et al., 2008; Tomanek et al., 2009). We borrow some of these metrics and propose several new ones.

To the best of our knowledge only one work exists that explores AL for coreference resolution. Gasperin (2009) experiments with an instance based approach in which batches of anaphoric pairs are selected on each iteration of AL. In these experiments, AL did not outperform the passive learning baseline, probably due to selecting batches of large size.

3 Active Learning Configurations

3.1 Instance Selection

The first active learning model we considered selects individual training instances – putatively coreferent mention pairs. This method is quite easy to simulate, and follows naturally from most of the theoretical active learning literature, but it has the drawback of being seemingly unrealistic as an annotation paradigm. That is, since coreference can span across an entire document, it is probably not practical to

[2]http://www.nlm.nih.gov/research/umls/

have a human expert annotate only a single instance at a time when a given instance may require many sentences of reading in order to contextualize the instance and properly label it. Moreover, even if such an annotation scheme proved viable, it may result in an annotated corpus that is only valuable for one type of coreference system architecture.

Nonetheless, active learning for coreference at the instance level is still useful. First, since this method most closely follows the successful active learning literature by using the smallest discrete problems, it can serve as a proof of concept for active learning in the coreference task – if it does not work well at this level, it probably will not work at the document level. Previous results (Gasperin, 2009) have shown that certain multiple instance methods do not work for coreference resolution, so testing on smaller selection sizes first can ensure that active learning is even viable at that scale. In addition, though instance selection may not be feasible for real world annotations, individual instances and metrics for selecting them are usually used as building blocks for more complex methods. In order for this to be possible it must be shown that the instances themselves have some value.

3.2 Document Selection

Active learning with document selection is a much more realistic representation of conventional annotation methods. Conventionally, a set of documents is selected, and each document is annotated exhaustively for coreference (Pradhan et al., 2011; Savova et al., 2011). Document selection fits into this workflow very naturally, by selecting the next document to annotate exhaustively based on some metric of which document has the best instances. In theory, this method can save annotation time by only annotating the most valuable documents.

Document selection is somewhat similar to the concept of *batch-mode active learning*, wherein multiple instances are selected at once, though batch-mode learning is usually intended to solve a different problem, that of an asymmetry between classifier training speed and annotation speed (Settles, 2010). A more important difference is that document selection requires that all of the instances in the batch must come from the same document. Thus, one might expect a priori that document selection for active learning will not perform as well as instance selection. However, it is possible that even smaller gains will be valuable for improving annotation time, and the more robust nature of a corpus annotated in such a way will make the long term benefits worthwhile.

In this work, we propose several metrics for selecting documents to annotate, all of which are based on instance level uncertainty. In the following descriptions, D is the set of documents, d is a single document, $\hat{d}$ is the selected document, $Instances(d)$ is a function which returns the set of pair instances in document d, i is an instance, $dist(i)$ is a function which returns the distance of instance i from the classification boundary, and I is the indicator function, which takes the value 1 if its argument is true and 0 otherwise. Note that high uncertainty occurs when $Abs(dist(i))$ approaches 0.

- Best instance – This method uses the uncertainty sampling criteria on instances, and selects the document containing the instance the classifier is least certain about. $\hat{d} = \underset{d \in D}{\text{argmin}}[\min_{i \in Instances(d)} Abs(dist(i))]$
- Highest average uncertainty – This method computes the average uncertainty of all instances in a document, and selects the document with the highest average uncertainty. $\hat{d} = \underset{d \in D}{\text{argmin}} \frac{1}{|Instances(d)|} \sum_{i \in Instances(d)} Abs(dist(i))$
- Least bad example – This method uses uncertainty sampling criteria to find the document whose most certain example is least certain, in other words the document whose most useless example is least useless. $\hat{d} = \underset{d \in D}{\text{argmin}} \max_{i \in Instances(d)} Abs(dist(i))$
- Narrow band – This method creates an uncertainty band around the discriminating boundary and selects the document with the most examples inside that narrow band. $\hat{d} = \underset{d \in D}{\text{argmax}} \sum_{i \in Instances(d)} I(Abs(dist(i) < 0.2))$
- Smallest spread – This method computes the distance between the least certain and most certain instances and selects the document minimizing that distance.

$\hat{d} = \underset{d \in D}{\text{argmin}}[\max_{i \in Instances(d)}(Abs(dist(i))) - \min_{i \in Instances(d)}(Abs(dist(i)))]$

- Most positives – This method totals the number of positive predicted instances in each document and selects the document with the most positive instances. $\hat{d} = \underset{d \in D}{\text{argmax}} \sum_{i \in Instances(d)} I(dist(i) > 0)$
- Positive ratio – This method calculates the percentage of positive predicted instances in each document and selects the document with the highest percentage. $\hat{d} = \underset{d \in D}{\text{argmax}} \frac{\sum_{i \in Instances(d)} I(dist(i)>0)}{|Instances(d)|}$

Many of these are straightforward adaptations of the instance uncertainty criteria, but others deserve a bit more explanation. The *most positives* and *positive ratio* metrics are based on the observation that the corpus is somewhat imbalanced – for every positive instance there are roughly 20 negative instances. These metrics try to account for the possibility that instance selection focuses on positive instances. The *average uncertainty* is an obvious attempt to turn instance metrics into document metrics, but *narrow band* and *smallest spread* metrics attempt to do the same thing while accounting for skew in the distribution of "good" and "bad" instances.

3.3 Document-Inertial Instance Selection

One of the biggest impracticalities of instance selection is that labeling any given instance may require reading a fair amount of the document, since the antecedent and anaphor can be quite far apart. Thus, any time savings accumulated by only annotating an instance is reduced since the reading time per instance is probably increased.

It is also possible that document selection goes too far in the other direction, and requires too many useless instances to be annotated to achieve gains. Therefore, we propose a hybrid method of document-inertial instance selection which attempts to combine aspects of instance selection and document selection.

This method uses instance selection criteria to select new instances, but will look inside the current document for a new instance within an uncertainty threshold rather than selecting the most uncertain instance in the entire training set. Sticking with the same document for several instances in a row can potentially solve the real world annotation problem that marking up each instance requires some knowledge of the document context. Instead, the context learned by selecting one instance can be retained if useful for annotating the next selected instance from the same document.

This also preserves one of the biggest advantages of instance selection, that of re-training the model after every selected instance. In batch-mode selection and document selection, many instances are selected according to criteria based on the same model starting point. As a result, the selected instances may be redundant and document scores based on accumulated instance scores may not reflect reality. Re-training the model between selected instances prevents redundant instances from being selected.

4 Evaluation

Evaluations of the active learning models described above took place in a simulation context. In active learning simulations, a labeled data set is used, and the unlabeled pool is simulated by ignoring or "covering" the labels for part of the data until the selection algorithm selects a new instance for annotation. After selection the next data point is simply put into the training data and its label is uncovered.

The data set used was the Ontology Development and Information Extraction (ODIE) corpus (Savova et al., 2011) used in the 2011 i2b2/VA Challenge on coreference resolution.[3] We used a set of 64 documents from the training set of the Mayo Clinic notes for our simulations.

Instances were created by using the training pipeline from the coreference system described in Section 2.1. As previously mentioned, this work uses the named entity anaphor classifier as it contains the most data points. This training set resulted in 6820 instances, with 311 positive instances and 6509 negative instances. Baseline ten-fold cross validation performance on this data set using an SVM with RBF kernel is an F-score of 0.48.

Simulations are performed using ten fold cross-validation. First, each data point is assigned to one

[3] https://www.i2b2.org/NLP/Coreference/

of ten folds (this is done randomly to avoid any autocorrelation issues). Then, for each iteration, one fold is made the seed data, another fold is the validation data, and the remainder are the unlabeled pool. Initially the labeled training data contains only the seed data set. The model is trained on the labeled training data, tested on the validation set, then used to select the next data point from the pool data set. The selected data point is then removed from the pool and added to the training data with its gold standard label(s), and the process repeats until the pool of unlabeled data is empty. Performance is averaged across folds to minimize the effects of randomness in seed and validation set selection. Typically, active learning is compared to a baseline of passive learning where the next data point to be labeled is selected from the unlabeled pool data set randomly.

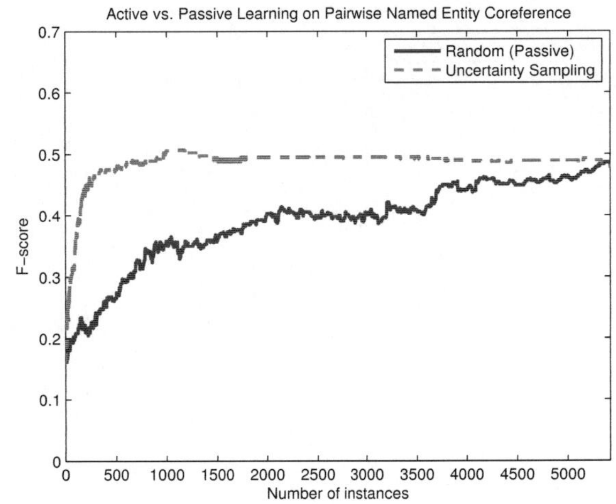

Figure 1: Instance selection simulation results. The x-axis is number of instances and the y-axis is ten-fold averaged f-score of the pairwise named entity classifier.

4.1 Instance Selection Experiments

Instance selection simulations follow the general template above, with each instance (representing a putative antecedent-anaphor pair) randomly assigned to a fold. After scoring on the validation set, uncertainty sampling is used to select a single instance from the unlabeled pool, and that instance is added to the training set.

Figure 1 shows the results of active learning using uncertainty selection on instances versus using passive learning (random selection). This makes it clear that if the classifier is allowed to choose the data, top performance can be achieved much faster than if the data is presented in random order. Specifically, the performance for uncertainty selection levels off at around 500 instances into the active learning, out of a pool set of around 5500 instances. In contrast, the passive learning baseline takes basically the entire dataset to reach the same performance.

This is essentially a proof of concept that there is such a thing as a "better" or "worse" instance when it comes to training a classifier for coreference. We take this as a validation for attempting a document selection experiment, with many metrics using instance uncertainty as a building block.

4.2 Document Selection Experiments

Document selection follows similarly to the instance selection above. The main difference is that instead of assigning pair vectors to folds, we assign documents to folds. To make a selection, each instance is labeled according to the model, document level metrics described in Section 3.2 are computed per document, and the document is selected which optimizes the metric being evaluated. All of that document's instances and labels are added to the training data, and the process repeats as before.

The results of these experiments are divided into two plots for visual clarity. Figure 2 shows the results of these experiments, roughly divided into those that work as well as a random baseline (left) and those that seem to work worse than a random baseline (right). The best performing metrics (on the left side of the figure) are *Positive Ratio*, *Least Worst*, *Highest Average*, and *Narrow Band*, although none of these performs noticeably better than random. The remaining metrics (on the right) seem to do worse than random, taking more instances to reach the peak performance near the end.

The performance of document selection suggests that it may not be a viable means of active learning. This may be due to a model of data distribution in which useful instances are distributed very uniformly throughout the corpus. In this case, an average document will only have 8–10 useful instances and many times as many that are not useful.

This was investigated by follow-up experiments on the instance selection which kept track of which

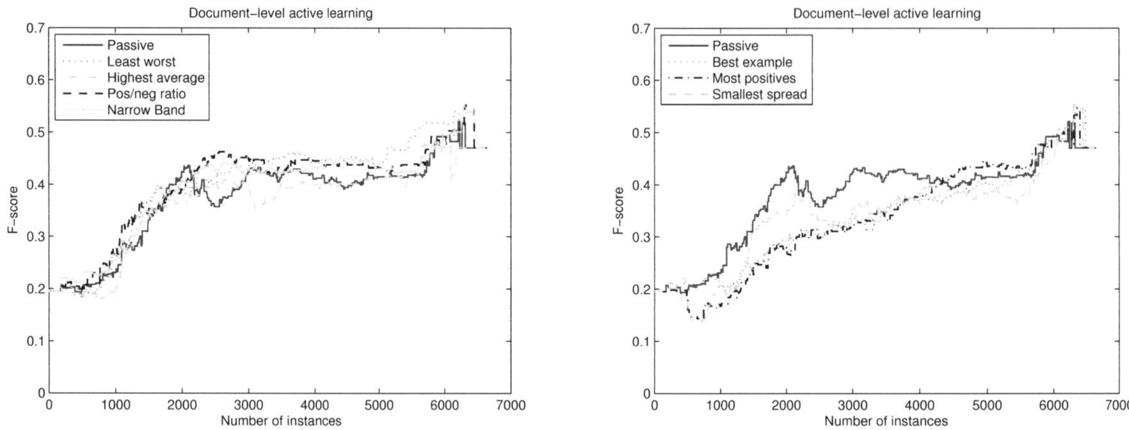

Figure 2: Two sets of document selection experiments.

document each instance came from. The experiments tracked the first 500 instances only, which is roughly the number of instances shown in Figure 1 to reach peak performance. Figure 3 (left) shows a histogram with document indices on the x-axis and normalized instance counts on the y-axis. The counts are normalized by total number of document vectors. In other words, we wanted to show whether there was a distinction between "good" documents containing lots of good instances and "bad" documents with few good instances.

The figure shows a few spikes, but most documents have approximately 10% of their instances sampled, and all but one document has at least one instance selected. Further investigation shows that the spikes in the figure are from shorter documents. Since shorter documents have few instances overall but always at least one positive instance, they will be biased to have a higher ratio of positive to negative instances. If positive instances are more uncertain (which may be the case due to the class imbalance), then shorter documents will have more selected instances per unit length.

We performed another follow-up experiment along these lines using the histogram as a measure of document value. In this experiment, we took the normalized histogram, selected documents from it in order of normalized number of items selected, and used that as a document selection technique. Obviously this would be "cheating" if used as a metric for document selection, but it can serve as a check on the viability of document selection. If the results are better than passive document selection, then there is some hope that a document level metric based on the uncertainty of its instances can be successful.

In fact, the right plot on Figure 3 shows that the "cheating" method of document selection still does not look any better than random document selection.

4.3 Document-Inertial Instance Selection Experiments

The experiments for document-inertial instance selection were patterned after the instance selection paradigm. However, each instance was bundled with metadata representing the document from which it came. In the first selection, the algorithm selects the most uncertain instance, and the document it comes from is recorded. For subsequent selections, the document which contained the previously selected instance is given priority when looking for a new instance. Specifically, each instance in that document is classified, and the confidence is compared against a threshold. If the document contains instances meeting the threshold, the most uncertain instance was selected. After each instance, the model is retrained as in normal instance selection, and the new model is used in the next iteration of the selection algorithm. For these experiments, the threshold is set at 0.75, where the distance between the classification boundary and the margin is 1.0.

Figure 4 shows the performance of this algorithm compared to passive and uncertainty sampling. Per-

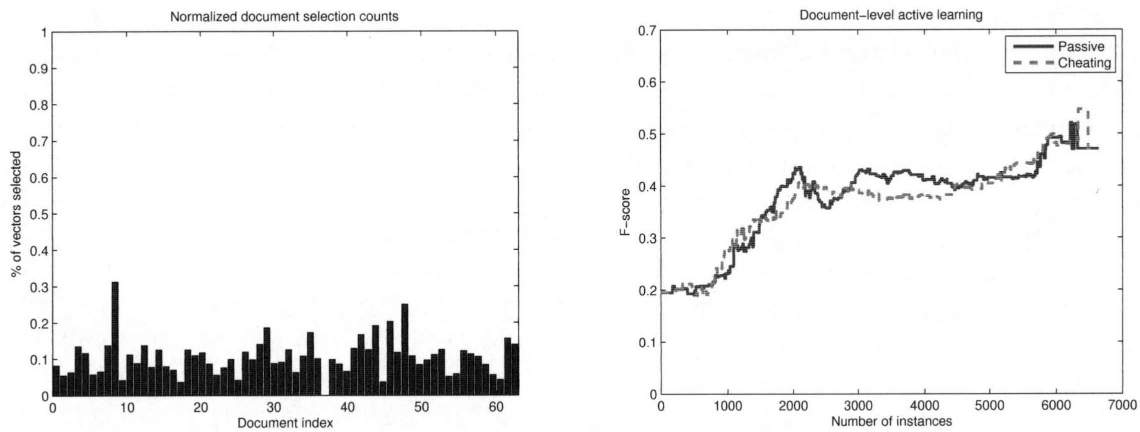

Figure 3: Left: Percentage of instances selected from each document. Right: Performance of a document selection algorithm that can 'cheat' and select the document with the highest proportion of good instances.

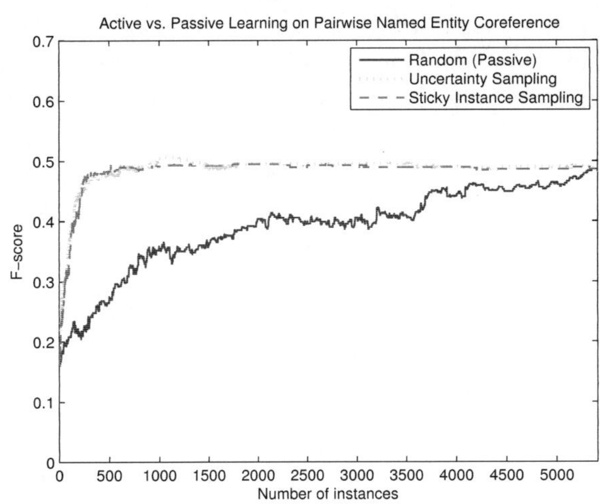

Figure 4: Document-inertial instance selection results.

formance using this algorithm is clearly better than passive learning and is similar to standard uncertainty selection ignoring document constraints.

5 Discussion and Conclusion

The results of these experiments paint a complex picture of the way active learning works for this domain and model combination. The first experiments with uncertainty selection indicate that the number of instances required to achieve classifier performance can be compressed. Selecting and training on all the good instances first leads to much faster convergence to the asymptotic performance of the classifier given the features and data set.

Attempting to extend this result to document selection met with mediocre results. Even the best performing of seven attempted algorithms seems to be about the same as random document selection. One can interpret these results in different ways.

The most pessimistic interpretation is that document selection simply requires too many useless instances to be annotated, good instances are spread too evenly, and so document selection will never be meaningfully faster than random selection. This interpretation seems to be supported by experiments showing that even if document selection uses a "cheating" algorithm to select the documents with the highest proportion of good instances it still does not beat a passive baseline.

One can also interpret these results to inspire further work, first by noting that all of the selection techniques attempt to build on the instance selection metrics. While our document selection metrics were more sophisticated than simply taking the n-best instances, Settles (2010) notes that some successful batch mode techniques explicitly account for diversity in the selections, which we do not. In addition, one could argue that our experiments were unduly constrained by the small number of documents available in the unlabeled pool, and that with a larger unlabeled pool, one would eventually encounter documents with many good instances. This may be true, but may be difficult in practice as clinical notes often need to be manually de-identified before any research use, and so it is not simply a matter of querying all records in an entire electronic medical record system.

The document-inertial instance selection showed that the increase in training speed can be maintained without switching documents for every instance. This suggests that while good training instances may be uniformly distributed, it is usually possible to find multiple *good enough* instances in the current document, and they can be found despite not selecting instances in the exact best order that plain instance selection would suggest.

Future work is mainly concerned with real world applicability. Document level active learning can probably be ruled out as being non-beneficial despite being the easiest to work into annotation work flows. Instance level selection is very efficient in achieving classifier performance but the least practical.

Document-inertial seems to provide some compromise. It does not completely solve the problems of instance selection, however, as annotation will still not be complete if done exactly as simulated here. In addition, the assumption of savings is based on a model that each instance takes a constant amount of time to annotate. This assumption is probably true for tasks like word sense disambiguation, where an annotator can be presented one instance at a time with little context. However, a better model of annotation for tasks like coreference is that there is a constant amount of time required for reading and understanding the context of a document, then a constant amount of time on top of that per instance.While modeling annotation time may provide some insight, it will probably be most effective to undertake empirical annotation experiments to investigate whether document-inertial instance selection actually provides a valuable time savings.

The final discussion point is that of producing complete document annotations. For coreference systems following the pairwise discriminative approach as in that described in Section 2.1, a corpus annotated instance by instance is useful. However, many recent approaches do some form of document-level clustering or explicit coreference chain building, and are not natively able to handle incompletely annotated documents.[4]

Future work will investigate this issue by quantifying the value of complete gold standard annotations versus the partial annotations that may be produced using document-inertial instance selection. One way of doing this is in simulation, by training a model on the 500 good instances that document-inertial instance selection selects, and then classifying the rest of the training instances using that model to create a "diluted" gold standard. Then, a model trained on the diluted gold standard will be used to classify the validation set and performance compared to the version trained on the full gold standard corpus. Similar experiments can be performed using other systems. The logic here is that if an instance was not in the top 10% of difficult instances it can be classified with high certainty. The fact that positive instances are rare and tend to be most uncertain is a point in favor of this approach – after all, high accuracy can be obtained by guessing in favor of negative once the positive instances are labeled. On the other hand, if document-inertial instance selection simply amounts to labeling of positive instances, it may not result in substantial time savings.

In conclusion, this work has shown that instance selection works for coreference resolution, introduced several metrics for document selection, and proposed a hybrid selection approach that preserves the benefits of instance selection while offering the potential of being applicable to real annotation. This work can benefit the natural language processing community by providing practical methods for increasing the speed of coreference annotation.

Acknowledgments

The project described was supported by award number NLM RC1LM010608, the Strategic Health IT Advanced Research Projects (SHARP) Program (90TR002) administered by the Office of the National Coordinator for Health Information Technology, and Integrating Informatics and Biology to the Bedside (i2b2) NCBO U54LM008748. The content is solely the responsibility of the authors and does not necessarily represent the official views of the NLM/NIH/ONC.

[4] Other recent unsupervised graphical model approaches using Gibbs sampling (Haghighi and Klein, 2007) may be able to incorporate partially annotated documents in semi-supervised training.

References

Caroline Gasperin. 2009. Active learning for anaphora resolution. In *Proceedings of the NAACL HLT Workshop on Active Learning for Natural Language Processing*, pages 1–8.

Aria Haghighi and Dan Klein. 2007. Unsupervised coreference resolution in a nonparametric bayesian model. In *Proceedings of the 45th Annual Meeting of the Association of Computational Linguistics*, pages 848–855.

David D. Lewis and William A. Gale. 1994. A sequential algorithm for training text classifiers. In *Proceedings of the ACM SIGIR Conference on Research and Development in Information Retrieval*, pages 3–12.

Vincent Ng and Claire Cardie. 2002. Improving machine learning approaches to coreference resolution. In *Proceedings of the 40th Annual Meeting of the Association for Computational Linguistics (ACL)*.

Vincent Ng. 2010. Supervised noun phrase coreference research: The first fifteen years. In *Proceedings of the 48th Annual Meeting of the Association for Computational Linguistics (ACL-10)*.

Sameer Pradhan, Lance Ramshaw, Mitchell Marcus, Martha Palmer, Ralph Weischedel, and Nianwen Xue. 2011. CoNLL-2011 shared task: Modeling unrestricted coreference in OntoNotes. In *Proceedings of the 15th Conference on Computational Natural Language Learning: Shared Task*, pages 1–27.

Guergana K. Savova, James J. Masanz, Philip V. Ogren, Jiaping Zheng, Sunghwan Sohn, Karin C. Kipper-Schuler, and Christopher G. Chute. 2010. Mayo clinical text analysis and knowledge extraction system (cTAKES): architecture, component evaluation and applications. *J Am Med Inform Assoc*, 17(5):507–513.

Guergana K. Savova, Wendy W. Chapman, Jiaping Zheng, and Rebecca S. Crowley. 2011. Anaphoric relations in the clinical narrative: corpus creation. *J Am Med Inform Assoc*, 18:459–465.

A.I. Schein and L.H. Ungar. 2007. Active learning for logistic regression: an evaluation. *Machine Learning*, 68(3):235–265.

B. Settles, M. Craven, and S. Ray. 2008. Multiple-instance active learning. *Advances in Neural Information Processing Systems (NIPS)*, 20:1289–1296.

Burr Settles. 2010. Active learning literature survey. Technical report, University of Wisconsin–Madison.

Katrin Tomanek, Florian Laws, Udo Hahn, and Hinrich Schütze. 2009. On proper unit selection in active learning: co-selection effects for named entity recognition. In *HLT '09: Proceedings of the NAACL HLT 2009 Workshop on Active Learning for Natural Language Processing*, pages 9–17, Morristown, NJ, USA. Association for Computational Linguistics.

S. Tong and D. Koller. 2002. Support vector machine active learning with applications to text classification. *The Journal of Machine Learning Research*, 2:45–66.

Jiaping Zheng, Wendy Webber Chapman, Rebecca S. Crowley, and Guergana K. Savova. 2011. Coreference resolution: A review of general methodologies and applications in the clinical domain. *Journal of Biomedical Informatics*, 44:1113–1122.

Jiaping Zheng, Wendy W Chapman, Timothy A Miller, Chen Lin, Rebecca S Crowley, and Guergana K Savova. 2012. A system for coreference resolution for the clinical narrative. *Journal of the American Medical Informatics Association*.

PubMed-Scale Event Extraction for Post-Translational Modifications, Epigenetics and Protein Structural Relations

Jari Björne [1,2], **Sofie Van Landeghem** [3,4], **Sampo Pyysalo** [5], **Tomoko Ohta** [5],
Filip Ginter [2], **Yves Van de Peer** [3,4], **Sophia Ananiadou** [5] and **Tapio Salakoski** [1,2]

[1]Turku Centre for Computer Science (TUCS), Joukahaisenkatu 3-5B, 20520 Turku, Finland
[2]Department of Information Technology, 20014 University of Turku, Finland
[3]Department of Plant Systems Biology, VIB, Technologiepark 927, 9052 Gent, Belgium
[4]Department of Plant Biotechnology and Bioinformatics, Ghent University, Gent, Belgium
[5]National Centre for Text Mining and University of Manchester,
Manchester Interdisciplinary Biocentre,131 Princess Street, Manchester, UK

Abstract

Recent efforts in biomolecular event extraction have mainly focused on core event types involving genes and proteins, such as gene expression, protein-protein interactions, and protein catabolism. The BioNLP'11 Shared Task extended the event extraction approach to sub-protein events and relations in the Epigenetics and Post-translational Modifications (EPI) and Protein Relations (REL) tasks. In this study, we apply the Turku Event Extraction System, the best-performing system for these tasks, to all PubMed abstracts and all available PMC full-text articles, extracting 1.4M EPI events and 2.2M REL relations from 21M abstracts and 372K articles. We introduce several entity normalization algorithms for genes, proteins, protein complexes and protein components, aiming to uniquely identify these biological entities. This normalization effort allows direct mapping of the extracted events and relations with post-translational modifications from UniProt, epigenetics from PubMeth, functional domains from InterPro and macromolecular structures from PDB. The extraction of such detailed protein information provides a unique text mining dataset, offering the opportunity to further deepen the information provided by existing PubMed-scale event extraction efforts. The methods and data introduced in this study are freely available from `bionlp.utu.fi`.

1 Introduction

Biomedical domain information extraction has in recent years seen a shift from focus on the extraction of simple pairwise relations (Pyysalo et al., 2008; Tikk et al., 2010) towards the extraction of *events*, represented as structured associations of arbitrary numbers of participants in specific roles (Ananiadou et al., 2010). Domain event extraction has been popularized in particular by the BioNLP Shared Task (ST) challenges in 2009 and 2011 (Kim et al., 2009; Kim et al., 2011). While the BioNLP ST'09 emphasized protein interactions and regulatory relationships, the expressive event formalism can also be applied to the extraction of statements regarding the properties of individual proteins. Accordingly, the EPI (Epigenetics and Post-Translational Modifications) subchallenge of the BioNLP ST'11 provided corpora and competitive evaluations for the detection of epigenetics and post-translational modification (PTM) events, while the REL (Entity Relations) subchallenge covers structural and complex membership relations of proteins (Ohta et al., 2011b; Pyysalo et al., 2011). The complex memberships and domains define the physical nature of an individual protein, which is closely linked to its function and biological activity. Post-translational modifications alter and regulate this activity via structural or chemical changes induced by the covalent attachment of small molecules to the protein. In epigenetic regulation, gene expression is controlled by the chemical modification of DNA and the histone proteins supporting chromosomal DNA. All of these aspects are important for defining the biological role of a protein, and thus the EPI and REL tasks enable the development of text mining systems that can extract a more complete picture of the biomolecular reactions and relations than previously possible (cf. Table 1). Furthermore, previous work has shown promising results for improving event extraction by integration of "static" entity relations (Pyysalo et al., 2009), in particular for the previously only available PTM event, phosphorylation (Van Landeghem et al., 2010).

Information on protein modifications is available in general-purpose protein databases such as UniProt, and there are also a number of dedicated database resources covering such protein modifications (Wu and others, 2003; Lee et al., 2006; Li et al., 2009). While the automatic extraction of PTMs from text has also been considered in a number of earlier studies, these have primarily involved single PTM reactions extracted with special-purpose methods (Hu et al., 2005; Yuan et al., 2006; Lee et al., 2008). The EPI task and associated work (Ohta et al., 2010) were the first to target numerous PTM reactions in a general framework using retrainable extraction methods. The automatic detection of modification statements using keyword matching-based methods has been applied also in support of DNA methylation DB curation (Ongenaert et al., 2008; Fang et al., 2011). However, as for PTM, the EPI task and its preparatory efforts (Ohta et al., 2011a) were the first to consider DNA methylation using the general event extraction approach. To the best of our knowledge, the present study is the first to extend the event extraction approach to PTM and DNA methylation event extraction to the scale of the entire available literature.

The Turku Event Extraction System (TEES), first introduced for the BioNLP ST'09 (Björne et al., 2009), was updated and generalized for participation in the BioNLP ST'11, in which it had the best performance on both the EPI and REL challenges (Björne and Salakoski, 2011). With an F-score of 53.33% for the EPI and 57.7% for the REL task, it performed over 16 pp better than the next best systems, making it well suited for our study. We apply this system to the extraction of EPI events and REL relations from all PubMed abstracts and all PMC open access articles, using a pipeline of open source text mining tools introduced in Björne et al. (2010).

We further process the result using a recently created bibliome-scale gene normalization dataset[1]. This normalization effort connects protein and gene mentions in text to their database IDs, a prerequisite for effective use of text mining results for most bioinformatics applications. In addition to protein names, the EPI and REL challenges refer to the protein substructures, modifications and complexes, which we also need to normalize in order to determine the biological context of these events. In this work, we develop a number of rule-based algorithms for the normalization of such non-protein entities.

With both proteins and other entities normalized, we can align the set of events extracted from the literature with biological databases containing annotations on protein features, such as UniProt. We can determine how many known and unknown features we have extracted from text, and what percentage of various protein feature annotations our text mining results cover. This association naturally also works in the other direction, as we can take a gene or protein and find yet unannotated post-translational modifications, domains, or other features from scientific articles, a promising use case for supporting biomedical database curation.

2 Methods

2.1 PMC preprocessing

PMC full texts are distributed in an XML format that TEES cannot use directly for event extraction. We convert this XML into a flat ASCII text format with a pipeline built on top of BioNLP ST'11 supporting resource tools (Stenetorp et al., 2011). This processing resolves embedded LaTeX expressions, separates blocks of text content (titles, sections, etc.) from others, maps non-ASCII characters to corresponding ASCII sequences, and normalizes whitespace. Resolving non-ASCII characters avoids increased error rates from NLP tools trained on ASCII-only data.

2.2 Event Extraction

We use the Turku Event Extraction System for extracting both REL relations and EPI events. TEES is a modular event extraction pipeline, that has recently been extended for all the subtasks of the BioNLP'11 ST, including EPI and REL (Björne and Salakoski, 2011). TEES performs all supported tasks using a shared graph scheme, which can represent both events and relations (Figure 1 D). The system also provides confidence scores enabling selection of the most likely correct predictions. Before event extrac-

[1]Data currently under review.

Event/relation type	Example
Hydroxylation	*HIF-alpha* proline **hydroxylation**
Phosphorylation	(D) siRNA-mediated ATM depletion blocks *p53* Serine-15 **phosphorylation**.
Ubiquitination	K5 **ubiquitinates** *BMPR-II* on a Membrane-proximal Lysine
DNA methylation	*RUNX3* is frequently inactivated by P2 **methylation** in solid tumors.
Glycosylation	Also, two asparagine residues in *alpha-hCG* were **glycosylated**.
Acetylation	This interaction was regulated by *Tat* **acetylation** at lysine 50.
Methylation	**Methylation** of lysine 37 of *histone H2B* is conserved.
Catalysis	*GRK2* **catalyzed** modest phosphorylation of BAC1.
Protein-Component	Three enhancer elements are located in the 40 kb intron of the *GDEP gene*.
Subunit-Complex	The most common form is a heterodimer composed of the *p65*/p50 subunits.

Table 1: Sentences with examples of the eight EPI event and two REL relation types, with highlighted **triggers**, and *protein* and site arguments. Relations have no trigger and Catalysis takes as an argument another event.

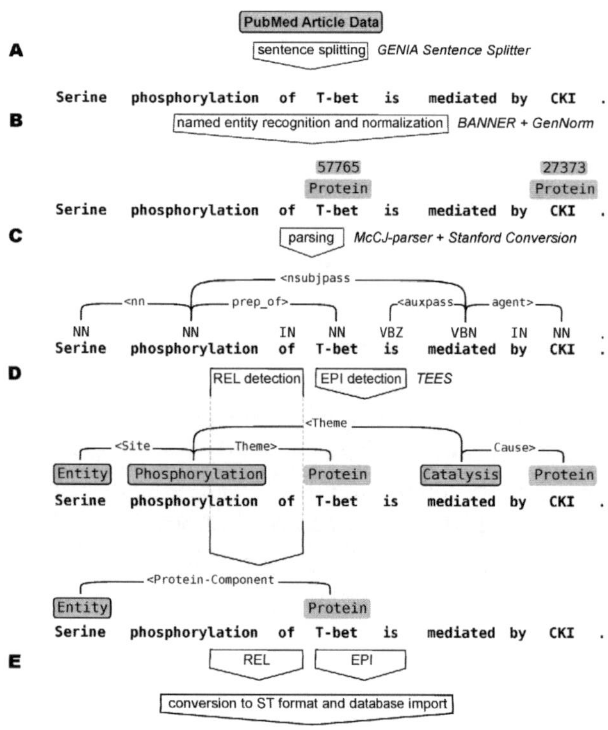

Figure 1: Event and relation extraction. Article text is split into sentences (A), where gene/protein entities are detected and normalized to their Entrez Gene IDs (B). Each sentence with at least one entity is then parsed (C). EPI events and REL relations are extracted from the parsed sentences (D) and following conversion to the BioNLP ST format are imported into a database (E). (Adapted from Björne and Salakoski (2011)).

tion, protein/gene names are detected and sentences are parsed. TEES handles all these preprocessing steps via a pipeline of tool wrappers for the GENIA Sentence Splitter (Kazama and Tsujii, 2003), the BANNER named entity recognizer (Leaman and Gonzalez, 2008), the McClosky-Charniak-Johnson (McCCJ) parser (Charniak and Johnson, 2005; McClosky, 2010) and the Stanford tools (de Marneffe et al., 2006). For a detailed description of TEES we refer to Björne and Salakoski (2011) and for the computational requirements of PubMed-scale event extraction to Björne et al. (2010).

2.3 Entity normalization

The extraction of events and relations as described in the previous sections is purely text-based and does not rely on any domain information from external resources. This ensures generalizability of the methods to new articles possibly describing novel interactions. However, practical use cases often require integration of text mining results with external resources. To enable such an integration, it is crucial to link the retrieved information to known gene/protein identifiers. In this section, we describe how we link text mining data to biomolecular databases by providing integration with Entrez Gene, UniProt, InterPro and the Protein Data Bank.

2.3.1 Protein annotations

A crucial step for integrating statements in domain text with data records is gene name normalization As part of a recent PubMed-scale effort,[2] gene

[2]Data currently under review.

normalizations were produced by the GenNorm system (Wei and Kao, 2011), assigning unique Entrez Gene identifiers (Sayers and others, 2010) to ambiguous gene/protein symbols. The GenNorm system represents the state-of-the-art in gene normalization, having achieved first rank by several evaluation criteria in the BioCreative III Challenge (Lu and others, 2011).

For practical applications, the Entrez Gene identifiers have been mapped to UniProt (The UniProt Consortium, 2011) through conversion tables provided by the NCBI. As Entrez Gene and UniProt are two of the most authoritative resources for gene and protein identification, these annotations ensure straightforward integration with other databases.

2.3.2 Complex annotations

The REL task Subunit-Complex relations all involve exactly one protein complex and one of its subunits, but the same complex may be involved in many different Subunit-Complex relations (Pyysalo et al., 2011). A key challenge for making use of these relations thus involves retrieving a unique identification of the correct complex. To identify protein complexes, we use the Protein Data Bank (PDB), an archive of structural data of biological macromolecules (Berman et al., 2000). This resource currently contains more than 80,000 3-D structures, and each polymer of a structure is annotated with its respective UniProt ID.

To assign a unique PDB ID to an entity involved in one or more Subunit-Complex relations, there is usually no other lexical context than the protein names in the sentence, e.g. "*the Rad9-Hus1-Rad1 complex*". Consequently, we rely on the normalized protein names (Section 2.3.1) to retrieve a list of plausible complexes, using data downloaded from UniProt to link proteins to PDB entries. Ambiguity is resolved by selecting the complex with the highest number of normalized proteins and giving preference to so-called representative chains. A list of representative chains is available at the PDB website, and they are determined by clustering similar protein chains[3] and taking the most confident ones based on resolution quality.

Each assignment of a PDB identifier is annotated with a confidence value between 0 and 1, expressing the percentage of proteins in the complex that could be retrieved and normalized in text. For example, even if one out of three UniProt identifiers is wrongly assigned for a mention, the correct complex might still be assigned with 0.66 confidence.

2.3.3 Domain annotations

Protein-Component relations define a relation between a gene/protein and one of its components, such as a gene promoter or a protein domain. To identify at least a substantial subset of these diverse relations, we have integrated domain knowledge extracted from InterPro. InterPro is a rich resource on protein families, domains and functional sites, integrating data from databases like PROSITE, PANTHER, Pfam, ProDom, SMART and TIGRFAMs (Hunter and others, 2012). Over 23,000 distinct InterPro entries were retrieved, linking to more than 16.5 million protein identifiers.

To assign an InterPro ID to an entity involved in one or more Protein-Component relations, a set of candidates is generated by inspecting the InterPro associations of each of the proteins annotated with that domain in text. For each such candidate, the description of the InterPro entry is matched against the lexical context around the entity by comparing the number of overlapping tokens, excluding general words, such as *domain*, and prepositions. The amount of overlap is normalized against the length of the InterPro description and expressed as a percentage, creating confidence values between 0 and 1.

Additionally, a simple pattern matching algorithm recognizes statements expressing an amino acid interval, e.g. "*aristaless domain (aa 527-542)*". When such expressions are found, the intervals as annotated in InterPro are matched against the retrieved interval from text, and the confidence values express the amount of overlap between the two intervals.

2.3.4 PTM site normalization

Six of the eight[4] EPI event types refer to post-translational modification of proteins. These events are *Hydroxylation*, *Phosphorylation*, *Ubiquitination*, *Glycosylation*, *Acetylation* and *(Protein) Methylation*. To evaluate the events predicted

[3]Requiring at least 40% sequence similarity.

[4]As we are interested in PTM sites, we make no distinction between "additive" PTMs such as *Acetylation* and their "reverse" reactions such as *Deacetylation*.

from text, we compare these to annotated post-translational modifications in UniProt. UniProt is one of the largest manually curated databases for protein knowledge, and contains annotations corresponding to each of the EPI PTM event types.

We use the reviewed and manually annotated UniProtKB/Swiss-Prot dataset (release 2012_02) in XML format. We take for each protein all *feature* elements of types *modified residue*, *cross-link* and *glycosylation site*. Each of these feature elements defines the site of the modification, either a single amino acid, or a sequence of amino acids. We select only annotations based on experimental findings, that is, features that do not have a non-experimental status (*potential*, *probable* or *by similarity*) to avoid e.g. features only inferred from the sequence.

The *modified residue* feature type covers the event types *Hydroxylation*, *Phosphorylation*, *Acetylation* and *Methylation*. We determine the class of the modification with the UniProt controlled vocabulary of post-translational modifications[5]. The *description* attribute is the ID attribute of an entry in the vocabulary, through which we can determine the more general keyword (KW) for that description, if defined. These keywords can then be connected to the corresponding event types in the case of *Hydroxylation*, *Phosphorylation*, *Acetylation* and *Methylation*. For *Ubiquitination* events, we look for the presence of the string "ubiquitin" in the *description* attribute of *cross-link* features. Finally, features corresponding to *Glycosylation* events are determined by their feature element having the type *glycosylation site*.

The result of this selection process is a list of individual modification features, which contain a type corresponding to one of the EPI PTM event types, the UniProt ID of the protein, and the position and amino acid(s) of the modification site. This data can be compared with extracted events, using their type, normalized protein arguments and modification site arguments. However, we also need to normalize the modification site arguments.

PTM sites are defined with a modification type and the numbered target amino acid residue. In EPI events, these residues are defined in the *site* argument target entities. To convert these into a form that can be aligned with UniProt, we apply a set of rules that try to determine whether a site is an amino acid. We start from the main site token, and check whether it is of the form *AA#*, where *AA* is an amino acid name, or a one or three letter code, and # an optional site number, which can also be in a token following the amino acid. For cases where the *site* entity is the word "residue" or "residues", we look for the amino acid definition in the preceding and following tokens. All strings are canonicalized with removal of punctuation, hyphens and parenthesis before applying the rules. In total, of the 177,994 events with a site argument, 75,131 could be normalized to an amino acid, and 60,622 of these to a specific residue number.

[5]http://www.uniprot.org/docs/ptmlist/

Event Type	Extracted	PMC (%)
Hydroxylation	14,555	34.17
Phosphorylation	726,757	44.00
Ubiquitination	74,027	70.46
DNA methylation	140,531	52.27
Glycosylation	154,523	42.31
Acetylation	114,585	69.40
Methylation	122,015	74.86
Catalysis	45,763	67.86
Total EPI	1,392,756	51.53
Protein-Component	1,613,170	52.59
Subunit-Complex	537,577	51.18
Total REL	2,150,747	52.23

Table 2: Total number of EPI events and REL relations extracted from PubMed abstracts and PMC full-text articles, with the fractions extracted from PMC.

3 Results

The source for extraction in this work is the set of 21 million PubMed abstracts and 372 thousand PMC open-access full-text articles. From this dataset, 1.4M EPI events and 2.2M REL relations were extracted (Table 2). For both tasks, about half of the results were extracted from PMC, confirming that full-text articles are an important source of information for these extraction targets. The total numbers of events and relations are considerably lower than e.g. the 21.3M events extracted for the GENIA task from PubMed abstracts (Björne et al., 2010; Van Landeghem et al., 2012), likely relating to the comparatively low frequency with which EPI and REL extraction targets are discussed with respect to the basic GENIA biomolecular reactions.

Event type	UniProt	Events	Match	Coverage	Events (site)	Match	Coverage
Hydroxylation	1,587	14,555	1,526	19	4,298	130	5
Phosphorylation	57,059	726,757	286,978	4,795	86,974	9,732	748
Ubiquitination	792	74,027	4,994	143	10,562	54	20
Glycosylation	6,708	154,523	18,592	897	22,846	68	31
Acetylation	6,522	114,585	15,470	764	25,689	158	30
Methylation	1,135	122,015	2,178	113	27,625	36	10
Total	73,803	1,206,462	329,738	6,731	177,994	10,178	844

Table 3: PTM events. PTMs that are not marked with non-experimental qualifiers are taken from UniProt. The *Events* column lists the total number of predicted events, and the *Events (site)* the number of events that also have a predicted site-argument. For these groups, *Match* is the number of events that matches a known PTM from UniProt, and *Coverage* the number of UniProt PTMs for which at least one match exists. For *Events* matching takes into account the PTM type and protein id, for *Events (site)* also the amino acid and position of the modified residue.

Event type	AA	UP	#	Highest confidence event	Article ID
Phosphorylation	S9	•	2	*p53* isolated from ML1, HCT116 and RKO cells, after short term genotoxic stress, were **phosphorylated** on Ser 6, Ser 9	PMC:2777442
Acetylation	S15		4	phosphorylated (Ser15), **acetylated** *p53*(Lys382)	PMC:2557062
Methylation	S15		1	**phosphorylation** of *p53* at serine 15 and acetylation	PM:10749144
Phosphorylation	S15	•	238	Chk2, as well as *p53* Ser(15) **phosphorylation** and its	PM:16731759
Phosphorylation	T18	•	12	*p53* stabilization and its **phosphorylation** in Thr18	PMC:3046209
Phosphorylation	S20	•	45	that **phosphorylation** of *p53* at Ser20 leads to	PMC:3050855
Phosphorylation	S33	•	14	**phosphorylation** of *p53* at serine 33 may be part of	PMC:35361
Phosphorylation	S37	•	20	serine 33 of *p53* in vitro when serine 37 is already	PMC:35361
Phosphorylation	S46	•	55	**phosphorylation** of *p53*, especially at Serine 46 by	PMC:2634840
Phosphorylation	T55	•	7	that **phosphorylation** of *p53* at Thr55 inhibits its	PMC:3050855
Phosphorylation	S99	•	0		
Phosphorylation	S183	•	0		
Phosphorylation	S269	•	0		
Phosphorylation	T284	•	0		
Ubiquitination	K291	•	0		
Acetylation	K292	•	0		
Ubiquitination	K292	•	0		
Acetylation	K305	•	0		
Phosphorylation	S313	•	1	**hyperphosphorylation** of *p53*, particularly of Ser313	PM:8649812
Phosphorylation	S314	•	0		
Phosphorylation	S315	•	6	to require **phosphorylation** of *p53* at serine 315 (35)	PMC:2532731
Methylation	K370	•	6	by **methylating** lysine 370 of *p53*	PMC:1636665
Acetylation	K372		1	for lysine 372 and 383 **acetylated** *p53* (Upstate,	PMC:1315280
Methylation	K372	•	5	**methylation** of *p53* by the KMT7(SET7/9) methyltransferase enzyme on Lys372	PMC:2794343
Acetylation	K373	•	16	p53 and **acetylated** *p53* (lysine-373 and lysine-382)	PMC:1208859
Methylation	K373	•	4	EHMT1-mediated *p53* **methylation** at K373	PM:20588255
Acetylation	K381	•	0		
Acetylation	K382	•	82	*p53* **acetylation** at lysine 382 was found not	PM:17898049
Methylation	K382	•	6	SET8 specifically **monomethylates** *p53* at lysine 382	PM:17707234
Methylation	K386	•	1	that **sumoylation** of *p53* at K386 blocks subsequent	PM:19339993
Phosphorylation	S392	•	35	and **phosphorylation** of *p53* at S392	PM:17237827

Table 4: Extracted and known PTM sites of *p53*. The type and site of the modification are in the first two columns. *UP* indicates whether the PTM is present in the UniProt annotation for *p53*. Column # shows the number of extracted events, followed by the event with the highest confidence score and the PubMed abstract or PMC full-text article it has been extracted from.

3.1 Extracted PTMs compared to UniProt

The EPI PTM events were compared to annotated PTMs from UniProt (Table 3). The majority of extracted PTM events (85%) have only a protein argument, and no information about the modification site, so these can only be compared by the protein id and PTM type. For the subset of proteins that also have a site, which can be normalized to an amino acid position, we can make a detailed comparison with UniProt. Finding a match for these normalized amino acids is more difficult, and for both categories, only a small fraction of proteins from UniProt is covered. In part this may be due to the limitations of the gene name normalization, as finding the exact species-specific protein ID remains a challenging task (Lu and others, 2011). However, even if the overall coverage is limited, well-known protein modifications can be assigned to specific residues, as we show in the next section.

3.2 Extracted PTMs for a single protein

For an in-depth example of PTM modifications, we study the protein *p53*, a central tumor suppressor protein that is the subject of many studies. *p53* is also among the proteins with the most UniProt PTM sites for which EPI events were predicted, making it a good example for a case study (see Table 4).

We take from UniProt all known *p53* PTMs corresponding to our EPI event types and list the number of predicted events for them (see Table 4). When the number of predicted events is high, the most confident prediction is usually a correctly extracted, clear statement about the PTM. All events for PTMs known in UniProt are correct except for the type of K386. For events not in UniProt, the two S15 ones are false positives, and K372 acetylation, while correctly extracted, is most likely a typo in the article. For the PTMs for which no event was extracted, we checked the reference article from UniProt annotation. K291, K292 ubiquitination, and K305 are from abstracts, and thus missed events. S183, S269 and T284 are from a non-open access PMC article, while S99, K292 acetylation, K305, S314 and K381 are from Excel or PDF format supplementary tables, sources outside our extraction input.

In total, we have extracted 561 PTM events related to *p53*, 554 of which correspond to a PTM annotated in UniProt. Of the 28 EPI-relevant PTMs on *p53*, 17 have at least one predicted event. The highest confidence events are about equally often from abstracts as from full texts.

Item	PubMeth	Extracted	Recall
PMID+UPID	2776	1698	61.2%
UPID	392	363	92.6%
PMID	1163	1120	96.3%

Table 5: Evaluation of DNA methylation event extraction recall against PubMeth.

3.3 DNA methylation analysis

Two recently introduced databases, PubMeth (Ongenaert et al., 2008) and MeInfoText (Fang et al., 2011) provide manually curated information on DNA methylation, primarily as it relates to cancer. To evaluate the coverage of DNA methylation event extraction, we focus here on PubMeth, as the full content of this database could be directly used. Each PubMeth DB record provides the primary name of the methylated gene and the PMID of the publication supporting the curation of the record. We used these two pieces of information to evaluate the recall [6] of DNA methylation event extraction.

We mapped PubMeth entries to UniProt identifiers (UPIDs), and extracted all unique (PMID, UPID) pairs from both PubMeth and the automatically extracted DNA methylation/demethylation events. The results of comparison of these sets of ID pairs are given in Table 5. We find that for over 60% of PubMeth entries, the system is able to recover the specific (document, gene) pair. This result is broadly in line with the recall of the system as evaluated in the BioNLP ST. However, if the matching constraint is relaxed, asking either 1) can the system extract the methylation of each gene in PubMeth *somewhere* in the literature or, inversely, 2) can the system detect *some* DNA methylation event in each document included in PubMeth as evidence, recall is over 90%. In particular, the evaluation indicates that the system shows very high recall for identifying documents discussing DNA methylation.

[6] As PubMeth does not aim for exhaustive coverage, precision cannot be directly estimated in this way. For example, PubMeth covers fewer than 2,000 documents and DNA methylation events were extracted from over 20,000, but due to differences in scope, this does not suggest precision is below 10%.

REL Type	Extracted	Match (p)	Match (e)
Prot-Cmp	1613.1K	561.8K	150.7K
SU-Cmplx	537.6K	226.5K	99.6K

Table 6: Numbers of extracted entity relations, with the protein (p) or both protein and entity (e) identified.

3.4 REL statistics

Table 6 presents the amount of extracted entity relations and the coverage of the normalization algorithms assigning protein, domain and complex identifiers. From a total of 537.6K Subunit-Complex relations, 226.5K (42%) involve a protein that could be unambiguously identified (Section 2.3.1). From this subset, 99.6K relations (44%) could be assigned to a PDB complex identifier (Section 2.3.2), accounting for 3800 representative 3D protein structures.

The Protein-Component relations are much more frequent in the data (1.6M relations) and here 35% of the relations (561.8K) involve a normalized protein mention. The assignment of InterPro domains to these Protein-Component relations (Section 2.3.3) further covers 150.7K relations in this subset (27%), identifying 5500 distinct functional domains. The vast majority of these annotations (99%) are produced by matching the lexical context against the InterPro descriptions, and only a few cases (200) matched against the amino-acid pattern.

4 Conclusions

We have combined state-of-the-art methods for gene/protein name normalization together with the best available methods for event-based extraction of protein post-translational modifications, reactions relating to the epigenetic control of gene expression, and part-of relations between genes/proteins, their components, and complexes. These methods were jointly applied to the entire available literature, both PubMed abstracts and PMC full-text documents, creating a text mining dataset unique in both scope and breadth of analysis. We further performed a comprehensive analysis of the results of this automatic extraction process against major biological database resources covering various aspects of the extracted information. This analysis indicated that text mining results for protein complexes, substructures and epigenetic DNA methylation provides already quite extensive coverage of relevant proteins. For post-translational modifications, we note that coverage still needs to be improved, but conclude that the extracted events already provide a valuable link to PTM related literature. In future work we hope to further extend the event types extracted by our PubMed-scale approach. The extraction methods as well as all data introduced in this study are freely available from `bionlp.utu.fi`.

Acknowledgments

We thank the Academy of Finland, the Research Foundation Flanders (FWO) and the UK BBSRC (reference number: BB/G013160/1) for funding, and CSC – IT Center for Science Ltd for computational resources.

References

Sophia Ananiadou, Sampo Pyysalo, Jun'ichi Tsujii, and Douglas B. Kell. 2010. Event extraction for systems biology by text mining the literature. *Trends in Biotechnology*, 28(7):381–390.

Helen M. Berman, John Westbrook, Zukang Feng, Gary Gilliland, T. N. Bhat, Helge Weissig, Ilya N. Shindyalov, and Philip E. Bourne. 2000. The protein data bank. *Nucleic Acids Research*, 28(1):235–242.

Jari Björne and Tapio Salakoski. 2011. Generalizing biomedical event extraction. In *Proceedings of the BioNLP Shared Task 2011 Workshop*, pages 183–191.

Jari Björne, Juho Heimonen, Filip Ginter, Antti Airola, Tapio Pahikkala, and Tapio Salakoski. 2009. Extracting complex biological events with rich graph-based feature sets. In *Proceedings of the BioNLP 2009 Workshop*, pages 10–18.

Jari Björne, Filip Ginter, Sampo Pyysalo, Jun'ichi Tsujii, and Tapio Salakoski. 2010. Scaling up biomedical event extraction to the entire PubMed. In *Proceedings of the BioNLP 2010 Workshop*, pages 28–36.

Eugene Charniak and Mark Johnson. 2005. Coarse-to-fine n-best parsing and MaxEnt discriminative reranking. In *Proceedings of the 43rd Annual Meeting of ACL*, pages 173–180.

Y.C. Fang, P.T. Lai, H.J. Dai, and W.L. Hsu. 2011. Meinfotext 2.0: gene methylation and cancer relation extraction from biomedical literature. *BMC bioinformatics*, 12(1):471.

Z. Z. Hu, M. Narayanaswamy, K. E. Ravikumar, K. Vijay-Shanker, and C. H. Wu. 2005. Literature mining and database annotation of protein phosphorylation using a rule-based system. *Bioinformatics*, 21(11):2759–2765.

Sarah Hunter et al. 2012. Interpro in 2011: new developments in the family and domain prediction database. *Nucleic Acids Research*, 40(D1):D306–D312.

Jun'ichi Kazama and Jun'ichi Tsujii. 2003. Evaluation and extension of maximum entropy models with inequality constraints. In *Proceedings of EMNLP 2003*, pages 137–144.

Jin-Dong Kim, Tomoko Ohta, Sampo Pyysalo, Yoshinobu Kano, and Jun'ichi Tsujii. 2009. Overview of BioNLP'09 shared task on event extraction. In *Proceedings of BioNLP 2009*, pages 1–9.

Jin-Dong Kim, Sampo Pyysalo, Tomoko Ohta, Robert Bossy, Ngan Nguyen, and Jun'ichi Tsujii. 2011. Overview of BioNLP Shared Task 2011. In *Proceedings of the BioNLP Shared Task 2011*, pages 1–6.

Robert Leaman and Graciela Gonzalez. 2008. BANNER: an executable survey of advances in biomedical named entity recognition. *Pacific Symposium on Biocomputing*, pages 652–663.

Tzong-Yi Lee, Hsien-Da Huang, Jui-Hung Hung, Hsi-Yuan Huang, Yuh-Shyong Yang, and Tzu-Hao Wang. 2006. dbPTM: an information repository of protein post-translational modification. *Nucleic acids research*, 34(suppl 1):D622–D627.

Hodong Lee, Gwan-Su Yi, and Jong C. Park. 2008. E3Miner: a text mining tool for ubiquitin-protein ligases. *Nucl. Acids Res.*, 36(suppl.2):W416–422.

Hong Li, Xiaobin Xing, Guohui Ding, Qingrun Li, Chuan Wang, Lu Xie, Rong Zeng, and Yixue Li. 2009. SysPTM: A Systematic Resource for Proteomic Research on Post-translational Modifications. *Molecular & Cellular Proteomics*, 8(8):1839–1849.

Zhiyong Lu et al. 2011. The gene normalization task in BioCreative III. *BMC Bioinformatics*, 12(Suppl 8):S2+.

Marie-Catherine de Marneffe, Bill MacCartney, and Christopher Manning. 2006. Generating typed dependency parses from phrase structure parses. In *Proceedings of LREC-06*, pages 449–454.

David McClosky. 2010. *Any domain parsing: automatic domain adaptation for natural language parsing*. Ph.D. thesis, Department of Computer Science, Brown University.

Tomoko Ohta, Sampo Pyysalo, Makoto Miwa, Jin-Dong Kim, and Jun'ichi Tsujii. 2010. Event extraction for post-translational modifications. In *Proceedings of BioNLP'10*, pages 19–27.

Tomoko Ohta, Sampo Pyysalo, Makoto Miwa, and Jun'ichi Tsujii. 2011a. Event extraction for DNA methylation. *Journal of Biomedical Semantics*, 2(Suppl 5):S2.

Tomoko Ohta, Sampo Pyysalo, and Jun'ichi Tsujii. 2011b. Overview of the epigenetics and post-translational modifications (EPI) task of BioNLP Shared Task 2011. In *Proceedings of BioNLP Shared Task 2011 Workshop*, pages 16–25.

Maté Ongenaert, Leander Van Neste, Tim De Meyer, Gerben Menschaert, Sofie Bekaert, and Wim Van Criekinge. 2008. PubMeth: a cancer methylation database combining text-mining and expert annotation. *Nucl. Acids Res.*, 36(suppl_1):D842–846.

Sampo Pyysalo, Antti Airola, Juho Heimonen, and Jari Björne. 2008. Comparative analysis of five protein-protein interaction corpora. *BMC Bioinformatics*, 9(Suppl. 3):S6.

Sampo Pyysalo, Tomoko Ohta, Jin-Dong Kim, and Jun'ichi Tsujii. 2009. Static relations: a piece in the biomedical information extraction puzzle. In *Proceedings of the BioNLP 2009 Workshop*, pages 1–9.

Sampo Pyysalo, Tomoko Ohta, and Jun'ichi Tsujii. 2011. Overview of the entity relations (REL) supporting task of BioNLP Shared Task 2011. In *Proceedings of the BioNLP Shared Task 2011 Workshop*, pages 83–88.

Eric W. Sayers et al. 2010. Database resources of the national center for biotechnology information. *Nucleic Acids Research*, 38(suppl 1):D5–D16.

Pontus Stenetorp, Goran Topić, Sampo Pyysalo, Tomoko Ohta, Jin-Dong Kim, and Jun'ichi Tsujii. 2011. Bionlp shared task 2011: Supporting resources. In *Proceedings of BioNLP Shared Task 2011 Workshop*, pages 112–120.

The UniProt Consortium. 2011. Ongoing and future developments at the universal protein resource. *Nucleic Acids Research*, 39(suppl 1):D214–D219.

Domonkos Tikk, Philippe Thomas, Peter Palaga, Jörg Hakenberg, and Ulf Leser. 2010. A comprehensive benchmark of kernel methods to extract protein-protein interactions from literature. *PLoS Comput Biol*, 6(7):e1000837, 07.

Sofie Van Landeghem, Sampo Pyysalo, Tomoko Ohta, and Yves Van de Peer. 2010. Integration of static relations to enhance event extraction from text. In *Proceedings of BioNLP'10*, pages 144–152.

Sofie Van Landeghem, Kai Hakala, Samuel Rönnqvist, Tapio Salakoski, Yves Van de Peer, and Filip Ginter. 2012. Exploring biomolecular literature with EVEX: Connecting genes through events, homology and indirect associations. *Advances in Bioinformatics*.

Chih-Hsuan Wei and Hung-Yu Kao. 2011. Cross-species gene normalization by species inference. *BMC bioinformatics*, 12(Suppl 8):S5.

Cathy H. Wu et al. 2003. The Protein Information Resource. *Nucl. Acids Res.*, 31(1):345–347.

X. Yuan, ZZ Hu, HT Wu, M. Torii, M. Narayanaswamy, KE Ravikumar, K. Vijay-Shanker, and CH Wu. 2006. An online literature mining tool for protein phosphorylation. *Bioinformatics*, 22(13):1668.

An improved corpus of disease mentions in PubMed citations

Rezarta Islamaj Doğan
National Center for Biotechnology Information
8600 Rockville Pike
Bethesda, MD 20894, USA
Rezarta.Islamaj@nih.gov

Zhiyong Lu
National Center for Biotechnology Information
8600 Rockville Pike
Bethesda, MD 20894, USA
Zhiyong.Lu@nih.gov

Abstract

The latest discoveries on diseases and their diagnosis/treatment are mostly disseminated in the form of scientific publications. However, with the rapid growth of the biomedical literature and a high level of variation and ambiguity in disease names, the task of retrieving disease-related articles becomes increasingly challenging using the traditional keyword-based approach. An important first step for any disease-related information extraction task in the biomedical literature is the disease mention recognition task. However, despite the strong interest, there has not been enough work done on disease name identification, perhaps because of the difficulty in obtaining adequate corpora. Towards this aim, we created a large-scale disease corpus consisting of 6900 disease mentions in 793 PubMed citations, derived from an earlier corpus. Our corpus contains rich annotations, was developed by a team of 12 annotators (two people per annotation) and covers all sentences in a PubMed abstract. Disease mentions are categorized into Specific Disease, Disease Class, Composite Mention and Modifier categories. When used as the gold standard data for a state-of-the-art machine-learning approach, significantly higher performance can be found on our corpus than the previous one. Such characteristics make this disease name corpus a valuable resource for mining disease-related information from biomedical text. The NCBI corpus is available for download at http://www.ncbi.nlm.nih.gov/CBBresearch/Fellows/Dogan/disease.html.

1 Introduction

Identification of biomedical entities has been an active area of research in recent years (Rinaldi et al., 2011, Smith et al., 2008, Yeh et al., 2005). Automatic systems, both lexically-based and machine learning-based, have been built to identify medically relevant concepts and/or their relationships. Biomedical entity recognition research covers not only gene/protein mention recognition (Tanabe et al., 2005, Campos et al., 2012), but also other medically relevant concepts such as disease names, chemical/drug names, treatments, procedures etc. Systems capable of achieving high performance on these tasks are highly desirable as entity recognition precedes all other information extraction and text mining tasks.

Disease information is sought very frequently in biomedical search engines. Previous PubMed log usage analysis (Islamaj Dogan et al., 2009) has shown that disease is the most frequent non-bibliographic information requested from PubMed users. Furthermore, disease information was often found to be queried together with Chemical/Drug or Gene/Protein information. Automatic recognition of disease mentions therefore, is essential not only for improving retrieval of relevant documents, but also for extraction of associations between diseases and genes or between diseases and drugs. However, prior research shows that automatic disease recognition is a challenging task due to variations and ambiguities in disease names (Leaman et al., 2009, Chowdhury and Lavelli 2010).

Lexically-based systems of disease name recognition, generally refer to the Unified Medical Language System (UMLS) (Burgun and Bodenreider

Table 1 AZDC corpus characteristics

Characteristics of the corpus	
Selected abstracts	793
Sentences	2,783
Sentences with disease mentions	1,757
Total disease mentions	3,224

2008). UMLS is a comprehensive resource of medically relevant concepts and relationships and METAMAP(Aronson and Lang 2010) is an example of a natural language processing (NLP) system that provides reliable mapping of the text of a biomedical document to UMLS concepts and their semantic types.

Machine learning systems, on the other hand, have been employed in order to benefit from the flexibility they allow over the rule-based and other statistical systems. However, machine learning systems are strongly dependent on the data available for their training; therefore a comprehensive corpus of examples representing as many variations as possible of the entity of interest is highly favorable.

To our best knowledge, there is one corpus of disease mentions in MEDLINE citations developed by Leaman et al., 2009. This corpus, AZDC corpus, was inspired by the work of Jimeno et al., 2008 and its overall characteristics are given in Table 1. This corpus has been the study of at least two different groups in building automatic systems for disease name recognition in biomedical literature (Leaman et al., 2009, Chowdhury and Lavelli, 2010). They both reported F-scores around 80% in 10-fold cross-validation experiments.

One common encountered difficulty in this domain is the fact that "disease" as a category has a very loose definition, and covers a wide range of concepts. "Disease" is a broadly-used term that refers to any condition that causes pain, suffering, distress, dysfunction, social problems, and/or death. In UMLS, the "disease" concept is covered by twelve different semantic types as shown in Table 2. The disease definition issue has been discussed extensively in other studies (Neveol et al., 2009, Neveol and Lu 2012).

Disease mentions are also heavily abbreviated in biomedical literature (Yeganova et al., 2010). These abbreviations are not always standard; the same abbreviated form may represent different defining strings in different documents. It is therefore, unclear whether these ambiguities could be resolved by an abbreviation look-up list from UMLS Metathesaurus and other available databases.

In this study, we present our efforts in improving the AZDC corpus by building a richer, broader and more complete disease name corpus. The NCBI corpus reflects a more representative view of what constitutes a disease name as it combines the decisions of twelve annotators. It also provides four different categories of disease mentions. Our work was motivated by the following observations:

- The need of a pool of experts:

  The AZDC corpus is the work of one annotator. While in terms of consistency this is generally a good thing, a pool of annotators guarantees a more representative view of the entity to be annotated and an agreement between annotators is preferred for categories with loose definitions such as "disease". Moreover, this would ensure that there would be fewer missed annotations within the corpus.
- The need of annotating all sentences in a document:

  The AZDC corpus has disease mention annotations of selected sentences in a collection of PubMed abstracts. In order to be able to perform higher level text mining tasks that explore relationships between diseases and other types of information such as genes or drugs, the disease name annotation has to include all sentences, as opposed to selected ones.

Our work is also related to other corpus annotation projects in the biomedical domain (Grouin et al., 2011, Tanabe at al., 2005, Thompson et al., 2009, Neveol at al., 2009, Chapman et al., 2012). These studies generally agree on the need of multiple experienced annotators for the project, the need of detailed annotation guidelines, and the need of large scale high-quality annotation corpora. The production of such annotated corpora facilitates the development and evaluation of entity recognition and information extraction systems.

2 Methods

Here we describe the NCBI corpus, and its annotation process. We discuss the annotation guidelines and how they evolved through the process.

2.1 The NCBI disease corpus

The AZDC corpus contains 2,783 sentences chosen from 793 PubMed abstracts. These selected

Table 2 The set of UMLS semantic types that collectively cover concepts of the "disease" category

UMLS sematic types	*Disease name example*
Acquired Abnormality	Hernia, Varicose Veins
Anatomical Abnormality	Bernheim aneurysm, Fistula of thoracic duct
Congenital Abnormality	Oppenheim's Disease, Ataxia Telangiectasia
Cell or Molecular Dysfunction	Uniparental disomy, Intestinal metaplasia
Disease or Syndrome	Acute pancreatitis, Rheumatoid Arthritis
Experimental Model of Disease	Collagen-Induced Arthritis, Jensen Sarcoma
Injury or Poisoning	Contusion and laceration of cerebrum
Mental or Behavioral Dysfunction	Schizophrenia, anxiety disorder, dementia
Neoplastic Process	Colorectal Carcinoma, Burkitt Lymphoma
Pathologic Function	Myocardial degeneration, Adipose Tissue Atrophy
Sign or Symptom	Back Pain, Seizures, Skeletal muscle paralysis
Finding	Abnormal or prolonged bleeding time

sentences were annotated for disease mentions, resulting in 1,202 unique mentions and 3,224 total mentions. The NCBI corpus starts with this original corpus; however, it is expanded to cover all the sentences in all the 793 PubMed abstracts.

2.2 Annotation guidelines

One fundamental problem in corpus annotation is the definition of what constitutes an entity to be tagged. Following the lead of the AZDC annotations, the group of annotators working on the NCBI corpus decided that a textual string would be annotated as a disease mention if it could be mapped to a unique concept in the UMLS Metathesaurus, if it corresponded to at least one of the semantic types listed in Table 2, and if it contained information that would be helpful to physicians and health care professionals.

Annotators were invited to use their common knowledge, use public resources of the National Library of Medicine such as UMLS or PubMed Health, Disease Ontology (Warren et al., 2006) and Wikipedia and consider the viewpoint of an average user trying to find information on diseases.

Initially, a set of 20 randomly chosen PubMed abstracts was used as a practice set for the development of annotation guidelines. After each annotator worked individually on the set, the results were shared and discussed among all annotators. The final annotation guidelines are summarized below and also made available at the corpus download website.

What to annotate?

1. *Annotate all specific disease mentions.*

A textual string referring to a disease name may refer to a Specific Disease, or a Disease Class. Disease mentions that could be described as a family of many specific diseases were annotated with an annotation category called Disease Class. The annotation category Specific Disease was used for those mentions which could be linked to one specific definition that does not include further categorization.

e.g. <Specific Disease> Diastrophic dysplasia </> is an <Disease Class> autosomal recessive disease</> characterized by short stature, very short limbs and joint problems that restrict mobility.

2. *Annotate contiguous text strings.*

A textual string may refer to two or more separate disease mentions. Such mentions are annotated with the Composite Mention category.

e.g. The text phrase "Duchenne and Becker muscular dystrophy" refers to two separate diseases. If this phrase is separated into two strings: "Duchenne" and "Becker muscular dystrophy", it results in information loss, because the word "Duchenne" on its own is not a disease mention.

3. *Annotate disease mentions that are used as modifiers for other concepts*

A textual string may refer to a disease name, but it may not be a noun phrase and this is better expressed with the Modifier annotation category.

e.g.: Although this mutation was initially detected in four of 33 <Modifier> colorectal cancer </> families analysed from eastern England, more extensive analysis has reduced the frequency to four of 52 English <Modifier> HNPCC </> kindreds analysed.

4. *Annotate duplicate mentions.*

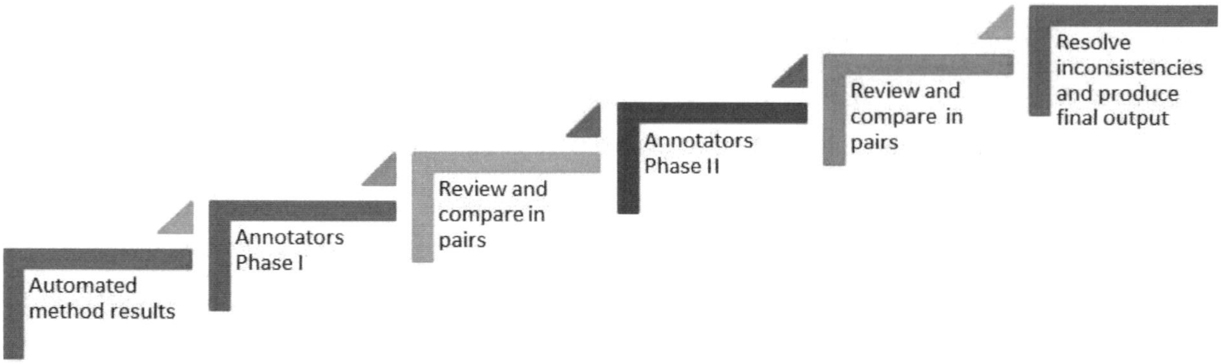

Figure 1. The annotation process

For each sentence in the PubMed abstract and title, the locations of all disease mentions are marked, including duplicates within the same sentence.

5. *Annotate minimum necessary span of text.*

The minimum span of text necessary to include all the tokens expressing the most specific form of the disease is preferred. For example, in case of the phrase "insulin-dependent diabetes mellitus", the disease mention including the whole phrase was preferred over its substrings such as "diabetes mellitus" or "diabetes".

6. *Annotate all synonymous mentions.*

Abbreviation definitions such as "Huntington disease" ("HD") are separated into two annotated mentions.

What not to annotate?

1. *Do not annotate organism names.*

Organism names such as "human" were excluded from the preferred mention. Viruses, bacteria, and other organism names were not annotated unless it was clear from the context that the disease caused by these organisms is discussed.

e.g. Studies of biopsied tissue for the presence of <Specific Disease> Epstein-Barr virus</> and <Specific Disease> cytomegalovirus </> were negative.

2. *Do not annotate gender.*

Tokens such as "male" and "female" were only included if they specifically identified a new form of the disease, for example "male breast cancer".

3. *Do not annotate overlapping mentions.*

For example, the phrase "von Hippel-Lindau (VHL) disease" was annotated as one single disease mention.

4. *Do not annotate general terms.*

Very general terms such as: disease, syndrome, deficiency, complications, abnormalities, etc. were excluded. However, the terms <u>cancer</u> and <u>tumor</u> were retained.

5. *Do not annotate references to biological processes.*

For example, terms corresponding to biological processes such as "tumorigenesis" or "cancerogenesis".

6. *Do not annotate disease mentions interrupted by nested mentions.*

Basically, do not break the contiguous text rule. E.g. WT1 dysfunction is implicated in both neoplastic (<u>Wilms tumor</u>, <u>mesothelioma</u>, <u>leukemia</u>, and <u>breast cancer</u>) and nonneoplastic (<u>glomerulosclerosis</u>) disease.

In this example, the list of all disease mentions includes: "neoplastic disease" and "nonneoplastic disease" in addition to the underlined mentions. However, they were not annotated in our corpus, because other tokens break up the phrase.

2.3 Annotators and the annotation process

The annotator group consisted of 12 people with background in biomedical informatics research and experience in biomedical text corpus annotation. The 793 PubMed citations were divided into sets of 25 PubMed citations each. Every annotator worked on 5 or 6 sets of 25 PubMed abstracts. The sets were divided randomly among annotators. Each set was shared by two people to annotate. To avoid annotator bias, pairs of annotators were chosen randomly for each set of 25 PubMed abstracts.

As illustrated in Figure 1, first, each abstract was pre-annotated using our in-house-developed CRF disease mention recognizer trained on the AZDC corpus. This process involved a 10-fold

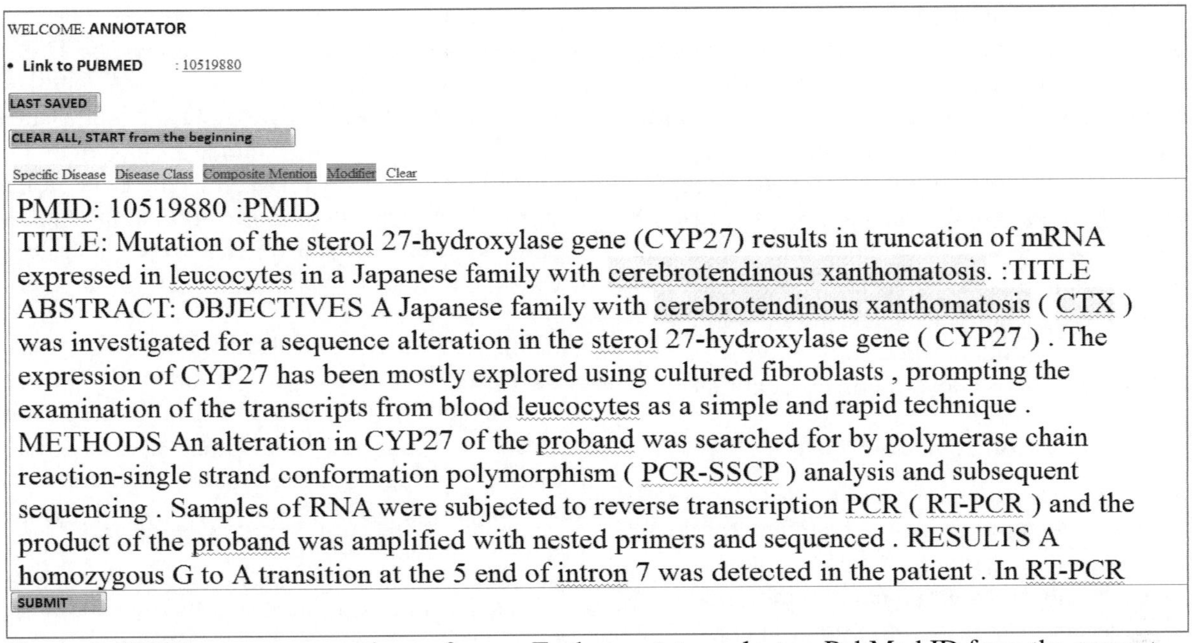

Figure 2. NCBI corpus annotation software. Each annotator selects a PubMed ID from the current working set, and is directed to this screen. Annotation categories are: Specific Disease (highlighted in yellow), Disease Class (green), Composite Mention (blue), or Modifier (purple). To annotate a disease mention in text, annotators highlight the phrase and click on the appropriate label on top of the editor screen. To delete a disease mention, annotators highlight the phrase and click on the Clear label on top of the editor. Annotators can retrieve the last saved version of their annotations for each particular document by clicking on "Last Saved" button. Annotators save their work by clicking on Submit button at the bottom of editor screen.

cross-validation scheme, where all sentences from the same PubMed abstract were assigned to the same split. The learning was performed on 9-folds and then, the PubMed abstracts assigned to the 10th fold were annotated for disease mentions on a sentence-by-sentence basis.

Annotation Phase I consisted of each pre-annotated abstract in the corpus being read and reviewed by two annotators working independently. Annotators could agree with the pre-annotation, remove it, or adjust its text span. Annotators could also add new annotations. After this initial round of annotations, a summary document was created highlighting the agreement and differences between two annotators in the annotations they produced for each abstract. This constituted the end of phase I. The pair of annotators working on the same set at this stage was given the summary document and their own annotations of Phase I.

In annotation Phase II, each annotator examined and edited his or her own annotations by reviewing the different annotations reported in the Phase I summary document. This resulted in a new set of annotations. After this round, a second summary document highlighting the agreement and differences between two annotators was created for each pair of annotators to review.

After phase II, each pair of annotators organized meetings where they reviewed, discussed and resolved their differences. After these meetings, a reconciled set of annotations was produced for each PubMed abstract. The final stage of the annotation process consisted of the first author going over all annotated segments and ensuring that annotations were consistent both in category and in text span across different abstracts and different annotation sets. For example if the phrase "classical galactosemia" was annotated in one abstract as a Specific Disease mention, all occurrences of that phrase throughout the corpus should receive consistent annotation. Identified hard cases were discussed at a meeting where all annotators were present and a final decision was made to reconcile differences. The final corpus is available at: http://www.ncbi.nlm.nih.gov/CBBresearch/Fellows/Dogan/disease.html

2.4 Annotation software

Annotation was done using a web interface (the prototype of PubTator (Wei et al., 2012)), as shown in Figure 2. Each annotator was able to log into the system and work independently. The system allowed flexibility to make annotations in the defined categories, modify annotations, correct the text span, delete as well as go back and review the process as often as needed. At the end of each annotation phase, annotators saved their work, and the annotation results were compared to find agreement and consistency among annotations.

2.5 Annotation evaluation metrics

We measured the annotators' agreement at phase I and II of the annotation process. One way to measure the agreement between two annotators is to measure their observed agreement on the sample of annotated items, as specified in Equation (1).

Agreement statistics are measured for each annotator pair, for each shared annotation set. Then, for each annotator pair the average agreement statistic is computed over all annotation sets shared between the pair of annotators. The final agreement statistic reflects the average and standard deviation computed over all annotator pairs. This is repeated for both phases.

Agreement between two annotators is measured on two levels: one, both annotators tag the same exact phrase based on character indices as a disease mention, and two, both annotators tag the same exact phrase based on character indices as a disease mention of the same category.

2.6 Application of the NCBI corpus

To compare the two disease corpora with regard to their intended primary use in training and testing machine learning algorithms, we performed a 10-fold cross validation experiment with BANNER (Leaman et al, 2009). We evaluated BANNER performance and compared Precision, Recall and F-score values for BANNER when trained and tested on AZDC corpus and the NCBI disease name corpus, respectively. In these experiments, disease mentions of all categories were included and are discussed in the Results section.

To compare the effect of improvement in disease name recognition, the different disease category annotations present in the NCBI corpus were

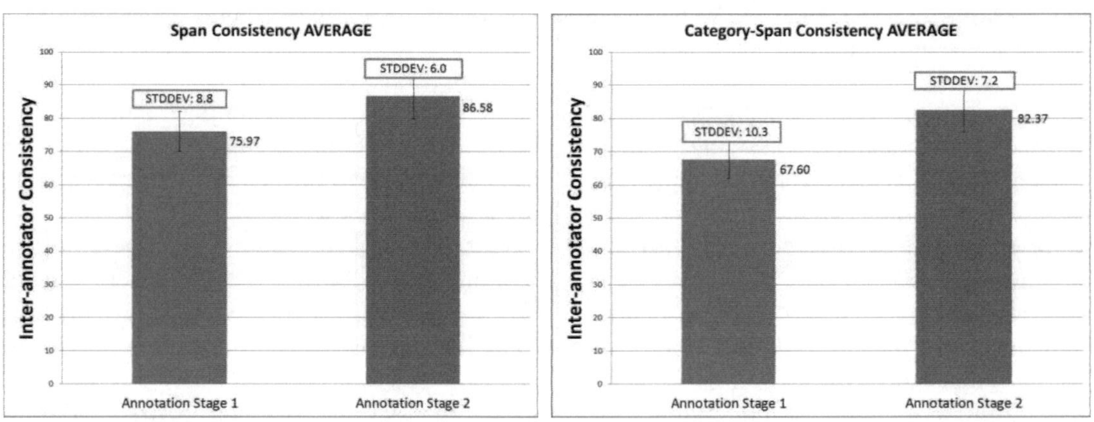

Figure 3 Inter-annotator annotation consistency measured at the span and span-category level

Table 3 The annotation results and corpus characteristics

Characteristics of the corpus	*NCBI corpus*	*AZDC*
Annotators	12	1
Annotated sentences in citation	ALL	Selected
PubMed Citations	793	793
Sentences	6,651	2,784
Sentences with disease annotations	3,752	1,757
Total disease mentions	6,900	3,228
Specific Disease	3,924	-
Disease Class	1029	-
Modifier	1,774	-
Composite Mention	173	-

(1)

$$Consistency = 100 * 2 * \frac{Agreement}{Annotations1 + \text{Annotations2}}$$

flattened into only one single category. This made the NCBI corpus compatible with the AZDC corpus.

3 Results and Discussion

3.1 Results of Inter-Annotator Agreement

Figure 3 shows the inter-annotator agreement results after Phase I and Phase II of the annotations. These statistics show a good agreement between annotators, especially after phase II of annotations. In particular, both span-consistency measure and span-category consistency measure is above 80% after phase II. These values show that our corpus reflects a high quality of annotations and that our two-stage annotation steps are effective in improving corpus consistency.

3.2 Agreement between automatic pre-annotation and final annotation results

In our previous work (Neveol et al, 2009) we have shown that automatic pre-annotation is found helpful by most annotators in assisting large-scale annotation projects with regard to speeding up the annotation time and improving annotation consistency while maintaining the high quality of the final annotations. Thus, we again used pre-annotation in this work. To demonstrate that human annotators were not biased towards the computer-generated pre-annotation, we compared the final annotation with the pre-annotation results. There are a total of 3295 pre-annotated disease mentions: 1750 were found also in the final corpus while the remaining 1545 were either modified or deleted. Furthermore, the final corpus consists of additional 3605 new annotations. Overall, the agreement between pre-annotation and final annotation results is only 35%.

3.3 Statistics of the NCBI disease corpus

After two rounds of annotation, several annotator meetings and resolving of inconsistencies, the NCBI corpus contains 793 fully annotated PubMed citations for disease mentions which are divided into these categories: Specific Disease, Disease Class, Composite Mention and Modifier. As shown in Table 3, the NCBI corpus contains more than 6K sentences, of which more than half contain disease mentions. There are 2,161 unique disease mentions total, which can be divided into these categories: 1,349 unique Specific Disease mentions, 608 unique Disease Class mentions, 121 unique Composite Disease mentions, and 356 unique Modifier disease mentions. The NCBI disease name corpus is available for download and can be used for development of disease name recognition tools, identification of Composite Disease Mentions, Disease Class or Modifier disease mention in biomedical text.

3.4 Characteristics of the NCBI corpus

This annotation task was initially undertaken for purposes of creating a larger, broader and more complete corpus for disease name recognition in biomedical literature.

The NCBI corpus addresses the inconsistencies of missed annotations by using a pool of experts for annotation and creating the annotation environment of multiple discussions and multiple rounds of annotation. The NCBI corpus addresses the problem of recognition of abbreviated disease mentions by delivering annotations for all sentences in the PubMed abstract. Processing all sentences in a document allows for recognition of an abbreviated form of a disease name. An abbreviated term could be tagged for later occurrences within the same document, if an abbreviation definition is recognized in one of the preceding sentences.

NCBI corpus provides a richer level of annotations characterized by four different categories of disease mentions: Specific Disease, Disease Class,

Table 4 NCBI corpus as training, development and testing sets for disease name recognition

Corpus Characteristics	*Training set*	*Development set*	*Test set*
PubMed Citations	593	100	100
Total disease mentions	5148	791	961
Specific Disease	2959	409	556
Disease Class	781	127	121
Modifier	1292	218	264
Composite Mention	116	37	20

Table 5 BANNER evaluation results on AZDC (original) corpus and on the NCBI corpus.

CRF-order	*Corpus*	*Precision*	*Recall*	*F-score*
1	AZDC	0.788	0.743	0.764
1	NCBI	**0.859**	**0.824**	**0.840**
2	AZDC	0.804	0.752	0.776
2	NCBI	**0.857**	**0.820**	**0.838**

Composite Mention and Modifier. Specific Disease mentions could be linked to one specific definition without further categorization, allowing for future normalization tasks. Composite Disease Mentions identify intricate lexical strings that express two or more disease mentions, allowing for future natural language processing tasks to look at them more closely. Modifier disease mentions identify non-noun phrase mentions, again useful for other text mining tasks.

Finally, the corpus can be downloaded and used for development and testing for disease name recognition and other tasks. To facilitate future work, we have divided the corpus into training, development and testing sets as shown in Table 4.

3.5 The NCBI corpus as training data for disease mention recognition

We replicated the BANNER experiments by comparing their cross-validation results on the original corpus (AZDC) and on the NCBI corpus. Our results reveal that BANNER achieves significantly better performance on the NCBI corpus: a 10% increase in F-score from 0.764 to 0.840. Table 5 shows detailed results for BANNER processing in precision, recall and F-score, for both corpora.

In addition, we performed BANNER experiments on the newly divided NCBI corpus with the following results: BANNER achieves an F-score of 0.845 on a 10 fold cross-validation experiment on the NCBI training set, an F-score of 0.819 when tested on the NCBI development set, after trained on the NCBI training set, and an F-score of 0.818 when tested on NCBI test set, after trained on NCBI training set.

3.6 Limitations of this work

The NCBI corpus was annotated manually, thus the tags assigned were judgment calls by human annotators. Annotation guidelines were established prior to the annotation process and they were refined during the annotation process, however grey areas still remained for which no explicit rules were formulated. In particular, inclusion of qualitative terms as part of the disease mention is a matter of further investigation as illustrated by the following example:

- Acute meningococcal pericarditis – Constitutes a disease mention and, exists as a separate concept in UMLS, however
- Acute Neisseria infection – May or may not include the descriptive adjective.

Similarly:

- Classical galactosemia – Includes the descriptive adjective, because it corresponds to a particular form of the disease.
- Inherited spinocerebellar ataxia – May or may not include the descriptive adjective.

Names containing conjunctions are difficult to tag. Although it might seem excessive to require a named entity recognizer to identify the whole expression for cases such as:

- Adenomatous polyps of the colon and rectum,
- Fibroepithelial or epithelial hyperplasias,
- Stage II or stage III colorectal cancer,

The NCBI disease name corpus rectifies this situation by annotating them as Composite Mention disease name category, thus, allowing for future NLP application to develop more precise methods in identifying these expressions.

Moreover, sentences which contained nested disease names require further attention, as the current annotation rule of annotating only contiguous phrases cannot select the outer mentions.

Finally, our current annotation guideline requires that only one of the four categories be assigned to each disease mention. This is not ideal because a disease mention may actually fit more than one category. For instance, a mention can be tagged as both "Modifier" and "Disease Class". In practice, for obtaining consistent annotations, the priority was given in the order of "Modifier", "Composite Mention", "Disease Class", and "Specific Disease" when more than one category deems appropriate. This aspect should be addressed at future work.

4 Conclusions

We have described the NCBI disease name corpus of tagged disease mentions in 793 PubMed titles and abstracts. The corpus was designed to capture disease mentions in the most common sense of the word, and is particularly relevant for biomedical information retrieval tasks that involve diseases. Annotations were performed for all sentences in a document, facilitating the future applications of complex information retrieval tasks connecting diseases to treatments, causes or other types of information. Annotation guidelines were designed with the goal of allowing flexible matching to UMLS concepts, while retaining true meaning of the tagged concept. A more detailed definition on what constitutes a disease name, accompanied with additional annotation rules, could help resolve some existing inconsistencies. The current corpus is reviewed several times by several annotators and describes a refined scale of annotation categories. It allows the separate definition and annotation of Composite mentions, Modifiers and distinguishes between Disease Class mentions versus Specific Diseases. The corpus is available for download[1].

Acknowledgments

Funding: This research was supported by the Intramural Research Program of the NIH, National Library of Medicine.
We sincerely thank Robert Leaman and Graciela Gonzalez for their help with BANNER, and the whole team of 12 annotators for their time and expertise on annotation of this corpus.

References

Aronson, A., Lang, F. 2010. An overview of MetaMap: historical perspective and recent advances. *J Am Med Inform Assoc*, 17(3): 229-236.

Burgun, A., Bodenreider, O. 2008. Accessing and integrating data and knowledge for biomedical research. *Yearb Med Inform*, 91-101.

Campos, D., Matos, S., Lewin, I., Oliveira, J., Rebholz-Schuhmann, D. 2012. Harmonisation of gene/protein annotations: towards a gold standard MEDLINE. *Bioinformatics,* 1;28(9):1253-61

Chapman, W.W., Savova, G.K., Zheng, J., Tharp, M., Crowley, R. 2012. Anaphoric reference in clinical reports: Characteristics of an annotated corpus. *J Biomed Inform*

Chowdhury, F.M., Lavelli, A. 2010. Disease mention recognition with specific features. *BioNLP*, 91-98.

Grouin, C., Rosset. S., Zweigenbaum, P., Fort, K., Galibert, O., Quintard, L. 2011. Proposal for an extension of traditional named entities: From guidelines to evalua-tion, an overview. *5th law workshop*, 92-100.

Islamaj Dogan, R., Murray, G. C., Neveol, A., Lu, Z. 2009. Understanding PubMed user search behavior through log analysis. *Database* (Oxford): bap018.

Jimeno,A., Jimnez-Ruiz, E., Lee, V., Gaudan, S., Berlanga,R., Reholz-Schuhmann, D.2008. Assessment of disease named entity recognition on a corpus of an-no-tated sentences. *BMC Bioinformatics*, 9(S-3).

Leaman, R., Miller, C., Gonzalez, G. 2009. Enabling Recognition of Diseases in Biomedical Text with Ma-chine Learning: Corpus and Benchmark. *Symposium on Languages in Biology and Medicine*, 82-89.

Neveol, A., Li, J., Lu, Z. 2012. Linking Multiple Disease-related resources through UMLS. *ACM International Health Informatics*.

Neveol, A., Islamaj Dogan, R., Lu, Z. 2011. Semi-automatic semantic annotation of PubMed Queries: a study on quality, efficiency, satisfaction. *J Biomed Inform*, 44(2):310-8.

Rinaldi, F., Kaljurand, K., Sætre, R. 2011. Terminological resources for text mining over biomedical scientific literature. *Artificial intelligence in medicine* 52(2)

Smith L., Tanabe L.K., Ando R.J., Kuo C.J., Chung I.F., Hsu C.N., Lin Y.S., Klinger R., Friedrich C.M., Ganchev K., Torii M., Liu H., Haddow B., Struble C.A., Povinelli R.J., Vlachos A., Baumgartner W.A. Jr., Hunter L., Carpenter B., Tsai R.T., Dai H.J., Liu F., Chen Y., Sun C., Katrenko S., Adriaans P., Blaschke C., Torres R., Neves M., Nakov P., Divoli A., Maña-López M., Mata J., Wilbur W.J. 2008.Overview of BioCreative II gene mention recognition. *Genome Biology*, 9 Suppl 2:S2.

Tanabe, L., Xie, N., Thom, L., Matten, W., Wilbur, W.J. 2005. GENETAG: a tagged corpus for gene /protein named entity recognition. *BMC Bioinformatics*, 6:S3.

Thompson, P., Iqbal, S.A., McNaught, J., Ananiadou, S. 2009. Construction of an annotated corpus to support biomedical information extraction. *BMC Bioinformatics*, 10:349.

Warren A., Kibbe J.D.O., Wolf W.A., Smith M.E., Zhu L., Lin S., Chisholm R., Disease Ontology. 2006

Wei C., Kao, H., Lu, Z., 2012. PubTator: A PubMed-like interactive curation system for document triage and literature Curation. *In proceedings of BioCreative* workshop, 145-150.

Yeganova, L., Comeau, D.C., Wilbur, W.J. 2011. Machine learning with naturally labeled data for identifying abbreviation definitions. *BMC Bioinformatics*. S3:S6

Yeh, A., Morgan, A., Colosime, M., Hirschman, L. 2005. BioCreAtIvE Task 1A: gene mention finding evaluation. *BMC Bioinformatics*, 6(Suppl 1):S2

[1] http://www.ncbi.nlm.nih.gov/CBBresearch/Fellows/Dogan/disease.html

New Resources and Perspectives for Biomedical Event Extraction

Sampo Pyysalo[1], Pontus Stenetorp[2], Tomoko Ohta[1], Jin-Dong Kim[3] and Sophia Ananiadou[1]

[1]National Centre for Text Mining and University of Manchester,
Manchester Interdisciplinary Biocentre, 131 Princess Street, Manchester, UK
[2]Tokyo University, 7-3-1 Hongo, Bunkyo-ku, Tokyo, Japan
[3]Database Center for Life Science, 2-11-16 Yayoi, Bunkyo-ku, Tokyo, Japan

Abstract

Event extraction is a major focus of recent work in biomedical information extraction. Despite substantial advances, many challenges still remain for reliable automatic extraction of events from text. We introduce a new biomedical event extraction resource consisting of analyses automatically created by systems participating in the recent BioNLP Shared Task (ST) 2011. In providing for the first time the outputs of a broad set of state-of-the-art event extraction systems, this resource opens many new opportunities for studying aspects of event extraction, from the identification of common errors to the study of effective approaches to combining the strengths of systems. We demonstrate these opportunities through a multi-system analysis on three BioNLP ST 2011 main tasks, focusing on events that none of the systems can successfully extract. We further argue for new perspectives to the performance evaluation of domain event extraction systems, considering a document-level, "off-the-page" representation and evaluation to complement the mention-level evaluations pursued in most recent work.

1 Introduction

Biomedical information extraction efforts are increasingly focusing on event extraction using structured representations that allow associations of arbitrary numbers of participants in specific roles (e.g. *Theme*, *Cause*) to be captured (Ananiadou et al., 2010). Domain event extraction has been advanced in particular by the BioNLP Shared Task (ST) events (Kim et al., 2011a; Kim et al., 2011b), which have introduced common task settings, datasets, and evaluation criteria for event extraction. Participants in these shared tasks have introduced dozens of systems for event extraction, and the resulting methods have been applied to automatically analyse the entire available domain literature (Björne et al., 2010) and applied in support of applications such as semantic literature search (Ohta et al., 2010; Van Landeghem et al., 2011b) and pathway curation support (Kemper et al., 2010).

It is possible to assess recent advances in event extraction through results for a task considered both in the BioNLP ST 2009 and 2011. By the primary evaluation criteria, the highest performance achieved in the 2009 task was 51.95% F-score, and a 57.46% F-score was reached in the comparable 2011 task (Kim et al., 2011b). These results demonstrate significant advances in event extraction methods, but also indicate that the task continues to hold substantial challenges. This has led to a call from task participants for further analysis of the data and results, accompanied by a proposal to release analyses from individual systems to facilitate such analysis (Quirk et al., 2011).

In this study, we explore new perspectives into the analyses and performance of event extraction methods. We build primarily on a new resource compiled with the support of the majority of groups participating in the BioNLP ST 2011, consisting of analyses from systems for the three main tasks sharing the text-bound event representation. We demonstrate the use of this resource through an evaluation focusing on events that cannot be extracted even by the union of combined systems, identifying particular remaining challenges for event extraction. We further propose and evaluate an alternate, document-level perspective to event extraction, demonstrating that when only unique events are considered for

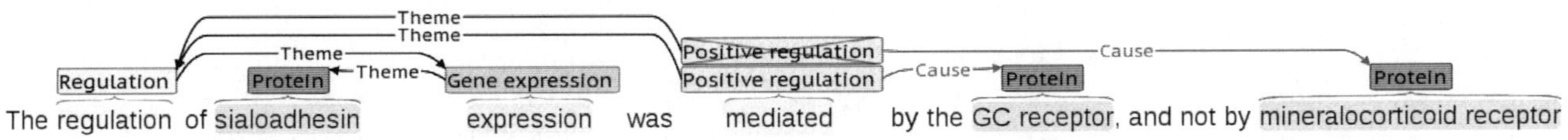

Figure 1: Example event annotations. The "crossed-out" event type identifies an event marked as negated. Event illustrations created using the STAV visualization tool (Stenetorp et al., 2011).

each document, the measured performance and even ranking of systems participating in the shared task is notably altered.

2 Background

In this work, we focus on the definition of the event extraction task first introduced in the BioNLP Shared Task 2009.[1] The task targets the extraction of *events*, represented as n-ary associations of participants (entities or other events), each marked as playing a specific *role* such as *Theme* or *Cause* in the event. Each event is assigned a *type* such as BINDING or PHOSPHORYLATION from a fixed, task-specific set. Events are further typically associated with specific *trigger* expressions that state their occurrence in text. As physical entities such as proteins are also identified in the setting with specific spans referring to the real-world entities in text, the overall task is "text-bound" in the sense of requiring not only the extraction of targeted statements from text, but also the identification of specific regions of text expressing each piece of extracted information. Events can further be marked with *modifiers* identifying additional features such as being explicitly negated or stated in a speculative context. Figure 1 shows an illustration of event annotations.

This BioNLP ST 2009 formulation of the event extraction task was followed also in three 2011 main tasks: the GE (Kim et al., 2011c), ID (Pyysalo et al., 2011a) and EPI (Ohta et al., 2011) tasks. A variant of this representation that omits event triggers was applied in the BioNLP ST 2011 bacteria track (Bossy et al., 2011), and simpler, binary relation-type representations were applied in three supporting tasks (Nguyen et al., 2011; Pyysalo et al., 2011b; Jourde et al., 2011). Due to the challenges of consistent evaluation and processing for tasks involving different representations, we focus in this work specifically on the three 2011 main tasks sharing a uniform representation: GE, ID and EPI.

[1]While far from the only formulation proposed in the literature, this specific task setting is the most frequently considered and arguably a *de facto* standard for domain event extraction.

3 New Resources for Event Extraction

In this section, we present the new collection of automatically created event analyses and demonstrate one use of the data through an evaluation of events that no system could successfully extract.

3.1 Data Compilation

Following the BioNLP ST 2011, the MSR-NLP group called for the release of outputs from various participating systems (Quirk et al., 2011) and made analyses of their system available.[2] Despite the obvious benefits of the availability of these resources, we are not aware of other groups following this example prior to the time of this publication.

To create the combined resource, we approached each group that participated in the three targeted BioNLP ST 2011 main tasks to ask for their support to the creation of a dataset including analyses from their event extraction systems. This suggestion met with the support of all but a few groups that were approached.[3] The groups providing analyses from their systems into this merged resource are summarized in Table 1, with references to descriptions of the systems used to create the included analyses. We compiled for each participant and each task both the final test set submission and a comparable submission for the separate development set.

As the gold annotations for the test set are only available for evaluation through an online interface (in order to avoid overfitting and assure the comparability of results), it is important to provide also development set analyses to permit direct comparison

[2]`http://research.microsoft.com/bionlp/`

[3]We have yet to hear back from a few groups, but none has yet explicitly denied the release of their data. Should any remaining group accept the release of their data, we will release a new, extended version of the resource.

	Task								System
Team	GE	EPI	ID	BB	BI	CO	REL	REN	description
UTurku	1	1	1	1	1	1	1	1	Björne and Salakoski (2011)
ConcordU	1	1	1			1	1	1	Kilicoglu and Bergler (2011)
UMass	1	1	1						Riedel and McCallum (2011)
Stanford	1	1	1						McClosky et al. (2011)
FAUST	1	1	1						Riedel et al. (2011)
MSR-NLP	1	1							Quirk et al. (2011)
CCP-BTMG	1	1							Liu et al. (2011)
BMI@ASU	1								Emadzadeh et al. (2011)
TM-SCS	1								Bui and Sloot (2011)
UWMadison	1								Vlachos and Craven (2011)
HCMUS	1						1		Le Minh et al. (2011)
PredX		1							-
VIBGhent							1		Van Landeghem et al. (2011a)

Table 1: BioNLP ST 2011 participants contributing to the combined resource.

	Events		
Task	Gold	FN	Recall
GE (task 1)	3250	1006	69.05%
EPI (CORE task)	601	129	78.54%
ID (CORE task)	691	183	73.52%

Table 2: Recall for the union of analyses from systems included in the combined dataset.

against gold annotations. The inclusion of both development and test set annotations also allows e.g. the study of system combination approaches where the combination parameters are estimated on development data for final testing on the test set (Kim et al., 2011a).

3.2 Evaluation

We demonstrate the use of the newly compiled dataset through a manual evaluation of GE, EPI and ID main task development set gold standard events that are not extracted by any of the systems for which analyses were available.[4] We perform evaluation on the GE subtask 1 and the EPI and ID task CORE subtasks, as all participating systems addressed the extraction targets of these subtasks.

We first evaluated each of the analyses against the development set of the respective task using the official shared task evaluation software, using options for the evaluation tools to list the sets of true positive (TP), false positive (FP) and false negative (FN) events. We then selected for each of the three tasks the set of events that were included in the FN list for all systems. This gives the results for the recall of the union of all systems shown in Table 2. The recall of the system union is approximately 30% points higher than that of any individual GE system (Kim et al., 2011c) and 25% points higher for EPI and ID (Ohta et al., 2011; Pyysalo et al., 2011a), suggesting potential remaining benefits from system combination. Nevertheless, a substantial fraction of the total set of gold events remains inaccessible also to this system union.

We then selected a random set of 100 events from each of the three sets of events that were not recovered by any system (i.e. 300 events in total) and performed a manual evaluation to identify frequent properties of these events that could contribute to extraction failures. In brief, we first performed a brief manual evaluation to identify common characteristics of these events, and then evaluated the 300 events individually to identify the set of these characteristics that apply to each event.

The results of the evaluation for common cases are shown in Table 3. We find that the most frequent property of the unrecoverable events is that they involve implicit arguments (Gerber and Chai, 2010), a difficult challenge that has not been extensively considered in domain event extraction. A closely related issue are events involving arguments in a sentence different from that containing the trigger ("cross-sentence"), connected either implicitly or through explicit coreference ("coreference"). Al-

[4] The final collection includes analyses from the systems of two groups that agreed to the release of their data after the completion of this analysis, but we expect the results to largely hold also for the final collection.

Type	GE	EPI	ID	Total
Implicit argument	18	33	15	66
Cross-sentence	14	40	4	58
Weak trigger	28	14	11	53
Coreference	12	20	18	50
Static Relation	6	28	6	40
Error in gold	17	4	9	30
Ambiguous type	2	9	11	22
Shared trigger	2	12	1	15

Table 3: Manual evaluation results for features of events that could not be recovered by any system.

though coreference was considered as as separate task in BioNLP ST 2011 (Nguyen et al., 2011), it is clear that it involves many remaining challenges for event extraction systems. Similarly, events where explicit arguments are connected to other arguments through "static" relations such as *part-of* (e.g. "*A binds the X domain of B*") represent a known challenge (Pyysalo et al., 2011b). These results suggest that further advances in event extraction performance could be gained by the integration of systems for the analysis of coreference and static relations, approaches for which some success has already been demonstrated in recent efforts (Van Landeghem et al., 2010; Yoshikawa et al., 2011; Miwa et al., 2012).

"Weak" trigger expressions that must be interpreted in context to determine whether they express an event, as well as a related class of events whose type must be disambiguated with reference to context ("ambiguous type") are comparatively frequent in the three tasks, while EPI in particular involves many cases where a trigger is shared between multiple events – an issue for approaches that assume each token can be assigned at most a single class. Finally, we noted a number of cases that we judged to be errors in the gold annotation; the number is broadly in line with the reported inter-annotator agreement for the data (see e.g. Ohta et al. (2011)).

While there is an unavoidable subjective component to evaluations such as this, we note that a similar evaluation performed following the BioNLP Shared Task 2009 using test set data reached broadly comparable results (Kim et al., 2011a). The newly compiled dataset represents the first opportunity for those without direct access to the test set data and submissions to directly assess the task results, as demonstrated here. We hope that this resource will encourage further exploration of both the data, the system analyses and remaining challenges in event extraction.

4 New Perspectives to Event Extraction

As discussed in Section 2, the BioNLP ST event extraction task is "text-bound": each entity and event annotation is associated with a specific span of text. Contrasted to the alternative approach where annotations are document-level only, this approach has a number of important benefits, such as allowing machine learning methods for event extraction to be directly trained on fully and specifically annotated data without the need to apply frequently error-prone heuristics (Mintz et al., 2009) or develop machine learning methods addressing the mapping between text expressions and document-level annotations (Riedel et al., 2010). Many of the most successful event extraction approaches involve direct training of machine learning methods using the text-bound annotations (Riedel and McCallum, 2011; Björne and Salakoski, 2011; McClosky et al., 2011). However, while the availability of text-bound annotations in data provided to task participants is clearly a benefit, there are drawbacks to the choice of exclusive focus on text-bound annotations in system output, including issues relating to evaluation and the applicability of methods to the task. In the following section, we discuss some of these issues and propose alternatives to representation and evaluation addressing them.

4.1 Evaluation

The evaluation of the BioNLP ST is instance-based and text-bound: each event in gold annotation and each event extracted by a system is considered independently, separating different mentions of the "same" real-world event. This is the most detailed (sensitive) evaluation setting permitted by the data, and from a technical perspective a reasonable choice for ranking systems performing the task.

However, from a practical perspective, this evaluation setting arguably places excessively strict demands on systems, and may result in poor correlation between measured performance and the practical value of systems. Our motivating observations are that specific real-world events tend to be mentioned multiple times in a single publication – especially the events that are of particular importance in the study – and that there are few practical applications for which it is necessary to find each such repeated mention. For example, in literature search for e.g. pathway or database curation support, one typical information need is to identify biomolecular reactions involving a specific protein. Event extraction can support such needs either by summarizing all events involving the protein that could be extracted from the literature (Van Landeghem et al., 2011b), or by retrieving documents (perhaps showing relevant text snippets) containing such events (Ohta et al., 2010). For the former to meet the information need, it may be sufficient that each different event is extracted once from the entire literature; for the latter, once from each relevant document. For uses such as these, there is no obvious need for, or, indeed, no very obvious benefit from the ability of extraction systems to separately enumerate every mention of every event in every publication. It is easy to envision other practical use cases where instance-level extraction performance is at best secondary and, we argue, difficult to identify ones where it is of critical importance.

For applications such as these, the important question is the reliability of the system at identifying events either on the level of documents or on the level of (a relevant subset of) the literature, rather than on the level of individual mentions. For a more complete and realistic picture of the practical value of event extraction methods, measures other than instance-level should thus also be considered.

4.2 Task setting

While applications can benefit from the ability of IE systems to identify a specific span of text supporting extracted information,[5] the requirement of the BioNLP ST setting that the output of event extraction systems must identify specific text spans for each entity and event makes it complex or impossible to address the task using a number of IE methods that might otherwise represent feasible approaches to event extraction.

[5]For example, for curation support tasks, this allows the human curator to easily check the correctness of extracted information and helps to select "evidence sentences", as included in many databases.

For example, Patwardhan and Riloff (2007) and Chambers and Jurafsky (2011) consider an IE approach where the extraction targets are MUC-4 style document-level templates (Sundheim, 1991), the former a supervised system and the latter fully unsupervised. These methods and many like them for tasks such as ACE (Doddington et al., 2004) work on the document level, and can thus not be readily applied or evaluated against the existing annotations for the BioNLP shared tasks. Enabling the application of such approaches to the BioNLP ST could bring valuable new perspectives to event extraction.

4.3 Alternative evaluation

We propose a new mode of evaluation that otherwise follows the primary BioNLP ST evaluation criteria, but incorporates the following two exceptions:

1. remove the requirement to match trigger spans
2. only require entity texts, not spans, to match

The first alternative criterion has also been previously considered in the GE task evaluation (Kim et al., 2011c); the latter has, to the best of our knowledge, not been previously considered in domain event extraction. We additionally propose to consider only the minimal set of events that are unique on the document level (under the evaluation criteria), thus eliminating effects from repeated mentions of a single event on evaluated performance. We created tools implementing this mode of evaluation with reference to the BioNLP ST 2011 evaluation tools.

While this type of evaluation has, to the best of our knowledge, not been previously applied specifically in biomedical event extraction, it is closely related (though not identical) to evaluation criteria applied in MUC, ACE, and the in-domain PPI relation extraction tasks in BioCreative (Krallinger et al., 2008).

4.4 Alternative representation

A true conversion to a document-level, "off the page" representation would require manual annotation efforts to identify the real-world entities and events referred to in text (Doddington et al., 2004). However, it is possible to reasonably approximate such a representation through an automatic heuristic conversion.

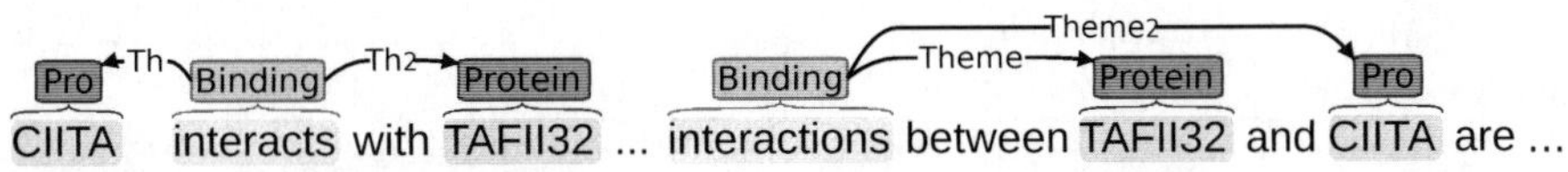

BioNLP Shared Task	*Document level*
T1 Protein 0 5 CIITA	T1 Protein CIITA
T2 Protein 21 28 TAFII32	T2 Protein TAFII32
T3 Binding 6 15 interacts	E1 Binding Theme:T1 Theme2:T2
E1 Binding:T3 Theme:T1 Theme2:T2	
T4 Protein 54 61 TAFII32	
T5 Protein 66 71 CIITA	
T6 Binding 33 45 interactions	
E2 Binding:T6 Theme:T4 Theme2:T5	

Figure 2: Illustration of BioNLP Shared Task annotation format and the proposed document-level ("off-the-page") format.

We first introduce a non-textbound annotation format that normalizes over differences in e.g. argument order and eliminates duplicate events. The format largely follows that of the shared task but removes any dependencies and references to text offsets (see Figure 2). The conversion process into this representation involves a number of steps. First, we merge duplicate pairs of surface strings and types, as different mentions of the same entity in different parts of the text are no longer distinguishable in the representation. In the original format, equivalence relations (Kim et al., 2011a) are annotated only for specific mentions. When "raising" the annotations to the document level, equivalence relations are reinterpreted to cover the full document by extending the equivalence to all mentions that share the surface form and type with members of existing equivalence classes. Finally, we implemented an event equivalence comparison to remove duplicate annotations from each document. The result of the conversion to this alternate representation is thus an "off-the-page" summary of the unique set of events in the document.

This data can then be used for training and comparison of methods analogously to the original annotations, but without the requirement that all analyses include text-bound annotations.

4.5 Experimental Results

We next present an evaluation using the alternative document-level event representation and evaluation, comparing its results to those for the primary shared task evaluation criteria. As comparatively few of the shared task participants attempted subtasks 2 and 3 for GE or the FULL task setting for EPI and ID, we consider only GE subtask 1 and the EPI and ID task CORE extraction targets in these experiments. We refer to the task overviews for the details of the subtasks and the primary evaluation criteria (Kim et al., 2011c; Pyysalo et al., 2011a; Ohta et al., 2011).

	Primary criteria			New criteria		
Group	Rec.	Prec.	F	Rec.	Prec.	F
FAUST	49.41	64.75	56.04	53.10	67.56	59.46
UMass	48.49	64.08	55.20	52.55	66.57	58.74
UTurku	49.56	57.65	53.30	54.23	60.11	57.02
MSR-NLP	48.64	54.71	51.50	53.55	58.24	55.80
ConcordU	43.55	59.58	50.32	47.42	60.85	53.30
UWMadison	42.56	61.21	50.21	46.09	62.50	53.06
Stanford	42.36	61.08	50.03	46.48	63.22	53.57
BMI@ASU	36.91	56.63	44.69	41.15	61.44	49.29
CCP-BTMG	31.57	58.99	41.13	34.82	66.89	45.80
TM-SCS	32.73	45.84	38.19	38.02	50.87	43.51
HCMUS	10.12	27.17	14.75	14.50	40.05	21.29

Table 4: Comparison of BioNLP ST 2011 GE task 1 results.

Tables 4, 5 and 6 present the results for the GE, EPI and ID tasks, respectively. For GE, we see consistently higher F-scores for the new criteria, in most cases reflecting primarily an increase in recall, but also involving increases in precision. The F-score differences range between 3-4% for most high-ranking systems, with more substantial increases for lower-ranking systems. Notable increases in precision are seen for some systems (e.g. HCMUS), indicating that the systems comparatively frequently extract correct information, but associated with the wrong spans of text.

Group	Primary criteria			New criteria		
	Rec.	Prec.	F	Rec.	Prec.	F
UTurku	68.51	69.20	68.86	74.20	69.14	71.58
FAUST	59.88	80.25	68.59	67.04	76.82	71.60
MSR-NLP	55.70	77.60	64.85	59.24	77.66	67.21
UMass	57.04	73.30	64.15	65.76	69.65	67.65
Stanford	56.87	70.22	62.84	62.74	67.12	64.86
CCP-BTMG	45.06	63.37	52.67	54.62	63.17	58.58
ConcordU	40.28	76.71	52.83	48.41	76.57	59.32

Table 5: Comparison of BioNLP ST 2011 EPI CORE task results.

Group	Primary criteria			New criteria		
	Rec.	Prec.	F	Rec.	Prec.	F
FAUST	50.84	66.35	57.57	50.11	65.33	56.72
UMass	49.67	62.39	55.31	49.34	60.98	54.55
Stanford	49.16	56.37	52.52	42.00	50.80	45.98
ConcordU	50.91	43.37	46.84	43.42	37.18	40.06
UTurku	39.23	49.91	43.93	48.03	51.84	49.86
PredX	23.67	35.18	28.30	20.94	30.69	24.90

Table 6: Comparison of BioNLP ST 2011 ID CORE task results.

For EPI (Table 5), we find comparable differences in F-score to those for GE, but there is a significant difference in the precision-recall balance: the majority of systems show over 5% points higher recall under the new criteria, but many show substantial losses in precision, while for GE precision was also systematically increased. This effect was not unexpected: we judge this to reflect primarily the increased number of opportunities to extract each unique event (higher recall) combined with the comparatively higher effect from errors from the reduction in the total number of unique correct extraction targets (lower precision). It is not clear from our analysis why a comparable effect was not seen for GE. Interestingly, most systems show a better precision/recall balance under the new criteria than the old, despite not optimizing for these criteria.

For ID (Table 6), we find a different effect also on F-score, with all but one system showing reduced performance under the new criteria, with some very clear drops in performance; the only system to benefit is UTurku. Analysis suggests that this effect traces primarily to a notable reduction in the number of simple PROCESS events that take no arguments[6] when considering unique events on the document level instead of each event mention independently.[7] Conversely, the Stanford system, which showed the highest instance-level performance in the extraction of PROCESS type events (see Pyysalo et al. (2011a)), shows a clear loss in precision.

The clear differences in performance and the many cases in which the system rankings under the two criteria differ demonstrate that the new evaluation criteria can have a decisive effect in which approaches to event extraction appear preferred. While there may be cases for which the original shared task criteria are preferred, there is at the very minimum a reasonable argument to be made that the emphasis these criteria place on the extraction of each instance of simple events is unlikely to reflect the needs of many practical applications of event extraction.

While these experimental results demonstrate that the new evaluation criteria emphasize substantially different aspects of the performance of the systems than the original criteria, they cannot *per se* serve as an argument in favor of one set of criteria over another. We hope that these results and the accompanying tools will encourage increased study and discussion of evaluation criteria for event extraction and more careful consideration of the needs of specific applications of the technology.

5 Discussion and Conclusions

We have presented a new resource combining analyses from the systems participating in the GE, ID and EPI main tasks of the BioNLP Shared Task 2011, compiled with the collaboration of groups participating in these tasks. We demonstrated one use of the resource through an evaluation of development set events that none of the participating systems could recover, finding that events involving implicit arguments, coreference and participants in more than once sentence continue to represent challenges to the event extraction systems that participated in these tasks.

We further argued in favor of new perspectives to the evaluation of domain event extraction systems,

[6]The ID task annotation criteria call for mentions of some high-level biological processes such as "infection" to be annotated as PROCESS even if no explicit participants are mentioned (Pyysalo et al., 2011a).

[7]It is interesting to note that there was an error in the UTurku system implementation causing it to fail to output any events without arguments (Jari Björne, personal communication), likely contributing to the effect seen here.

emphasizing in particular the need for document-level, "off-the-page" representations and evaluation to complement the text-bound, instance-level evaluation criteria that have so far been applied in the shared task evaluation. We proposed a variant of the shared task standoff representation for supporting such evaluation, and introduced evaluation tools implementing the proposed criteria. An evaluation supported by the introduced resources demonstrated that the new criteria can in cases provide substantially different results and rankings of the systems, confirming that the proposed evaluation can serve as an informative complementary perspective into event extraction performance.

In future work, we hope to further extend the coverage of the provided system outputs as well as their analysis to cover all participants of all tasks in the BioNLP Shared Task 2011. We also aim to use the compiled resource in further study of appropriate criteria for the evaluation of event extraction methods and deeper analysis of the remaining challenges in event extraction.

To encourage further study of all aspects of event extraction, all resources and tools introduced in this study are provided freely to the community from `http://2011.bionlp-st.org`.

Acknowledgments

We wish to thank the members of all groups contributing to the combined resource, and in particular the members of the MSR-NLP group for providing both the initial suggestion for its creation as well as the first publicly released analyses from their system. We would also like to thank the anonymous reviewers for their many insightful comments.

This work was funded in part by UK Biotechnology and Biological Sciences Research Council (BBSRC) under project Automated Biological Event Extraction from the Literature for Drug Discovery (reference number: BB/G013160/1), by the Ministry of Education, Culture, Sports, Science and Technology of Japan under the Integrated Database Project and by the Swedish Royal Academy of Sciences.

References

Sophia Ananiadou, Sampo Pyysalo, Jun'ichi Tsujii, and Douglas B. Kell. 2010. Event extraction for systems biology by text mining the literature. *Trends in Biotechnology*, 28(7):381–390.

Jari Björne and Tapio Salakoski. 2011. Generalizing biomedical event extraction. In *Proceedings of the BioNLP Shared Task 2011 Workshop*.

Jari Björne, Filip Ginter, Sampo Pyysalo, Jun'ichi Tsujii, and Tapio Salakoski. 2010. Complex event extraction at PubMed scale. *Bioinformatics*, 26(12):i382–390.

Robert Bossy, Julien Jourde, Philippe Bessières, Maarten van de Guchte, and Claire Nédellec. 2011. BioNLP Shared Task 2011 - Bacteria Biotope. In *Proceedings of BioNLP Shared Task 2011 Workshop*, pages 56–64.

Quoc-Chinh Bui and Peter. M.A. Sloot. 2011. Extracting biological events from text using simple syntactic patterns. In *Proceedings of BioNLP Shared Task 2011 Workshop*, pages 143–146.

Nathanael Chambers and Dan Jurafsky. 2011. Template-based information extraction without the templates. In *Proceedings of the ACL-HLT 2011*, pages 976–986.

George Doddington, Alexis Mitchell, Mark Przybocki, Lance Ramshaw, Stephanie Strassel, and Ralph Weischedel. 2004. The automatic content extraction (ACE) program–tasks, data, and evaluation. In *Proceedings of LREC*, volume 4, pages 837–840.

Ehsan Emadzadeh, Azadeh Nikfarjam, and Graciela Gonzalez. 2011. Double layered learning for biological event extraction from text. In *Proceedings of BioNLP Shared Task 2011 Workshop*, pages 153–154.

Matthew Gerber and Joyce Chai. 2010. Beyond nombank: A study of implicit arguments for nominal predicates. In *Proceedings of ACL 2010*, pages 1583–1592.

Julien Jourde, Alain-Pierre Manine, Philippe Veber, Karën Fort, Robert Bossy, Erick Alphonse, and Philippe Bessières. 2011. BioNLP Shared Task 2011 – Bacteria gene interactions and renaming. In *Proceedings of BioNLP Shared Task 2011 Workshop*, pages 65–73.

Brian Kemper, Takuya Matsuzaki, Yukiko Matsuoka, Yoshimasa Tsuruoka, Hiroaki Kitano, Sophia Ananiadou, and Jun'ichi Tsujii. 2010. PathText: a text mining integrator for biological pathway visualizations. *Bioinformatics*, 26(12):i374–i381.

Halil Kilicoglu and Sabine Bergler. 2011. Adapting a general semantic interpretation approach to biological event extraction. In *Proceedings of the BioNLP Shared Task 2011 Workshop*.

Jin-Dong Kim, Tomoko Ohta, Sampo Pyysalo, Yoshinobu Kano, and Jun'ichi Tsujii. 2011a. Extracting bio-molecular events from literature - the BioNLP'09 shared task. *Computational Intelligence*, 27(4):513–540.

Jin-Dong Kim, Sampo Pyysalo, Tomoko Ohta, Robert Bossy, Ngan Nguyen, and Jun'ichi Tsujii. 2011b.

Overview of BioNLP Shared Task 2011. In *Proceedings of BioNLP Shared Task*, pages 1–6.

Jin-Dong Kim, Yue Wang, Toshihisa Takagi, and Akinori Yonezawa. 2011c. Overview of the Genia Event task in BioNLP Shared Task 2011. In *Proceedings of the BioNLP Shared Task 2011 Workshop*.

Martin Krallinger, Florian Leitner, Carlos Rodriguez-Penagos, Alfonso Valencia, et al. 2008. Overview of the protein-protein interaction annotation extraction task of BioCreative II. *Genome Biology*, 9(Suppl 2):S4.

Quang Le Minh, Son Nguyen Truong, and Quoc Ho Bao. 2011. A pattern approach for biomedical event annotation. In *Proceedings of BioNLP Shared Task 2011 Workshop*, pages 149–150.

Haibin Liu, Ravikumar Komandur, and Karin Verspoor. 2011. From graphs to events: A subgraph matching approach for information extraction from biomedical text. In *Proceedings of the BioNLP Shared Task 2011 Workshop*.

David McClosky, Mihai Surdeanu, and Christopher Manning. 2011. Event extraction as dependency parsing. In *Proceedings of ACL-HLT 2011*, pages 1626–1635.

Mike Mintz, Steven Bills, Rion Snow, and Dan Jurafsky. 2009. Distant supervision for relation extraction without labeled data. In *Proceedings of ACL-IJCNLP 2009*, pages 1003–1011.

Makoto Miwa, Paul Thompson, and Sophia Ananiadou. 2012. Boosting automatic event extraction from the literature using domain adaptation and coreference resolution. *Bioinformatics*.

Ngan Nguyen, Jin-Dong Kim, and Jun'ichi Tsujii. 2011. Overview of BioNLP 2011 Protein Coreference Shared Task. In *Proceedings of BioNLP Shared Task 2011 Workshop*, pages 74–82.

Tomoko Ohta, Takuya Matsuzaki, Naoaki Okazaki, Makoto Miwa, Rune Sætre, Sampo Pyysalo, and Jun'ichi Tsujii. 2010. Medie and info-pubmed: 2010 update. *BMC Bioinformatics*, 11(Suppl 5):P7.

Tomoko Ohta, Sampo Pyysalo, and Jun'ichi Tsujii. 2011. Overview of the Epigenetics and Post-translational Modifications (EPI) task of BioNLP Shared Task 2011. In *Proceedings of the BioNLP Shared Task 2011 Workshop*.

Siddharth Patwardhan and Ellen Riloff. 2007. Effective information extraction with semantic affinity patterns and relevant regions. In *Proceedings of EMNLP-CoNLL 2007*, pages 717–727.

Sampo Pyysalo, Tomoko Ohta, Rafal Rak, Dan Sullivan, Chunhong Mao, Chunxia Wang, Bruno Sobral, Jun'ichi Tsujii, and Sophia Ananiadou. 2011a. Overview of the Infectious Diseases (ID) task of BioNLP Shared Task 2011. In *Proceedings of the BioNLP Shared Task 2011 Workshop*.

Sampo Pyysalo, Tomoko Ohta, and Jun'ichi Tsujii. 2011b. Overview of the entity relations (REL) supporting task of BioNLP Shared Task 2011. In *Proceedings of BioNLP Shared Task 2011 Workshop*, pages 83–88.

Chris Quirk, Pallavi Choudhury, Michael Gamon, and Lucy Vanderwende. 2011. MSR-NLP entry in BioNLP Shared Task 2011. In *Proceedings of BioNLP Shared Task 2011 Workshop*, pages 155–163.

Sebastian Riedel and Andrew McCallum. 2011. Fast and robust joint models for biomedical event extraction. In *Proceedings of EMNLP 2011*, pages 1–12.

Sebastian Riedel, Limin Yao, and Andrew McCallum. 2010. Modeling relations and their mentions without labeled text. *Machine Learning and Knowledge Discovery in Databases*, pages 148–163.

Sebastian Riedel, David McClosky, Mihai Surdeanu, Andrew McCallum, and Chris Manning. 2011. Model combination for event extraction in BioNLP 2011. In *Proceedings of the BioNLP Shared Task 2011 Workshop*.

Pontus Stenetorp, Goran Topić, Sampo Pyysalo, Tomoko Ohta, Jin-Dong Kim, and Jun'ichi Tsujii. 2011. BioNLP Shared Task 2011: Supporting Resources. In *Proceedings of the BioNLP Shared Task 2011 Workshop*.

Beth M. Sundheim. 1991. Third message understanding evaluation and conference (MUC-3): Phase 1 status report. In *Proceedings of the Speech and Natural Language Workshop*, pages 301–305.

Sofie Van Landeghem, Sampo Pyysalo, Tomoko Ohta, and Yves Van de Peer. 2010. Integration of static relations to enhance event extraction from text. In *Proceedings of BioNLP 2010*, pages 144–152.

Sofie Van Landeghem, Thomas Abeel, Bernard De Baets, and Yves Van de Peer. 2011a. Detecting entity relations as a supporting task for bio-molecular event extraction. In *Proceedings of BioNLP Shared Task 2011 Workshop*, pages 147–148.

Sofie Van Landeghem, Filip Ginter, Yves Van de Peer, and Tapio Salakoski. 2011b. Evex: a pubmed-scale resource for homology-based generalization of text mining predictions. In *Proceedings of BioNLP 2011 Workshop*, pages 28–37.

Andreas Vlachos and Mark Craven. 2011. Biomedical event extraction from abstracts and full papers using search-based structured prediction. In *Proceedings of BioNLP Shared Task 2011 Workshop*, pages 36–40.

Katsumasa Yoshikawa, Sebastian Riedel, Tsutomu Hirao, Masayuki Asahara, and Yuji Matsumoto. 2011. Coreference based event-argument relation extraction on biomedical text. *Journal of Biomedical Semantics*, 2(Suppl 5):S6.

Combining Compositionality and Pagerank for the Identification of Semantic Relations between Biomedical Words

Thierry Hamon
LIM&BIO UFR SMBH
Université Paris 13, France
thierry.hamon@univ-paris13.fr

Christopher Engström
Division of Applied Mathematics
Mälardalen University
Västerås, Sweden

Mounira Manser
LIM&BIO UFR SMBH
Université Paris 13, France

Zina Badji and Natalia Grabar
CNRS UMR 8163 STL
Université Lille 1&3
59653 Villeneuve d'Ascq, France
natalia.grabar@univ-lille3.fr

Sergei Silvestrov
Division of Applied Mathematics
Mälardalen University
Västerås, Sweden

Abstract

The acquisition of semantic resources and relations is an important task for several applications, such as query expansion, information retrieval and extraction, machine translation. However, their validity should also be computed and indicated, especially for automatic systems and applications. We exploit the compositionality based methods for the acquisition of synonymy relations and of indicators of these synonyms. We then apply pagerank-derived algorithm to the obtained semantic graph in order to filter out the acquired synonyms. Evaluation performed with two independent experts indicates that the quality of synonyms is systematically improved by 10 to 15% after their filtering.

1 Introduction

Natural languages have extremely rich means to express or to hide semantic relations: these can be more or less explicit. Nevertheless, the semantic relations are important to various NLP tasks within general or specialized languages (*i.e.*, query expansions, information retrieval and extraction, text mining or machine translation) and their deciphering must be tackled by automatic approaches. We focus in this work on synonymy relations. Thus, it is important to be able to decide whether two terms (*i.e.*, *anabolism* and *acetone anabolism*, *acetone anabolism* and *acetone biosynthesis*, *replication of mitochondrial DNA* and *mtDNA replication*) convey the same, close or different meanings. According to the ability of an automatic system to decipher such relations, the answers of the system will be more or less exhaustive. Several solutions may be exploited when deciphering the synonymy relations:

1. Exploitation of the existing resources in which the synonyms are already encoded. However, in the biomedical domain, such resources are not well described. If the morphological description is the most complete (NLM, 2007; Schulz et al., 1999; Zweigenbaum et al., 2003), little or no freely available synonym resources can be found, while the existing terminologies often lack the synonyms.
2. Exploitation and adaptation of the existing methods (Grefenstette, 1994; Hamon et al., 1998; Jacquemin et al., 1997; Shimizu et al., 2008; Wang and Hirst, 2011).
3. Proposition of new methods specifically adapted to the processed data.

Due to the lack of resources, we propose to exploit the solutions 2 and 3. In either of these situations, the question arises about the robustness and the validity of the acquired relations. For instance, (Hamon and Grabar, 2008) face two problems: (1) contextual character of synonymy relations (Cruse, 1986), *i.e.*, two words are considered as synonyms if they can occur within the same context, which makes this relation more or less broad depending on the usage; (2) ability of automatic tools to detect and characterize these relations, *i.e.*, two words taken out of their context can convey different relations than the one expected. Our objective is to assess the reliability of synonymy resources. We propose to weight and to filter the synonym relations with the pagerank-derived algorithm (Brin and Page, 1998). When

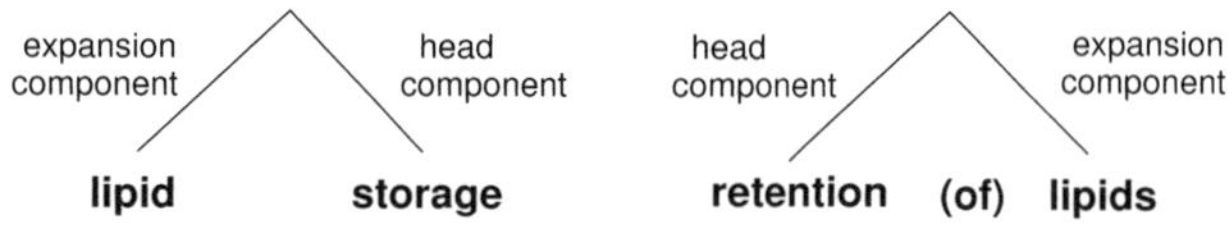

Figure 1: Parsing tree of the terms *lipid storage* and *retention of lipids*

processing textual data, this algorithm has been previously applied in different contexts such as semantic disambiguation (Mihalcea et al., 2004; Sinha and Mihalcea, 2007; Agirre and Soroa, 2009), summarization (Fernandez et al., 2009) and, more recently, for the identification of synonyms (Sinha and Mihalcea, 2011). This last work takes into account the usage of a given word in corpora and its known synonyms from lexical resources. Other related works propose also the exploitation of the random walk algorithm for the detection of semantic relatedness of words (Gaume, 2006; Hughes and Ramage, 2007) and of documents (Hassan et al., 2007). Our work is different from the previous work in several ways: (1) the acquisition of synonymy is done on resources provided by a specialized domain; (2) the pagerank algorithm is exploited for the filtering of semantic relations generated with linguistically-based approaches; (3) the pagerank algorithm is adapted to the small size of the processed data.

In the following of this paper, we present first the material (section 2), then the method we propose (section 3). We then describe the experiments performed and the results (section 4), as well as their evaluation and discussion (section 5). Finally, we conclude and indicate some perspectives (section 6).

2 Material

We use the *Gene Ontology* (*GO*) as the original resource from which synonym lexicon (or elementary synonym relations) are induced. The goal of the *GO* is to produce a structured vocabulary for describing the roles of genes and their products in any organism. *GO* terms are structured with four types of relations: subsumption `is-a`, meronymy `part-of`, `synonymy` and `regulates`. The version used in the current work is issued from the UMLS 2011AA. It provides 54,453 concepts and their 94,161 terms. The generated pairs of terms have 119,430 `is-a` and 101,254 `synonymy` relations.

3 Methods

Our method has several steps: preprocessing of *GO* terms (section 3.1), induction of elementary synonyms (section 3.2) and their characterization with lexical and linguistic indicators (section 3.3), analysis of the synonymy graph, its weighting thanks to the pagerank algorithm and its filtering (section 3.4). We also perform an evaluation of the generated and filtered synonymy relations (section 3.5).

In the following, we call *original synonyms* those synonyms which are provided by *GO*, and we call *elementary synonyms* those synonyms which are induced by the compositionality based approach.

3.1 Preprocessing the *GO* terms: Ogmios NLP platform

The aim of terminology preprocessing step is to provide syntactic analysis of terms for computing their syntactic dependency relations. We use the `Ogmios` platform[1] and perform: segmentation into words and sentences; POS-tagging and lemmatization (Tsuruoka et al., 2005); and syntactic analysis[2]. Syntactic dependencies between term components are computed according to assigned POS tags and shallow parsing rules. Each term is considered as a syntactic binary tree composed of two elements: head component and expansion component. For instance, *lipid* is the head component of the two terms analyzed on figure 1.

3.2 Compositionality based induction of synonyms

GO terms present compositional structure (Verspoor et al., 2003; Mungall, 2004; Ogren et al., 2005). In the example below (concept GO:0009073) the compositionality can be observed through the substitution of one of the components (underlined):

> *aromatic amino acid family* <u>*biosynthesis*</u>
> *aromatic amino acid family* <u>*anabolism*</u>
> *aromatic amino acid family* <u>*formation*</u>
> *aromatic amino acid family* <u>*synthesis*</u>

We propose to exploit the compositionality for induction of synonym resources (*i.e.*, *biosynthesis*, *anabolism*, *formation*, *synthesis* in the given example).

[1] http://search.cpan.org/~thhamon/Alvis-NLPPlatform/
[2] http://search.cpan.org/~thhamon/Lingua-YaTeA/

While the cited works are based on the string matching, our approach exploits their syntactic analysis, which makes it independent on their surface graphical form (like examples on figure 1).

Compositionality assumes that the meaning of a complex expression is fully determined by its syntactic structure, the meaning of its parts and the composition function (Partee, 1984). This assumption is very often true in specialized langages, which are known to be compositional. On the basis of syntactically analysed terms, we apply a set of compositional rules: if the meaning $\mathcal{M}$ of two complex terms $A\ rel\ B$ and $A'\ rel\ B$, where A is its head and B its expansion components, is given as following:

$$\mathcal{M}(A\ rel\ B) = f(\mathcal{M}(A), \mathcal{M}(B), \mathcal{M}(rel))$$

$$\mathcal{M}(A'\ rel\ B) = f(\mathcal{M}(A'), \mathcal{M}(B), \mathcal{M}(rel))$$

for a given composition function f, if $A\ rel\ B$ and $A'\ rel\ B$ are complex synonym terms and if B components are identical (such as *acetone* within *acetone catabolism* and *acetone breakdown*), then the synonymy relation between components A and A' $\{$*catabolism*, *breakdown*$\}$ can be induced. The modification is also accepted on expansion component B: from terms *replication of mitochondrial DNA* and *mtDNA replication* (fig. 1), we can induce synonymy between *mitochondrial DNA* and *mtDNA*. Finally, the modification is also accepted for both components $A\ rel\ B$ and $A'\ rel\ B'$, such as in *nicotinamide adenine dinucleotide catabolism* and *NAD breakdown*, where one pair, *i.e.* $\{$*catabolism*, *breakdown*$\}$, can be known from previously processed synonyms and allow to induce the new pair $\{$*nicotinamide adenine dinucleotide*, *NAD*$\}$. It should noticed that *rel* depends on the original relations: if the original terms are synonyms then the elementary terms are also synonyms, if the original terms are hierarchically related then the elementary terms are also hierarchically related, etc.

3.3 Lexically-based profiling of the induced elementary synonyms

In order to test and improve the quality of the induced synonymy relations, we confront these synonyms with approaches which allow to acquire the hyperonymy relations. All these resources are endogeneously acquired from the same terminology *GO*:

- Each induced pair of synonyms is controlled for the lexical inclusion (Kleiber and Tamba, 1990; Bodenreider et al., 2001). If the test is positive, like in the pair $\{$*DNA binding*, *binding*$\}$ this would suggest that this pair may convey a hierarchical relation. Indeed, it has been observed that lexical subsumption marks often a hierarchical subsumption. Thus, in the pair $\{$*DNA binding*, *binding*$\}$, *binding* is the hierarchical parent of *DNA binding*, while *DNA binding* has a more specific meaning than *binding*. One can assume that the cooccurrence of synonymy with the lexical subsumption makes the synonymy less reliable;
- The same compositional method, as described in the previous section, is applied to original *GO* term pairs related through `is-a` relations. In this way, we can also infer `is-a` elementary relations. Thus, if a pair of induced synonyms is also induced through `is-a` relations, *i.e.* $\{$*binding*, *DNA binding*$\}$, this also makes the synonymy relations less reliable.

In summary, an induced synonymy relation is considered to be less reliable when it cooccurs with a lexical inclusion or with `is-a` relation. For instance, several edges from figure 2 present the cooccurrence of synonymy relations with the `is-a` relations (such as, $\{$*holding*, *retention*$\}$, $\{$*retention*, *storage*$\}$ or $\{$*retention*, *sequestering*$\}$).

3.4 Pagerank-derived filtering of the induced elementary synonyms

The induced semantic relations can be represented as graphs where the nodes correspond to words and the edges to one or more relations between given two words. An example of what it can look like can be seen on figure 2: the induced synonymy relations may indeed cooccur with non-synonymy relations, like the hierarchical relations `is-a`. We propose to use a pagerank approach (Brin and Page, 1998) in order to separate a given graph of synonym relations into subsets (or groups) within which all the words are considered as synonyms with each other but not with any other word outside their subset. In order not to influence the results by the varying size of the graphs, we exploit a non-normalized version of pagerank (Engström, 2011). Thus, given the usual

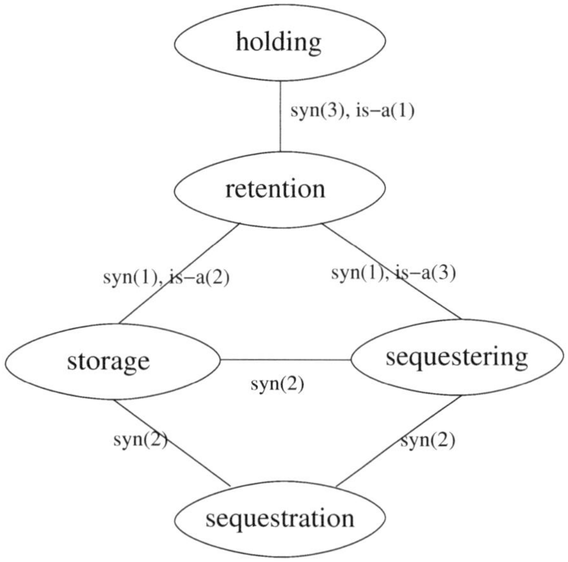

Figure 2: An example of graph generated thanks to the induced semantic relations: pairs related with synonymy relations `syn` may also be related with non-synonymy relations (like hierarchical relation `is-a`)

normalized version $P_{S_i}^{(1)}$ of pagerank:

Definition 1 $P_S^{(1)}$ *for system S is defined as the eigenvector with eigenvalue one to the matrix*

$$M = c(A + gv^T)^T + (1-c)v1^T$$

where g is a $n \times 1$ vector with zeros for nodes with outgoing nodes and 1 for all dangling nodes, $0 < c < 1$, A is the linkmatrix with sum of every row equal to one, v is a non-negative weightvector with sum one.

As we mentioned, with the processed data we have to use the non-normalized version of pagerank:

Definition 2 $P_S^{(2)}$ *for system S is defined as:*

$$P_S^{(2)} = \frac{P_S^{(1)}||V||_1}{d}, \text{with } d = 1 - \sum cA^T P_S^{(1)}$$

where V is the part of a global weightvector corresponding to the system S. We let V be the one vector such that all words are weighted equally.

Looking at the example from figure 2, we start from any node and then randomly either stop by a probability c or choose (possibly weighted by edge-weights) a new node by the probability $1-c$ from any of those linked to the chosen node. The pagerank of a node can then be seen as the sum of the probabilities of all paths to the node in question (starting in every node once including itself).

Usually A is a two-dimensional matrix in which the sum of every row is equal to one and all non-zero elements are equal between them. In order to use different types of relations and different weights on these relations we calculate cA. Given B, where B contains the weights of different edges and their type, we calculate A as:

$$A_{i,j} = (B_{i,j,SYN}/(B_{i,j,OTHER}+1))/n_i$$

where n_i is the total number of edges connected to node i. We treat all relations as symmetric relations for the filtering algorithm when creating B. While some relations aren't symmetric it seems reasonable to assume they affect the likelihood of synonyms in both directions. We also do not distinguish non-synonym relations among them. However, we try a few variations on how to weight A such as assigning different weights to synonym and non-synonym relations or using a logarithmic scale to decrease the effect of very different weights in B.

Further to the weighting, the rows of A do not necessarily sum to one. We propose then not to choose a specific value for c, but to threshold the sum of every row in cA to 0.95. This means that for most of the rows we set $c_{row} = 1/\sum A_{row} \cdot 0.95$, but for rows with a low sum we don't increase the strength of the links but rather keep them as they are ($c_{row} = 1$). Choosing the threshold can be seen as choosing c in the ordinary pagerank formulation. A low threshold means that only the immediate surrounding of a node may impact its pagerank, while a high threshold means that distant nodes may also have an impact. Higher threshold is also useful to separate the pagerank of nodes and to make slower the convergence when calculating the pagerank. When the sum of all rows is less than one and all non-zero elements are positive we can guarantee that the pagerank algorithm converges (Bryan and Leise, 2006). We also use the *Power Method* modified for the non-normalized version of pagerank (Engström, 2011). On the basis of these elements, we apply the following algorithm for segmenting the graph into groups of nodes:

1. Calculate weighted linkmatrix;
2. Calculate pagerank from uniform weightvector v_i;

3. Select the node with the highest pagerank;
4. Calculate pagerank from non-uniform weightvector (zero vector with a single 1 for the selected node);
5. Nodes with $P^{(2)} > cutoff$ are selected as synonyms with selected node and each other;
6. Remove the found synonym nodes from the graph;
7. If the graph is non empty, restart from step 1;
8. Otherwise end: words belonging to the same group are considered as synonyms.

We present the application of the algorithm on the example from figure 2 using the *cutoff* $= 1.5$. We start by calculating the weights on the links (weighted linkmatrix). For instance, given the relation from *storage* to *retention* we have: $A_{i,j} = (B_{i,j,SYN}/(B_{i,j,OTHER} + 1))/n_i = (1/(2+1))/3 = 1/9$. After computing the weights for all the relations and thresholding the sum of rows to 0.95, when the sum of weights out of a node is larger than 0.95, we obtain figure 3. This gives the pagerank from uniform vector $[4.8590, 7.7182, 16.4029, 16.1573, 15.4152]$, in which we select the node *storage* with the highest pagerank. Pagerank from non-uniform weightvector is then $[0.5490, 1.0970, 4.7875, 4.0467, 3.9079]$, in which we select the nodes with rank larger than *cutoff* $= 1.5$ (*storage*, *sequestration*, *sequestering*) as synonyms. After removing these nodes, we recalculate the weight matrix and repeate the algorithm: the two remaining nodes are found to belong to the same group. We then terminate the algorithm.

3.5 Evaluation protocol

The evaluation is performed against the manually validated synonymy relations. This validation has been done by two independent experts with the background in biology. They were asked to validate the induced synonyms acquired as the step 3.2 of the method. The inter-expert Cohen's kappa is 0.75. On the basis of this evaluation, we compute the precision: percentage of relations which allow to correctly group terms within the connected components and the groups. We compute two kinds of precision (Sebastiani, 2002): micro-precision which is the classical conception of this measure obtained at the level of the relations, and macro-precision which corresponds to the mean of the precisions obtained at the level of connected components or groups. The evaluation is done with the induced synonyms and also after their filtering with the pagerank-derived algorithm. This last evaluation leads to a better observation of the efficiency of the pagerank algorithm.

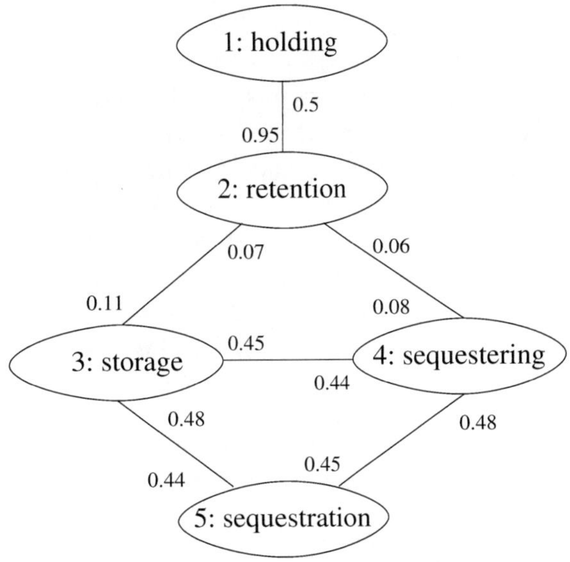

Figure 3: Example from figure 2 with weighted links

4 Experiments and Results

The *GO* terms have been fully processed with the NLP tools (POS-tagging and syntactic analysis) in order to prepare the next step, during which the elementary relations and the indicators are acquired.

4.1 Application of the lexical NLP methods

We applied the NLP method to the *GO* terms. The application of the compositionality approach to original synonymy and hierarchical relations generated 3,707 and 10,068 elementary relations, synonymous and hierarchical respectivelly. Depending on the syntactic structure of the original terms, the synonymy relations are induced between simple or complex terms, but also between their abbreviated and full forms, between the morpho-syntactic variants, etc. Very few of these synonyms exist within *GO* or within the WordNet resource (Fellbaum, 1998). We also detected 1,608 lexical inclusions. The lexical inclusions and the `is-a` relations are preserved only if they cooccur with induced synonymy relations. All these relations are then grouped into connected components (figure 2): the synonymy relations correspond to edges, term components correspond to nodes, while the information on is-a relations and on lexical inclusions appears as reliability indicators of the synonymy edges. A total of 2,017 connected components are generated. The biggest connected component contains 140 nodes and 183 edges. At this step, the connected components are evaluated against the reference data: we compute the precision.

4.2 Filtering of the induced synonyms with the pagerank-derived algorithm

We the apply the pagerank-derived algorithm to the induced synonyms, but also to the combinations of these synonyms with is-a relations and/or with lexical inclusions. The objective is then to filter the induced synonyms and to improve their reliability. We perform seven experiments, in which the synonymy and the indicators may receive the same importance or may be weighted:

1. syn: only the elementary synonymy relations are considered;
2. syn-isa: combination of synonymy and hierarchical is-a relations;
3. syn-incl: combination of synonymy relations with lexical inclusions;
4. syn-isa-incl: combination of synonymy and hierarchical relations with lexical inclusions;
5. syn-isa(535): combination of synonymy relations with lexical inclusions, using different weights: $(A_{i,j} = 5B_{i,j,SYN}/(3B_{i,j,OTHER} + 5))/n_i$;
6. syn-isa(353): combination of synonymy relations with lexical inclusions, using different weights: $(A_{i,j} = 3B_{i,j,SYN}/(5B_{i,j,OTHER} + 3))/n_i$.
7. syn-isa(log): combination of synonymy relations with lexical inclusions, using logarithmic weights: $(A_{i,j} = ((1/ln(2))ln(B_{i,j,SYN} + 1)/((1/ln(2))ln(B_{i,j,OTHER} + 2)))/n_i$.

According to the method described in section 3.4, the connected components of the synonymy relations obtained in section 3.2 are segmented again into one or more smaller and more homogeneous groups. The number of groups varies between 745 and 1,798 across the experiments. Moreover, around 25% of the synonymy relations may be removed by pagerank. These connected components and groups can also be evaluated against the reference data and we can compute the precision.

5 Evaluation and Discussion

The evaluation has been done by two independent experts, with the Cohen's kappa inter-expert agreement 0.75. We exploit the reference data of the two experts separately (we distinguish expert_1 and expert_2) and in common. We also distinguish macro-precision and micro-precision. Finally, the precision is first evaluated after the induction step with the NLP methods, and then after the processing of the acquired synonymy relations through the pagerank-derived algorithm and their filtering.

For the weighting of the non-synonymy and synonymy relations, we tested and applied several coefficients: 5, 3 and 5 in experiment 5 (*syn-isa535*); 3, 5 and 3 in experiment 6 (*syn-isa353*), etc. Different weights have been tested ranging from 1 to 7, as well as the log variations. On the whole, these variations have no significant impact on the results. But then, it is very important to respect the dependence among these coefficients and not to set them randomly.

The filtering of the synonymy relations has to control two factors: (1) the first is related to the fact that the removed relations are to be true negatives and that among them there should be no or a small number of correct relations; while (2) the second is related to the fact that the remaining relations are to be true positives and that among them there should be no or a small number of wrong relations.

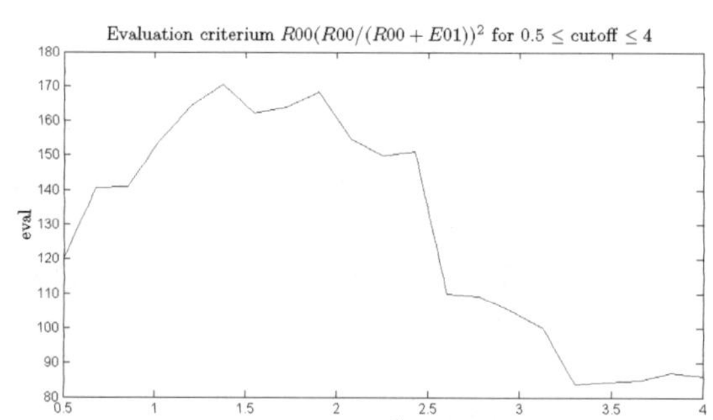

Figure 4: Impact of the cutoff values on the filtering of synonymy relations

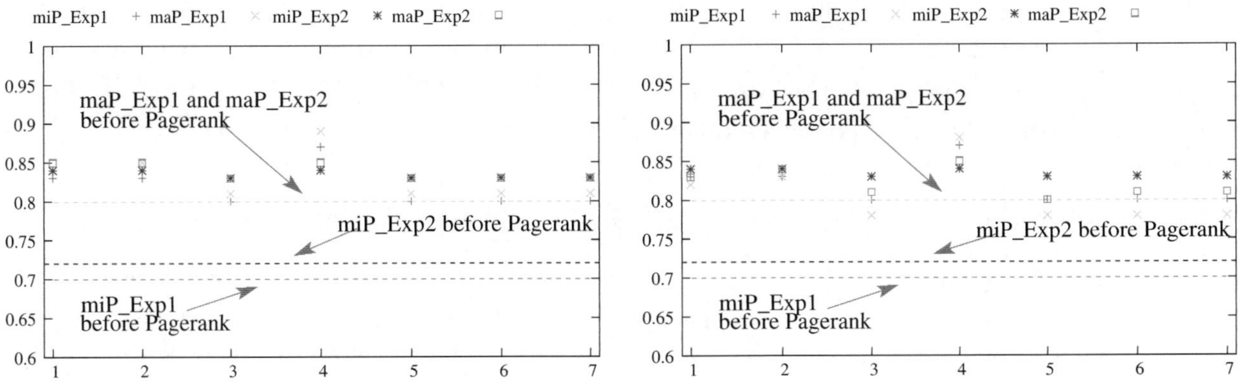

(a) Connected components with terms

(b) Groups of terms

Figure 5: Evaluation of the results in terms of micro-precision miP and of macro-precision maP for connected components and for groups of terms (performed according to the reference data provided by two experts)

On figure 4, we present the impact of the cutoff values on the selection and filtering of the synonyms. Like with other parameters, we have tested several values between 0.5 and 4. This figure illustrates the distribution of the correctly removed relations. The cutoff values have an important impact on the results: we can observe that the optimal cutoff values are set between 1.5 and 2 because they allow to remove the highest number of the wrong relations. We have set the cutoff value to 1.5. The choice of cutoff is an important factor for the definition of the amount of the links that are to be removed: the higher the *cutoff* the higher the number of clusters. On the data processed in this work, the cutoff value has been defined experimentally thanks to the observation of the processed data. For the generalization of this method to new unknown but similar linguistic data (new terminology, new langage, new domain...), the *cutoff* will be either set in order to remove a certain predefined number of links or will be defined from a typical sample of the data.

Contrary to the cutoff values, the choice of threshold doesn't greatly impact the results, although using a lower threshold makes it harder to choose a good cutoff values since the ranking of different nodes will be closer to each other.

As for the analysis of the precision and of the relations which are correctly kept within the connected components, let's observe figure 5. On this figure, we present the evaluation results performed within the connected components with induced synonyms (figure 5(a)) and within the groups of filtered synonyms (figure 5(b)). On the y-axis we indicate the precision values, and on the x-axis, we indicate the different experiments performed as mentioned above: 1 in which only synonyms are exploited, 2 in which synonyms are combined with hierarchical `is-a` relations, 3 in which synonyms are combined with lexical inclusions, etc. Horizontal lines correspond to the precision obtained before the application of the pagerank: they remain the same whatever the experiment. These lines correspond to three reference data provided by the $expert_1$, the $expert_2$ and by their common data. As for the points, they indicate the precision obtained further to the pagerank: it varies according to experiments and experts. On the basis of figure 5, we can observe that:

- the difference between the expert evaluations is very low (0.02);
- the pagerank allows to increase the precision (between 0.10 and 0.15 for micro-precision, while macro-precision varies by 0.05);
- the consideration of synonymy alone provides performant results;
- the consideration of `is-a` relations improves the results but lexical inclusions decrease them;
- the increased weight of some of the quality indicators has no effect on the evaluation;
- macro-precision is superior to micro-precision because our data contain mainly small groups,

while the few large connected components have a very low precision;

- there is but a small difference between connected components (figure 5(a)) and groups (figure 5(b));
- the consideration of `is-a` relations and of lexical inclusions provides the best precision but the amount of the remaining synonyms is then the lowest. As we explained, it is important to keep the highest number of the correct relations, although when a lot of relations is removed, it is logical to obtain a higher precision. This means that the combination of `is-a` relations and of lexical inclusions is not suitable because it removes too much of synonyms.

In relation with this last observation, is should be noted that the balance between the removed and the remaining relations is a subtle parameter.

The obtained results indicate that the pagerank is indeed useful for the filtering of synonyms, although the parameters exploited by this algorithm must be defined accurately. Thus, it appears that synonymy alone may be sufficient for this filtering. When the quality indicators are considered, `is-a` relations are suitable for this filtering because very often they propose true hierarchical relations. However, the lexical inclusions have a negative effect of the filtering. We assume this is due to the fact that the lexical inclusions are ambiguous: they may convey hierarchical relations but also equivalence relations (Haralambous and Lavagnino, 2011). Indeed, contextually some terms may be shortened or may be subject to an elision while their meaning is not impacted.

Currently, the pagerank is limited by the fact that it is applied to a relatively small set of data while it is designed to process very large data. Then, it can be interesting to enrich the model and to be able to take into account other quality indicators, such as frequencies, productivity or other semantic relations proposed within *GO* (`part-of` and `regulates`). Moreover, we can also give a lesser weight to some indicators (such as lexical inclusions) with penalties and keep the strong weight for other indicators. In the current model of the pagerank, we threshold rows to < 0.95. However, we assume that the algorithm may have problems with very large and very connected graphs: the pagerank may spread out in the graph too much and possibly allow the first words with the highest pagerank to make groups with only one word. This can be corrected if an additional calculation is added and when the group contains only one word at step 5.

6 Conclusion and Perspectives

We propose an original approach for inducing synonyms from terminologies and for their filtering. The methods exploit the NLP methods, compositionality principle and pagerank-derived algorithm. This work is motivated by the fact that synonymy is a contextual relation and its validity and universality are not guaranteed. We assume the semantic cohesiveness of synonymy relations should be qualified and quantified. The compositionality and NLP methods allow to acquire endogeneously the synonymy relations and the quality indicators, while the pagerank-derived algorithm leads to the filtering of the acquired synonyms. Its functionning is based upon the synonymy relations and also upon the acquired indicators (*is-a* relations and lexical inclusions). It appears that the synonymy relations alone provide good clues for their filtering. The *is-a* relations are also fruitful, while the use of the lexical inclusions appears not to be suitable.

In the future, we plan to add and test other indicators. Other experiments will also be done with the pagerank approach. For instance, it will be interesting to propose a model which takes into account that, within a cluster, words may be synonym with some cluster words but not with all the words of the cluster. This method can be adapted for the processing of corpora and also applied to terms from other terminologies. The acquired and filtered synonymy relations will be exploited within the NLP applications in order to test the efficiency of these resources and also the usefulness and efficiency of their filtering. Moreover, the compositionality approach can be adapted and exploited for the paraphrasing of the biomedical terms and for the improvement of their understanding by non expert people.

References

E Agirre and A Soroa. 2009. Personalizing PageRank for word sense disambiguation. In *EACL 2009*, pages 33–41, Athens, Greece, March.

O Bodenreider, A Burgun, and TC Rindflesch. 2001. Lexically-suggested hyponymic relations among medical terms and their representation in the UMLS. In URI INIST CNRS, editor, *Terminologie et Intelligence artificielle (TIA)*, pages 11–21, Nancy.

S Brin and L Page. 1998. The anatomy of a large-scale hypertextual Web search engine. *Computer Networks and ISDN Systems*, 30(1–7):107–117.

K Bryan and T Leise. 2006. The $\$25,000,000,000$ eigenvector: the linear algebra behind google. *SIAM Rev.*, 48(3):569–581.

David A. Cruse. 1986. *Lexical Semantics*. Cambridge University Press, Cambridge.

C Engström. 2011. Pagerank as a solution to a linear system, pagerank in changing systems and non-normalized versions of pagerank. Master's thesis, Mathematics, Centre for Mathematical sciences, Lund University. LUTFMA-3220-2011.

C Fellbaum. 1998. A semantic network of english: the mother of all WordNets. *Computers and Humanities. EuroWordNet: a multilingual database with lexical semantic network*, 32(2-3):209–220.

S Fernandez, E SanJuan, and JM Torres-Moreno. 2009. Résumés de texte par extraction de phrases, algorithmes de graphe et énergie textuelle. In *Société Francophone de Classification*, pages 101–104.

B Gaume. 2006. Cartographier la forme du sens dans les petits mondes lexicaux. In *JADT*.

G Grefenstette. 1994. *Explorations in automatic thesaurus discovery*. Kluwer Academic Publishers.

T Hamon and N Grabar. 2008. Acquisition of elementary synonym relations from biological structured terminology. In *Computational Linguistics and Intelligent Text Processing (5th International Conference on NLP, 2006)*, number 4919 in LNCS, pages 40–51. Springer.

T Hamon, A Nazarenko, and C Gros. 1998. A step towards the detection of semantic variants of terms in technical documents. In *COLING-ACL'98*, pages 498–504.

Y Haralambous and E Lavagnino. 2011. La réduction de termes complexes dans les langues de spćialité. *TAL*, 52(1):37–68.

S Hassan, R Mihalcea, and C Banea. 2007. Random-walk term weighting for improved text classification. In *ICSC*, pages 242–249.

T Hughes and D Ramage. 2007. Lexical semantic relatedness with random graph walks. In *EMNLP-CoNLL*, pages 581–589. Association for Computational Linguistics.

C Jacquemin, JL Klavans, and E Tzoukerman. 1997. Expansion of multi-word terms for indexing and retrieval using morphology and syntax. In *ACL/EACL 97)*, pages 24–31, Barcelona, Spain.

G Kleiber and I Tamba. 1990. L'hyperonymie revisitée : inclusion et hiérarchie. *Langages*, 98:7–32, juin.

R Mihalcea, P Tarau, and E Figa. 2004. Pagerank on semantic networks, with application to word sense disambiguation. In *COLING*, pages 1126–1132.

CJ Mungall. 2004. Obol: integrating language and meaning in bio-ontologies. *Comparative and Functional Genomics*, 5(6-7):509–520.

NLM, 2007. *UMLS Knowledge Sources Manual*. National Library of Medicine, Bethesda, Maryland. www.nlm.nih.gov/research/umls/.

PV Ogren, KB Cohen, and L Hunter. 2005. Implications of compositionality in the Gene Ontology for its curation and usage. In *Pacific Symposium of Biocomputing*, pages 174–185.

BH Partee, 1984. *Compositionality*. F Landman and F Veltman.

S Schulz, M Romacker, P Franz, A Zaiss, R Klar, and U Hahn. 1999. Towards a multilingual morpheme thesaurus for medical free-text retrieval. In *Medical Informatics in Europe (MIE)*, pages 891–4.

F Sebastiani. 2002. Machine learning in automated text categorization. *ACM Computing Surveys*, 34(1):1–47.

N Shimizu, M Hagiwara, Y Ogawa, K Toyama, and H Nakagawa. 2008. Metric learning for synonym acquisition. In *COLING*, pages 793–800.

R Sinha and R Mihalcea. 2007. Unsupervised graph-based word sense disambiguation using measures of word semantic similarity. In *IEEE International Conference on Semantic Computing (ICSC 2007)*, pages 363–369.

RS Sinha and RF Mihalcea. 2011. Using centrality algorithms on directed graphs for synonym expansion. In *FLAIRS*.

Y Tsuruoka, Y Tateishi, JD Kim, T Ohta, J McNaught, S Ananiadou, and J Tsujii. 2005. Developing a robust part-of-speech tagger for biomedical text. *LNCS*, 3746:382–392.

CM Verspoor, C Joslyn, and GJ Papcun. 2003. The Gene Ontology as a source of lexical semantic knowledge for a biological natural language processing application. In *SIGIR workshop on Text Analysis and Search for Bioinformatics*, pages 51–56.

T Wang and G Hirst. 2011. Exploring patterns in dictionary definitions for synonym extraction. *Natural Language Engineering*, 17.

P Zweigenbaum, R Baud, A Burgun, F Namer, É Jarrousse, N Grabar, P Ruch, F Le Duff, B Thirion, and S Darmoni. 2003. Towards a Unified Medical Lexicon for French. In *Medical Informatics in Europe (MIE)*, pages 415–20.

Domain Adaptation of Coreference Resolution for Radiology Reports

Emilia Apostolova, Noriko Tomuro, Pattanasak Mongkolwat*, Dina Demner-Fushman∓
College of Computing and Digital Media, DePaul University, Chicago, IL
*Department of Radiology, Northwestern University Medical School, Chicago, IL
∓Communications Engineering Branch, National Library of Medicine, Bethesda, MD
emilia.aposto@gmail.com, tomuro@cs.depaul.edu,
p-mongkolwat@northwestern.edu, ddemner@mail.nih.gov

Abstract

In this paper we explore the applicability of existing coreference resolution systems to a biomedical genre: radiology reports. Analysis revealed that, due to the idiosyncrasies of the domain, both the formulation of the problem of coreference resolution and its solution need significant domain adaptation work. We reformulated the task and developed an unsupervised algorithm based on heuristics for coreference resolution in radiology reports. The algorithm is shown to perform well on a test dataset of 150 manually annotated radiology reports.

1 Introduction

Coreference resolution is the process of determining whether two expressions in natural language refer to the same entity in the world. General purpose coreference resolution systems typically cluster all mentions (usually noun phrases) in a document into coreference chains according to the underlying reference entity. A number of coreference resolution algorithms have been developed for general texts. To name a few, Soon et al. (2001) employed machine learning on the task and achieved an F-score of 62.6 and 60.4 on the MUC-6 (1995) and MUC-7 (1997) coreference corpora respectively. Ng et al. (2002) improved this learning framework and achieved F-scores of 70.4 and 63.4 respectively on the same datasets.

There are also a number of freely available off-the-shelf coreference resolution modules developed for the general domain. For example, BART (Versley et al., 2008) is an open source coreference resolution system which provides an implementation of the Soon et al. algorithm (2001). The Stanford Deterministic Coreference Resolution System (Raghunathan et al., 2010) uses an unsupervised sieve-like approach to coreference resolution. Similarly, the GATE Information Extraction system (Cunningham et al., 2002) includes a rule-based coreference resolution module consisting of orthography-based patterns and a pronominal coreferencer (matching pronouns to the most recent referent).

While coreference resolution is a universal discourse problem, both the scope of the problem and its solution could vary significantly across domains and text genres. Newswire coreference resolution corpora (such as the MUC corpus) and general purpose tools do not always fit the needs of specific domains such as the biomedical domain well.

The importance and distinctive characteristics of coreference resolution for biomedical articles has been recognized, for example (Castano et al., 2002; Gasperin, 2006; Gasperin et al., 2007; Su et al., 2008). Within the biomedical field, clinical texts have been noted as a genre that needs specialized coreference corpora and methodologies (Zheng et al., 2011). The importance of the task for the clinical domain has been attested by the 2011 i2b2 NLP shared task (Informatics for Integrating Biology and the Bedside[1]) which provided an evaluation platform for coreference resolution for clinical texts.

However, even within the clinical domain, coreference in different sub-genres could vary signifi-

[1]https://www.i2b2.org/NLP/Coreference/

cantly. In this paper we demonstrate the idiosyncrasies of the task of coreference resolution in a clinical domain sub-genre, radiology reports, and describe an unsupervised system developed for the task.

2 Coreference Resolution for Radiology Reports

Radiology reports have some unique characteristics that preclude the use of coreference resolution modules or algorithms developed for the general biomedical domain or even for other types of clinical texts. The radiology report is a clinical text used to communicate medical image findings and observations to referring physicians. Typically, radiology reports are produced by radiologists after examining medical images and are used to describe the findings and observations present in the accompanied images.

The radiology report accompanies an imaging study and frequently refers to artifacts present in the image. In radiology reports, artifacts present in the image exhibit *discourse salience*, and as a result are often introduced with definite pronouns and articles. For example, consider the sentence *The pericardial space is clear.* The definite noun phrase *the pericardial space* does not represent an anaphoric (or cataphoric) discourse entity and has no antecedent. In contrast, coreference resolution in general texts typically considers definite noun phrases to be anaphoric discourse entities and attempts to find their antecedents.

Another important distinction between general purpose coreference resolution and the coreference resolution module needed by an NLP system for clinical texts is the scope of the task. General purpose coreference resolution systems typically cluster all mentions in a document into coreference chains. Such comprehensive mention clustering is often not necessary for the purposes of clinical text NLP systems. Biomedical Information Extraction systems typically first identify named entities (medical concepts) and map them to unambiguous biomedical standard vocabularies (e.g. UMLS[2] or RadLex[3] in the radiological domain). While multiple mentions of the same named entity could exist in a document, in most cases these mentions were previously assigned to the same medical concept. For example, multiple report mentions of '*the heart*' or '*the lung*' will normally be mapped to the same medical concept and clustering of these mentions into coreference chains is typically not needed.

3 Task Definition

Analysis revealed that the coreference resolution task could be simplified and still meet the needs of most Information Extraction tasks relevant to the radiological domain. Due to their nature, texts describing medical image finding and observations do not contain most pronominal references typically targeted by coreference resolution systems. For example, no occurrence of personal pronouns (e.g. *he, I*), possessive pronouns (e.g. *his, my*), and indefinite pronouns (e.g. *anyone, nobody*) was found in the validation dataset. Demonstrative pronouns and non-pleonastic 'it' mentions were the only pronominal references observed in the dataset[4]. The following examples demonstrate the use of demonstrative pronouns and the non-pleonastic 'it' pronoun (shown in bold):

There is prominent soft tissue swelling involving the premaxillary tissues. ***This*** *measures approximately 15 mm in thickness and extends to the inferior aspect of the nose.*

There is a foreign object in the proximal left mainstem bronchus on series 11 image 17 that was not present on the prior study. ***It*** *has a somewhat ovoid to linear configuration.*

Following these observations, the coreference resolution task has been simplified as follows. Coreference chains are assigned only for demonstrative pronouns and 'it' noun phrases. The coreference resolution task then involves selecting for each mention a single best antecedent among previously annotated named entities (medical concepts) or the NULL antecedent.

4 Dataset

A total of 300 radiology reports were set aside for validation and testing purposes. The dataset consists

[2] http://www.nlm.nih.gov/research/umls/

[3] http://www.radlex.org/

[4] Pleonastic 'it' refers to its use as a 'dummy' pronoun, e.g. *It is raining*, while non-pleonastic use of the pronoun refers to a specific entity.

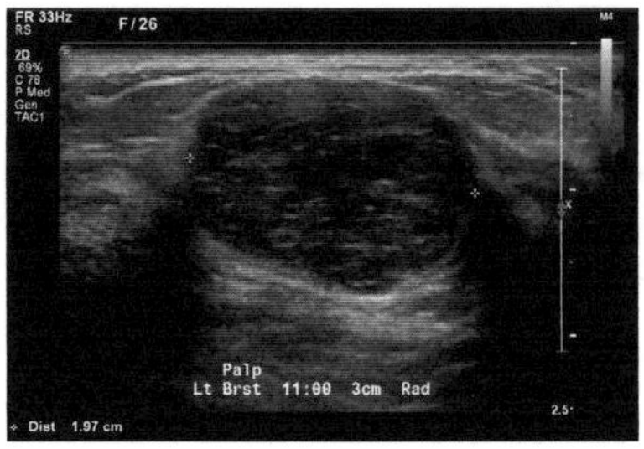

Figure 1: A sample DICOM image from an imaging study described by the following radiology report snippet: ... *FINDINGS: Targeted sonography of the upper inner left breast was performed. At the site of palpable abnormality, at the 11 o'clock position 3 cm from the nipple, there is an oval circumscribed, benign-appearing hypoechoic mass measuring 2.0 x 1.6 x 1.4 cm. There is mild internal blood flow. It is surrounded by normal appearing glandular breast tissue....*

of 100 Computed Tomography Chest reports, 100 Ultrasound Breast reports, and 100 Magnetic Resonance Brain reports, all randomly selected based on their report types from a dataset of more than 100,000 de-identified reports spanning a period of 9 years[5]. These three types of reports represent a diverse dataset covering representative imaging modalities and body regions. Figure 1 shows a sample Breast Ultrasound DICOM[6] image and its associated radiology report.

The reports were previously tagged (using an automated system) with medical concepts and their semantic types (e.g. anatomical entity, disorder, imaging observation, etc.). Half of the dataset (150 reports) was manually annotated with coreference chains using the simplified task definition described above. The other half of the dataset was used for validation of the system described next.

5 Method and Results

The coreference resolution task involves selecting for each mention a single best antecedent among previously annotated named entities or the NULL antecedent. Mentions are demonstrative pronoun phrases or definite noun phrases containing previously annotated named entities.

We implemented an algorithm for the task described above which was inspired by the work of Haghighi and Klein (2009). The algorithm first identifies mentions within each report and orders them linearly according to the position of the mention head. Then it selects the antecedent (or the NULL antecedent) for each mention as follows:

1. The possible antecedent candidates are first filtered based on a distance constraint. Only mentions of interest belonging to the preceding two sentences are considered. The rationale for this filtering step is that radiology reports are typically very concise and less cohesive than general texts. Paragraphs often describe multiple observations and anatomical entities sequentially and rarely refer to mentions more distant than the preceding two sentences.

2. The remaining antecedent candidates are then filtered based on a syntactic constraint: the co-referent mentions must agree in number (singular or plural based on the noun phrase head).

3. The remaining antecedent candidates are then filtered based on a semantic constraint. If the two mentions refer to named entities, the named entities need to have the same semantic category[7].

4. After filtering, the closest mention from the set of remaining possible antecedents is selected. If the set is empty, the NULL antecedent is selected.

Pairwise coreference decisions are considered transitive and antecedent matches are propagated transitively to all paired co-referents.

The algorithm was evaluated on the manually annotated test dataset. Results (Table 1) were computed using the pairwise F1-score measure: precision, recall, and F1-score were computed over all pairs of mentions in the same coreference cluster.

Precision	Recall	F1-score
74.90	48.22	**58.66**

Table 1: Pairwise coreference resolution results.

The system performance is within the range of state-of-the-art supervised and unsupervised coreference resolution systems[8]. F1-scores could range

[5]The collection is a proprietary dataset belonging to Northwestern University Medical School.

[6]Digital Imaging and Communications in Medicine, © The National Electrical Manufacturers Association.

[7]The same semantic type in the case of UMLS concepts or the same parent in the case of RadLex concepts.

[8]Source code for the described system will be made available upon request.

between 39.8 and 67.3 for various methods and test sets (Haghighi and Klein, 2009). The simplification of the coreference resolution problem described above allowed us to focus only on coreference chains of interest to clinical text Information Extraction tasks and positively influenced the outcome. In addition, our goal was to focus on high precision results as opposed to optimizing the overall F1-score. This guarantees that coreference resolution errors will result in mostly omissions of coreference pairs and will not introduce information extraction inaccuracies.

6 Conclusion

In this paper, we presented some of the challenges involved in the task of adapting coreference resolution for the domain of clinical radiology. We presented a domain-specific definition of the coreference resolution task. The task was reformulated and simplified in a practical manner that ensures that the needs of biomedical information extraction systems are still met. We developed an unsupervised approach to the task of coreference resolution of radiology reports and demonstrate state-of-the-art precision and reasonable recall results. The developed system is made publicly available to the NLP research community.

References

J. Castano, J. Zhang, and J. Pustejovsky. 2002. Anaphora resolution in biomedical literature. In *International Symposium on Reference Resolution*. Citeseer.

D.H. Cunningham, D.D. Maynard, D.K. Bontcheva, and M.V. Tablan. 2002. GATE: A Framework and Graphical Development Environment for Robust NLP Tools and Applications.

C. Gasperin, N. Karamanis, and R. Seal. 2007. Annotation of anaphoric relations in biomedical full-text articles using a domain-relevant scheme. In *Proceedings of DAARC*, volume 2007. Citeseer.

C. Gasperin. 2006. Semi-supervised anaphora resolution in biomedical texts. In *Proceedings of the Workshop on Linking Natural Language Processing and Biology: Towards Deeper Biological Literature Analysis*, pages 96–103. Association for Computational Linguistics.

A. Haghighi and D. Klein. 2009. Simple coreference resolution with rich syntactic and semantic features. In *Proceedings of the 2009 Conference on Empirical Methods in Natural Language Processing: Volume 3-Volume 3*, pages 1152–1161. Association for Computational Linguistics.

V. Ng and C. Cardie. 2002. Improving machine learning approaches to coreference resolution. In *Proceedings of the 40th Annual Meeting of the Association for Computational Linguistics*, pages 104–111.

K. Raghunathan, H. Lee, S. Rangarajan, N. Chambers, M. Surdeanu, D. Jurafsky, and C. Manning. 2010. A multi-pass sieve for coreference resolution. In *Proceedings of the 2010 Conference on Empirical Methods in Natural Language Processing*, pages 492–501. Association for Computational Linguistics.

W.M. Soon, H.T. Ng, and D.C.Y. Lim. 2001. A machine learning approach to coreference resolution of noun phrases. *Computational Linguistics*, 27(4):521–544.

J. Su, X. Yang, H. Hong, Y. Tateisi, J. Tsujii, M. Ashburner, U. Leser, and D. Rebholz-Schuhmann. 2008. Coreference resolution in biomedical texts: a machine learning approach. *Ontologies and Text Mining for Life Sciences 08*.

Y. Versley, S.P. Ponzetto, M. Poesio, V. Eidelman, A. Jern, J. Smith, X. Yang, and A. Moschitti. 2008. Bart: A modular toolkit for coreference resolution. In *Proceedings of the 46th Annual Meeting of the Association for Computational Linguistics on Human Language Technologies: Demo Session*, pages 9–12. Association for Computational Linguistics.

J. Zheng, W.W. Chapman, R.S. Crowley, and G.K. Savova. 2011. Coreference resolution: A review of general methodologies and applications in the clinical domain. *Journal of biomedical informatics*.

What can NLP tell us about BioNLP?

Attapol Thamrongrattanarit, Michael Shafir, Michael Crivaro, Bensiin Borukhov, Marie Meteer
Department of Computer Science
Brandeis University
Waltham, MA 02453, USA
{tet, mshafir, mcrivaro, bborukhov, mmeteer}@brandeis.edu

Abstract

The goal of this work is to apply NLP techniques to the field of BioNLP in order to gain a better insight into the field and show connections and trends that might not otherwise be apparent. The data we analyzed was the proceedings from last decade of BioNLP workshops. Our findings reveal the prominent research problems and techniques in the field, their progression over time, the approaches that researchers are using to solve those problems, insightful ways to categorize works in the field, and the prominent researchers and groups whose works are influencing the field.

1 Introduction

Thanks to improving technology and the discovery of stronger statistical methods, natural language processing techniques have more power than ever to give us insights into real datasets too large for humans to efficiently process. In the field of BioNLP, we see that natural language processing has a wide range of applications within the medical domain from analysis of clinical data to literature. With the increasing amount of publications in this growing field, building a classification structure is helpful both for categorizing papers in a sensible way and for recognizing the trends that brought the field to where it is today. Understanding the current nature of the field can show us where the most effort is needed, while taking a look at where the field has been can highlight successes and even unanswered questions.

As the use of NLP in the medical domain has expanded in recent years so has the amount of freely-available online research. With this wealth of information comes a problem, however, as it is not truly feasible for humans to read through all the research out there and classify it in a way that will capture the less-obvious trends and the finer relationships between seemingly-disconnected works. Instead, we propose that statistical methods can help us discover both the most reasonable way to partition the field and also see how the research has changed over the past decade. The longer term goal for the work is to contribute to a "map" of the field that can be a community resource, such as www.medlingmap.org, described in Meteer, et al. (2012).

Schuemie et al. (2009) used clustering techniques to analyze the domain of Medical Informatics. They processed a large number of Medline abstracts to find a subset of the journals classified as "Medical Informatics" whose content was sufficiently related to constitute a basis for the field. Using hierarchical clustering, they determined that such a group of journals exists and, as we might expect, the rest of the journals were largely disconnected. They also used this cluster of journals as the basis for a topic modeling task. Analyzing the articles from their new basis of journals, they found three very strong, topic-based clusters, each comprised of three sub-clusters. Overall, Schuemie et al. (2009) demonstrated how it is possible to gain a great deal of insight into the nature of a field by using statistical methods over that field's literature. More recently, Gupta and Manning (2011) used automatic methods to tag documents for "focus," "technique," and "domain" by examining over 15,000 ACL abstracts. This level of categorization is useful because it expands beyond the simple notion of the "topic" to implicitly show if a work, for example, is about an application of named-entity recognition or if it simply uses NER to achieve a greater task. The techniques demonstrated by Gupta and Manning could be very enlightening if applied to the BioNLP proceedings, though in this paper we refrain from drawing conclusions about individual papers. Instead, we will relate them through the topics extracted from the full-text proceedings.

For our task, we look to the ACL and NAACL-associated workshops on NLP applications in the medical domain. Entering its 11th year, the BioNLP workshop (under a variety of names) has given us ten rich and varied proceedings in addition to a pair of more focused shared tasks. All in all, the workshops have produced over 270 unique papers. Our data of 270 documents was small relative to (Schuemie et al., 2009) 6.3 million documents; therefore, we chose to expand our analysis to the full text of the documents instead of just the abstracts. Additionally, using the full papers allowed us to capture information about document content that abstracts alone could not provide.

2 Methods and Results

2.1 Pipeline Architecture

We implemented a document processing pipeline that would allow our approaches to be generalizable, easily reproducible, and extendable. Each of our analytic processes was integrated into this pipeline and parameterized to allow us proper flexibility for empirical experimentation. The pipeline works by managing the interaction between a configurable set of data layers and a configurable set of processing stages over those layers. It supports saving and loading its internal state between stages. In addition, layers and stages follow specific templates that reduce the amount of code to write and maintain. The ordering and activation of each stage is also parameterized. This pipeline allowed us to quickly and efficiently experiment with various approaches and combine them. The sample implementation of this pipeline is available publicly at github.com/attapol/mapping_bionlp.

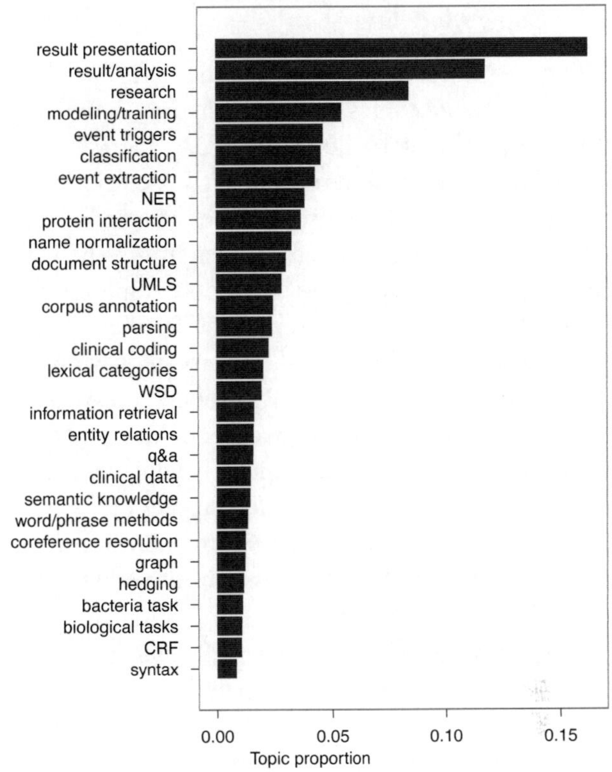

Figure 1: Average topic proportion across all the documents output by the LDA model

2.2 Preprocessing

The papers from the BioNLP workshop are all available freely from the ACL Anthology Archive [1]. We first extracted the text from the PDF files using pdf2text unix tool and then tagged them all for title, authors, places of origin, abstract, content, and references. In all cases, the abstract, content, and references were separated automatically using a script, and the places had to be hand-annotated. Papers from 2004 onward (starting with the first BioLINK workshop) have complete BibTeX entries that allowed us to automatically extract the titles and authors, but for 2002 and 2003 this work had to be done manually. Since we wanted to perform our analysis solely on the prose of the papers, and not on any of the numerical data, we filtered out portions of the text containing elements such as tables, graphs, footnotes, and URLs. We also filtered out stopwords (as defined by the NLTK package (Bird and Loper, 2004) for Python).

[1] aclweb.org

2.3 Topic Modeling

Using the Mallet toolkit (McCallum, 2002), we were able to generate topics from our cleaned data using the Latent Dirichlet Allocation (LDA) model. This approach allows us to represent each document as a vector of topic proportions instead of a bag of words, which prevents the problem of sparsity. When we set the number of topics to 30, the system output a set of distinct topics that seem to describe a range of tasks and methods within the domain of BioNLP. The topics generated by the LDA model reflect areas of study that are being pursued, techniques that are being applied, and resources that are being consulted in the field. A list of the generated topics along with the associated keywords is shown in Table 1 and the distributions of the topics across the entire document set is displayed in Figure 1.

Additionally, we found that the topics generated by LDA were more informative about the full content of a work than those generated by TF-IDF as TF-IDF would often give too much weight to specific examples over general concepts. For example, TF-IDF tended to select specific names of resources and ontologies rather than general terms. For example, it selected "Frame-net" instead "ontology" and "RadLex" instead of "lexicon". We concluded that, while interesting, TF-IDF results were not strongly suited for capturing an overall glimpse of the field. However, we think that TF-IDF can be much more useful in its more traditional capacity of finding document-specific keywords; we aim to use these indices to partially automate keyword generation for MedlingMap (Meteer et al., 2012), which is our accompanying project.

2.4 Topic Correlation

While looking at the topic proportions for each of our LDA topics overall can help us paint a broad picture of the field, it can also help to look at the relationship between these topics as they occur in the documents. Some topics appear highly ranked in nearly all papers, such as the topic that is characterized by terms such as "system" and "results", and the topic that includes "precision" and "recall" because they reflect the performance evaluation convention in the field. However, most topics are only dominant in a small subset of the papers. Some topics refer to tasks (e.g. named-entity recognition, hedging) and others refer to techniques (e.g. CRFs, parsing). We can look at how often pairs of task-related topic and method-related topic co-occur to see if researchers in the community are using certain techniques in conjunction with solving certain problems. We first turned a topic proportion vector into a binary vector where each element indicates which topic is discussed more extensively than average. Then, we counted the co-occurrences of tasks and methods of interest. To demonstrate this, we computed the number of papers that substantially discuss event extraction in conjunction with parsing, graph, lexical categories, or semantic knowledge (Figure 2). This topic comparison method provides a means of visualizing how researchers in the field are approaching BioNLP problems. It reveals that parsing and graph-based methods are commonly used in biological event extraction, while lexical categories and semantic knowledge are not as central to many of the approaches to this task. Moving forward, tracking how these correlations change over time will provide an insightful reflection of the field's progress on the task in a more meaningful way than evaluation scores alone. While a deeper analysis of all of such trends is beyond the scope of this paper, it certainly warrants further investigation.

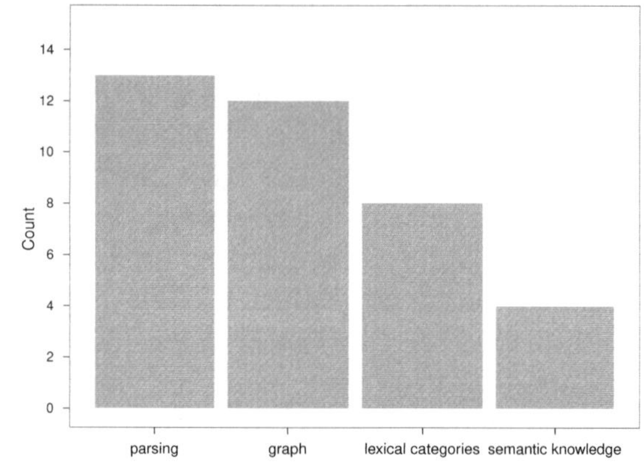

Figure 2: The bar plot shows the frequency of the co-occurrences between the event extraction topic and some of the method-related topics.

Table 1: The resulting topics and their associated keywords generated by LDA model with 30 topics

Topic Name	Keywords
Event Extraction	event, task, extraction, types, data, annotation
Coreference Resolution	anaphora, resolution, referring, links, antecedent
Graph	graph, relationships, nodes, edges, path, constraint, semanics
Clinical Coding	medical, data, codes, patients, notes, reports
Hedging	negation, scope, cues, speculative, hedge, lexical
Clinical Data	condition, historical, clinical, temporal, reports, context
Bacteria Task	bacteria, names, location, organisms, taxonomic, host, roles, type
Entity Relations	relations, entities, feature, static, renaming, annotated, pairs
Document Structure Analysis	rst, classification, abstracts, identification, data, terms
Q&A	question, answer, structure, passage, evidence, purpose
Event Triggers	triggers, dependency, binding, type, training, token, detection
Semantic Knowledge	semantic, frame, structures, argument, patterns, domain, types
Protein Interaction	protein, patterns, interaction, extraction, biological
Parsing	dependency, parser, tree, syntactic, structures, grammar, link
Name Normalization	gene, names, dictionary, normalization, protein, database, synonyms
Named Entity Recognition	entity, named, word, recognition, features, class, protein
Information Retrieval	search, queries, interface, text, retrieval, document
Corpus Annotation	corpus, annotation, guidelines, agreement, papers
Lexical Categories	semantic, categories, resources, simstring, lexical, gazetteer, features
Research	text, figure, knowledge, domain, research, complex, processing
CRF	crf, skip, chain, linear, dependency, words, edges, sentence
Result Discussion	system, based, results, set, table, test, shown, approach
Biological Tasks	species, disease, mutation, mentions, features, entities, acronym
UMLS	terms, semantic, phrases, umls, concepts, ontology, corpus
Word/Phrase Methods	words, measures, morphological, tag, token, chunking, form
WSD	disambiguation, sense, word, semantic, wsd, ambiguous
Result Analysis	found, number, precision, recall, cases, high, related, results
Classification	features, training, data, classification, set, learning, svm
Modeling/Training	training, data, model, tagger, performance, corpus, annotated
Syntax	attachment, pps, np, fragments, pp, noun, vp, nos, pattern

2.5 Trends within the subdisciplines in Biomedical NLP Literature

Our analysis of temporal trends builds on the idea proposed by (Hall et al., 2008) in their analysis of the changing trends in the field of computational linguistics over time. In their approach, they attempted, among other things, to analyze which topics were up and coming in the field and which were becoming less popular. Given their sound results, we decided to perform the same kind of trend analysis over the BioNLP topics. For many of our 30 topics, there was little change in the topic frequency over time. Considering the relative youth of the BioNLP field, this result is not entirely surprising. We did, however, find a few topics that have undergone notable changes in these past ten years, as observable in Figure 3. In particular, we found that two topics have seen surges of activity in recent years, whereas there were three topics that started out strong in the early years but that have since petered off. The two topics that have gained popularity in the past few years both involve biomedical events. Specifically, one such topic is primarily about event extraction tasks, and the other is about event triggers and the more fine-grained roles one needs to tag to categorize such events. The popularity of these two tasks is hardly surprising, given that they were the focus of the 2009 and 2011 shared tasks which were about working with events in both general and detailed ways. We do notice, however, that the growing trends continue in 2010 as well, when there was no shared task, and so we can see that events are of great interest in the field at present even without the added incentive of the shared tasks. It is reasonable to suggest that the 2009 BioNLP Shared Task in event extraction generated interest in the topic that continued through 2010 and 2011. Two more topics originally saw their popularity rise in the early years, but have

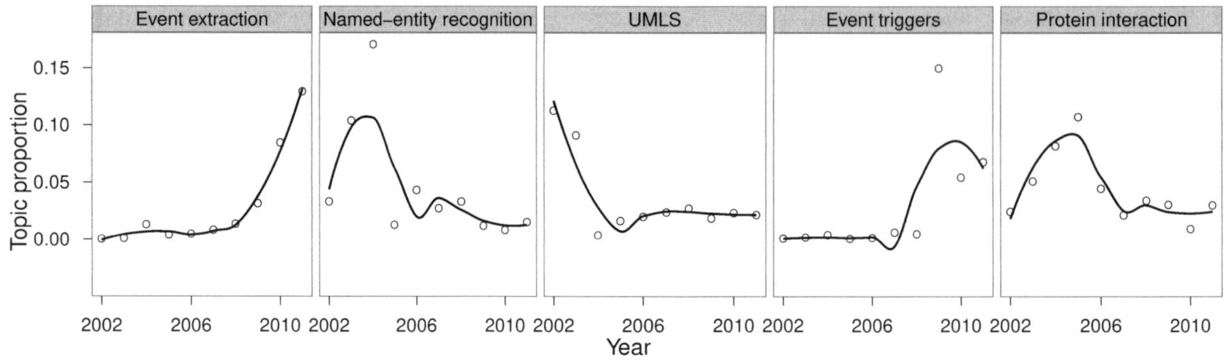

Figure 3: Topic proportions for some topics have gone through dramatic changes, which reflect how research interest and methodology evolve over time.

since seen it fade. Each of these is a specific task: named-entity recognition, which dropped off after 2004, and protein interaction, which saw a sharp decline after 2005. Although a detailed causal analysis is beyond the scope of this paper, we might wonder what accounts for these drops in topic proportion. The explanation that seems most likely is that great strides were made in these areas early on, but we have since reached a plateau in advancements. As such, the research has moved elsewhere. The only topic to see a steady decrease from the start was the topic associated with the Unified Medical Language System. In general, we can view a trend associated with a resource differently from one associated with a task. Above, when discussing tasks, we saw where the research currently has been heading and where it has been. With a resource, we could consider an upward trend to represent either an increased number of applications to a task or perhaps an expansion of the resource itself. In the case of UMLS, the downward trend likely suggests that the field has moved away from this particular resource, either because it does not apply as well to newer tasks or because it has been replaced with something more powerful.

2.6 Cluster Analysis

Our next step with the LDA-generated topics was to run a k-means clustering algorithm. We used the same topic proportion vector and a Euclidean metric to create the feature space for clustering. We used the standard k-means function in the statistical language R (R Development Core Team, 2010). The assumption of the LDA model biases each topic proportion vector to be sparse (Blei et al., 2003), and this turns out to be true in our data set. Therefore, we chose the number of clusters to match the number of topics so that the document space can be partitioned proportionally to its dimensionality. This clustering provides us with a useful schema for document classification within the domain of BioNLP. We can use the clusters as a guide for how to organize the current papers, and we can also view the clusters as a guide for how to select relevant research to build future work on. Clusters bring together related papers from different research groups and multiple workshops, such as those shown in Table 2. In all of these examples, the selection of these sets of papers simply based on keyword search would be very difficult, since many of the key terms are going to be present in a much larger set of documents.

2.7 Author Relation Analysis

As an additional task, we investigated the connections between authors in the BioNLP proceedings. Eggers et al. (2005) used a graph to visualize who was being cited by whom in ISI publications. There, the hope was to identify which authors worked within the same subdisciplines by examining clusters within the citation graph. By examining who cited whom in the BioNLP publications, we hoped instead to uncover the authors of the most influential papers, both within our own clusters and outside the scope of the BioNLP workshops. In our model, which can be viewed in Figure 4, we constructed a

List of papers assigned to the cluster where the most discussed topic is parsing (44.74% on average)
A Comparative Study of Syntactic Parsers for Event Extraction
Analysis of Link Grammar on Biomedical Dependency Corpus Targeted at Protein-Protein Interactions
On the unification of syntactic annotations under the Stanford dependency scheme
A Transformational-based Learner for Dependency Grammars in Discharge Summaries
A Study on Dependency Tree Kernels for Automatic Extraction of Protein-Protein Interaction

List of papers assigned to the cluster where the most discussed topic is clinical data (48.74% on average)
Applying the TARSQI Toolkit to Augment Text Mining of EHRs
Temporal Annotation of Clinical Text
Extracting Distinctive Features of Swine (H1N1) Flu through Data Mining Clinical Documents
ConText: An Algorithm for Identifying Contextual Features from Clinical Text
Distinguishing Historical from Current Problems in Clinical Reports – Which Textual Features Help?

Table 2: Two sample clusters from running k-means clustering algorithm on the corpus

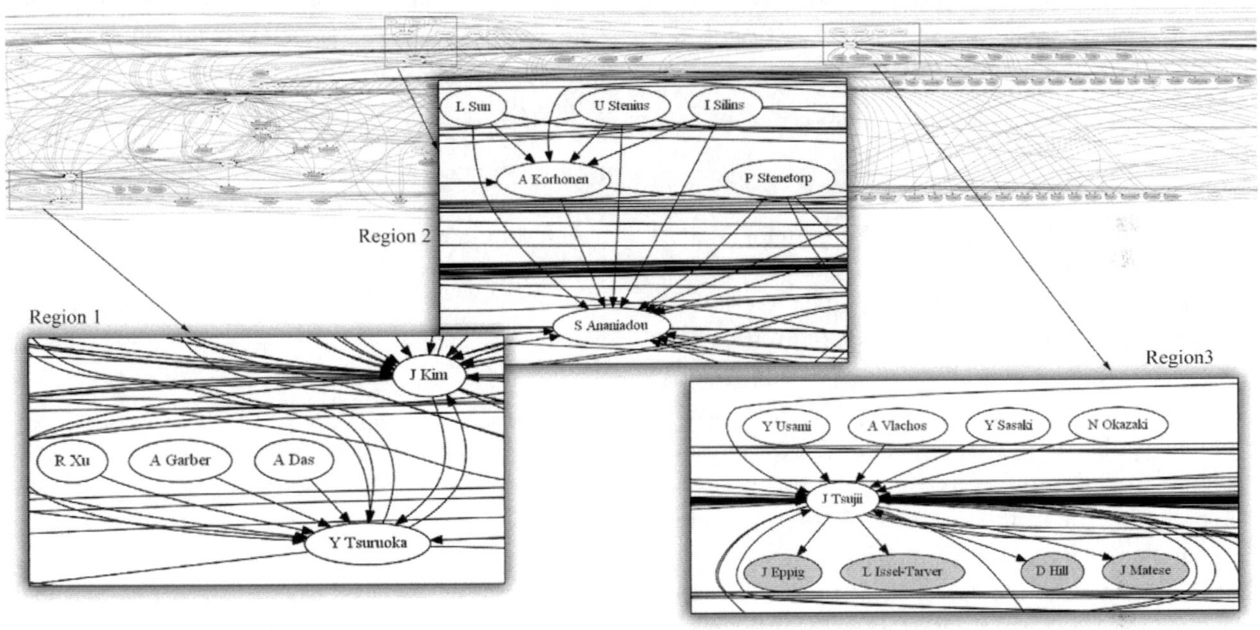

Figure 4: Citation relation graph. Each node represents an author whose papers are either published in the BioNLP proceedings or are cited by one of the papers in the proceedings. Each edge represents a citation activity.

directed graph of author citations from the BioNLP workshops and shared tasks. We disregarded the author ordering within each paper and gave the same weights for all authors whose names appear on the paper. In this graph, a node points to another node if that author cited the other author at least three times. Additionally, a white node signifies an author who published in the BioNLP workshop between 2008 and 2011, whereas a grey node is someone who did not, but was cited in papers during that time span. As can be seen in Figure 4 above, which is itself only a piece of the complete graph, this graph is rather large and complex, showing us a large degree of interconnectedness and interdependence in the field. Simply from the density of the lines, we can find some of the most influential figures, such as Jun'ichi Tsujii, shown in Region 3 and Yoshimasa Tsuruoka, shown in Region 2. Unsurprisingly, Tsujii's node is bustling with activity, as a very large number of authors cite works with Tsujii as an author, and his own prolific authorship (or co-authorship) naturally has him citing a variety of authors. The white nodes near his own show the authors who published BioNLP papers and primarily referenced his works, whereas the grey nodes near his show people who didn't publish, but who Tsujii cited in the proceedings multiple times. Thus, proximity can also be very telling in a graph like this. Since nodes with a heavier reliance on one another tend to end up closer to one another, we can also observe something of a "citation hierarchy" in sections of the graph. Region 2 is a prime example of this notion. We observe Ananiadou at the bottom with a large number of incoming edges. Above her node, we see Korhonen, who cites Ananiadou but is also cited by a number of other authors herself. Finally, above Korhonen there are a series of single nodes who cite her (and Ananiadou) but are without incoming edges of their own. We can think of this as something of a "local hierarchy", consisting of authors who are closely connected, with the more heavily-cited (and heavily-citing) easy to pick out.

3 Next Steps

The work described here provides a snapshot into the field. Underlying the work is a toolset able to reproduce the results on new sets of data to continue tracking the trends, topics, and collaborations. However, to be really useful to the research community, the results need to be captured in a way that can facilitate searches in this domain and support ongoing research. In order to do this, we are in the process of incorporating the results presented here in a content management system, MedLingMap (Meteer et al., 2012), which supports faceted indexing. Research in search interface design has shown that techniques which can create hierarchical faceted metadata stuctures of a domain significantly increase the ability of users to efficiently access documents in the collection (Stoica et al., 2005). The techniques described here can be fed into MedLingMap to create much of the metadata required to efficiently navigate the space.

4 Conclusion

In this report, we have outlined a variety of methods that can be used to gain a better understanding of BioNLP as a field. Our use of topic modeling demonstrates that the field already has several well-defined tasks, techniques, and resources, and we showed that we can use these topics to gain insight into the major research areas in the field and how those efforts areas are progressing. We put forth that this analysis could be powerful in recognizing when a problem has been effectively "solved", when a technique falls out of favor, and when a resource grows outdated. At the same time, we can see rising trends, such as how the 2009 shared task spurred an obvious 2010 interest in event extraction, and the correlations in the field between certain approaches and certain tasks. Through clustering, we were able to show that these topics also can help us separate the documents from the field into distinctive groups with a common theme, which can aid in building a database for current documents and classifying future ones. Finally, we ended with an analysis of author relations based on citation frequency and demonstrated how such a structure can be useful in identifying influential figures through their works.

As a further benefit of this work, we propose to use it to create a more lasting resource for the community that makes these results available to support search and and navigation in the bio-medical NLP field.

References

Andrew McCallum. 2002. MALLET: A Machine Learning for Language Toolkit. http://mallet.cs.umass.edu.

David Blei, Andrew Ng, and Michael Jordan. 2003. Latent Dirichlet allocation. *Journal of Machine Learning Research*, 3:993-1022

David Hall, Dan Jurafsky, and Christopher D. Manning. 2008. Studying the history of ideas using topic models. *In EMNLP.*

MJ Schuemie, JL Talmon, PW Moorman, and JA Kors 2009. Mapping the domain of medical informatics. *Methods Inf Med* 48:76-83.

Marie Meteer, Bensiin Borukhov, Michael Crivaro, Michael Shafir, and Attapol Thamrongrattanarit 2012. MedLingMap: Growing a resource for the Bio-Medical NLP field.

R Development Core Team 2010. R: A language and environment for statistical computing. http://www.R-project.org.

S. Eggers, Z. Huang, H. Chen, L. Yan, C. Larson, A. Rashid, M. Chau, and C. Lin. 2005 Mapping Medical Informatics Research Medical Informatics: Knowledge Management and Data Mining in BioMedicine. Springer Science+Business Media, Inc.

S Gupta, and C. Manning 2011. Analyzing the dynamics of research by extracting key aspects of scientific papers. Proceedings of IJCNLP.

Steven Bird, Edward Loper, and Ewan Klein. 2009. Natural Language Processing with Python. OReilly Media Inc.

Emilia Stoica, Marti A. Hearst, and Megan Richardson. 2007. Automating creation of hierarchical faceted metadata structure. Human Language Technologies: The Annual Conference of the North American Chapter of the Association for Computational Linguistics (NAACL-HLT 2007).

A Prototype Tool Set to Support Machine-Assisted Annotation

Brett R. South[1,2], Shuying Shen[1,2], Jianwei Leng[2], Tyler B. Forbush[4], Scott L. DuVall[3,4], Wendy W. Chapman[5]

Departments of [1]Biomedical Informatics, [2]Internal Medicine, and [3]Radiology University of Utah, Salt Lake City, Utah, USA

[4]IDEAS Center SLCVA Healthcare System, Salt Lake City, Utah, USA

[5]University of California, San Diego, Division of Biomedical Informatics, La Jolla, California, USA

brett.south@hsc.utah.edu, shuying.shen@hsc.utah.edu, jianwei.leng@utah.edu, tyler.forbush@utah.edu, scott.duvall@utah.edu, wendy.w.chapman@gmail.com

Abstract

Manually annotating clinical document corpora to generate reference standards for Natural Language Processing (NLP) systems or Machine Learning (ML) is a time-consuming and labor-intensive endeavor. Although a variety of open source annotation tools currently exist, there is a clear opportunity to develop new tools and assess functionalities that introduce efficiencies into the process of generating reference standards. These features include: management of document corpora and batch assignment, integration of machine-assisted verification functions, semi-automated curation of annotated information, and support of machine-assisted pre-annotation. The goals of reducing annotator workload and improving the quality of reference standards are important considerations for development of new tools. An infrastructure is also needed that will support large-scale but secure annotation of sensitive clinical data as well as crowdsourcing which has proven successful for a variety of annotation tasks. We introduce the Extensible Human Oracle Suite of Tools (eHOST) http://code.google.com/p/ehost that provides such functionalities that when coupled with server integration offer an end-to-end solution to carry out small or large scale as well as crowd sourced annotation projects.

1 Introduction

Supervised learning methods benefit from a reference standard that is used to train and evaluate the performance of Natural Language Processing (NLP) or Machine Learning (ML) systems for information extraction and classification. Ideally, generating a reference standard involves the review of more than one annotator with an accompanying adjudication step to resolve discrepancies (Roberts et al., 2007; Roberts et al., 2009). However, manual annotation of clinical, texts is time-consuming, expensive, and requires considerable effort. Reducing the time and costs required for manual annotation could be achieved by developing new tools that integrate methods to more efficiently annotate clinical texts and integrate a management interface that allows administration of large or small scale annotation projects. Such a tool could also integrate methods to pre-annotate entities such as noun phrases or clinical concepts mapped to a standard vocabulary. Efficiencies could be realized via reduction in human workload, modification of annotation tasks that could include crowd sourcing, and implementation of machine-assisted approaches.

Typically annotation of clinical texts requires human reviewers to identify information classes of interest called *"markables"*. These tasks may also require reviewers to assign attributes to those information classes and build relations between spans of annotated text. For each annotation task there may be one or many types of markables and each markable class may be associated with one or more spans of text and may include single or even multiple tokens. These tasks may occur simultaneously, or may also be done in different steps and by multiple reviewers. Furthermore, these activities require written guidelines that clearly explicate what information to annotate, specifics about each markable class, such as how much information to include in annotated spans, or syntactic rules to provide further guidance on annotated spans. Annotation tasks may benefit by incorporating rules or guidelines as part of the annotation task itself in the form of machine-assisted verification.

There are many annotation tools available, and the majority of them were designed for linguistic or gene annotation. Linguistic annotation tools such as Callisto and WordFreak are standalone clients suitable for small to medium scale tasks where collaborative effort is not emphasized. Functionality integrated with eHOST was inspired by existing features of these tools with the intent of providing a more efficient means of reference standard generation in a large collaborative environment. One annotation tool called Knowtator, a plug-in for Protégé (Musen, M.A., et al, 1995) developed by Ogren (2006) has been widely used to annotate clinical texts and generate reference standards. However, no standalone system exists that can provide end users with the ability to manually or semi-automatically edit, curate, and easily navigate annotated information. There are also specific functionalities that are missing from open source annotation tools in the clinical and biomedical domains that would introduce efficiencies into manual annotation tasks. These functionalities include: annotation of clinical texts along with database storage of stand-off annotations, the ability to interactively annotate texts in a way that allows users to react to either pre-annotations imported from NLP or ML systems or use exact string matching across an active corpus to identify similar spans of text to those already annotated. Additionally, these systems do not generally support crowd sourcing, machine-assisted pre-annotation or verification approaches integrated directly with the annotation tool.

This paper discusses development of a prototype open source system designed to provide functionality that supports these activities and offers an end-to-end solution when coupled with server integration to reduce both annotator and administrative workload associated with reference standard. We introduce the Extensible Human Oracle Suite of Tools (eHOST) created with these expectations in mind.

2 Background

Our goal for these development efforts was to build a prototype open source system that improves upon existing tools by including new functions and refining capabilities available in other annotation tools. The resulting GUI interface provides a means of visually representing annotated information, its attributes, and relations between annotated mentions. These efforts also focused integrating various machine-assisted approaches that can be used to easily curate and navigate annotated information within a document corpus, pre-annotate information, and also verify annotations based on rules checks that correspond with annotation guidelines or linguistic and syntactic cues.

The eHOST provides basic functionality including manual annotation of information representing markable classes and assignment of information attributes and relationships between markable classes. Annotations exported from eHOST are written using the XML format as Knowtator thus allowing integration of inputs and outputs to and from Knowtator and indirectly to Protégé 3.3.1. Coupling eHOST with an integrated server package such as the one under development by the VA Informatics and Computing Infrastructure (VINCI) called the Chart Administration Server for Patient Review (CASPR) provides one method of increasing efficiencies for small or large-scale annotation efforts that could also include crowd sourcing.

2.1 System Features Development

In the domains of computational linguistics and biomedical informatics various approaches that can be used to improve annotation efficiencies have been evaluated for a variety of tasks including information extraction and classification. While several methods may help reduce the time and costs required to create reference standards, one of the simplest approaches may include integrating machine-assisted methods to pre-annotate relevant spans of text allowing the annotator to add missing annotations, modify spans, or delete spurious annotations. Neveol (2011) evaluated use of automatic semantic pre-annotation of PubMed queries. This study showed a significant reduction in the number of required annotations when using pre-annotations, reduction in annotation time with higher inter-annotator agreement. Pre-annotation using simple approaches such as regular expressions coupled with dictionaries (South et al., 2010a) based on the UMLS as a source of lexical knowledge (Friedman, 2001) and pre-annotation of information representing protected health information (South et al., 2010b). In both cases finding that annotators preferred particular types of pre-annotation over others, but improvements in reference standard quality occur when pre-annotation was provided. Others have explored the use of third party tools for the pre-annotation task for UMLS concepts (Savova, 2008) and pre-annotation using an algorithmic approach (Chapman, et al., 2007) combined with domain expert annotations reused for temporal relation annotation (Mowery, 2008). Savova (2008) suggests limited utility when a third party tool is used for pre-annotation and Mowery (2008) suggest that even with domain expert pre-annotations, additional features are required to discern temporality. Finally, Fort and Sagot (2008) evaluated using pre-annotation for part-of-speech tagging on the Penn Tree bank corpus and demonstrate a gain in quality and annotation speed even with a not so accurate tagger.

Semi-automated curation has been explored as a means to build custom dictionaries for information extraction tasks (Riloff, 1993). More recently this approach was spurred on by the BioCreative II competition (Yeh et al., 2003). Alex et al., (2008), explored the use of NLP-assisted text mining to speed up curation of biomedical texts. Settles et al., (2008) estimates true labeling costs and provides a review of active and interactive learning approaches as a means of providing labels and reducing the cost of obtaining training data (Settles, 2010). Although eHOST does not yet include an active learning module it does provide one means of interactive annotation so these are important considerations for future development efforts.

In the biomedical informatics domain crowd sourcing has been evaluated as part of the 2009 i2b2 Medication Challenge (Uzuner, 2010). Nowak and Ruger (2010) provide estimates of annotation reliability from crowd sourcing of image annotation. Hsueh et al., (2009) provide estimates of the quality of crowd sourcing for sentiment classification using both experts and non-expert annotators. In all three cases the resulting annotation set was of comparable quality to that derived from expert annotators. Wang et al., (2008) make general recommendations for best approaches to crowd sourcing that include closer interactions between human and machine methods in ways that more efficiently connect domain expertise with the annotation task.

Subsequent sections in this paper walk the reader through the various basic and advanced features eHOST provides. These features have been developed in a way that provides flexibility to add additional modules that support improvements in annotation workflow and efficiency for a variety of annotation scenarios applicable to computational linguistics and biomedical informatics. Some of these features may be useful for crowd-sourced efforts whereas others may simply represent an improvement in the way annotation is visualized or how manual effort can be reduced. Figures in this paper use a set of synthetic clinical documents and a demonstration annotation project based on the 2010 and 2011 i2b2/VA annotation tasks as examples available from http://code.google.com/p/ehost.

2.2 Systems Architecture

The eHOST is a client application that can run on most operating systems that supports Java including, most Microsoft Windows x86/x64 platforms, Apple Mac OS X, Sun Solaris, and Linux. The application uses standardized formats including a file folder system, and structured XML inputs and outputs. These capabilities also support integration with other open source tools for annotation and knowledge management including Knowtator and Protégé. An Extract-Transform-Load process (ETL) is used by the system to import concept information from different sources, such as XML or Protégé PINS files. These inputs sources are normalized for loading into eHOST. All data that exists in the data pool can be transformed into various output formats. Raw input data documents in a single text file or sequential text files in a file folder system.

Information representing an annotation including concept attributes such as the annotated span, attributes, and relationships between annotations are inserted into a common data pool using a dynamic structured storage space. The data pool ensures that eHOST has capabilities to add new functions easily without making major changes to system architecture.

2.3 Annotation Project Workspace

In eHOST each project has its own user assigned workspace that includes an annotation schema and document corpus. Annotation schema can also be imported from an existing Protégé PINS file. Project settings can be inherited from existing projects for similar annotations tasks using eHOST. Other workspace functions include quickly switching between up to five of the most recently used workspaces. A workspace can be assigned for each annotation layer or document batch. In these situations, an annotator would receive a pre-compiled project that specifies all settings including any text documents and the annotation schema. Defining a workspace is a particularly useful function in situations where annotations may be crowd sourced and there may be multiple layers of annotation that are potentially fielded to many annotators.

2.3.1 Corpus Management

For any annotation task, the end user must manage the document corpus, which can originate from a server or a file folder system that contains individual text files. Using the stand-alone eHOST client tool, corpus management is accomplished via the current workspace (Figure 1). When the user initializes a new project, documents are placed in a "corpus" folder that is associated with the newly created annotation project. All text files, are copied to the "corpus" folder at the time of workspace assignment. Therefore, there is no risk of deleting the original documents associated with each new annotation project. This feature makes distribution of projects easier, because of the consistency between the workspace, corpus assignment and annotation output folders. For crowd-sourced projects eHOST can be integrated with a backend server via web services using an administrative module called CASPR.

Figure 1. eHOST corpus management

2.3.2 Viewer/Editor Panels

Figure 2 shows an annotation for *"full body pain"*, (shown with black bar above and below the active annotation) and information for that annotation including the annotated span, the class assignment and an assertion for the 2010 and 2011 i2b2/VA Challenge annotation tasks (Uzuner et al., 2011 and Uzuner et al., 2012). The result editor tab and its associated panels serve as the central place for basic annotation features. These functionalities include: assigning an annotator, creating new annotations or adjusting annotated spans of text and assignment of attributes or creating relationships between annotated spans of text. Other functions in the results editor tab include navigation between documents in the active corpus, resizing the text displayed in the document viewer, and *"save"* and *"save as"* functions that assigns a path for XML output files. The end user can easily remove all annotations in a document or remove specific kinds of annotations by deleting a "markable" class as well as remove attributes, and relationships between all annotations.

From the navigator screen in the stand-alone eHOST client tool a user can build annotation schema specifying markable classes, their associated attributes, and any allowed relationships. The navigator interface allows the user to review all annotated spans either within the current document or across the entire document corpus, toggle the view of each class on or off, see counts for all unique annotations and all annotations for each class, and choose a class for a fast annotate mode.

An annotation editor panel allows the user to view more detailed information for each selected annotation. This includes the time stamp of when the annotation was created, annotator assignment, comments on the annotation and class, attribute and relationship information.

Annotations can be created using several approaches from the result editor. In the normal mode, a class assignment window appears when the user selects a span of text, new annotations are generated by selecting any one of the markable classes. Activating a *"one click annotate"* mode is possible by checking the box next to a class of markables. Under this mode, any text selected is automatically annotated as that markable class. This feature improves task efficiencies when categories of markables are low or annotations of the same category cluster in small sections. Keyboard shortcuts have also been integrated with eHOST to reduce annotator click burden and dependence on a mouse. These shortcuts are available for tasks such as modification of spans, deletion of annotations, and navigation between annotations.

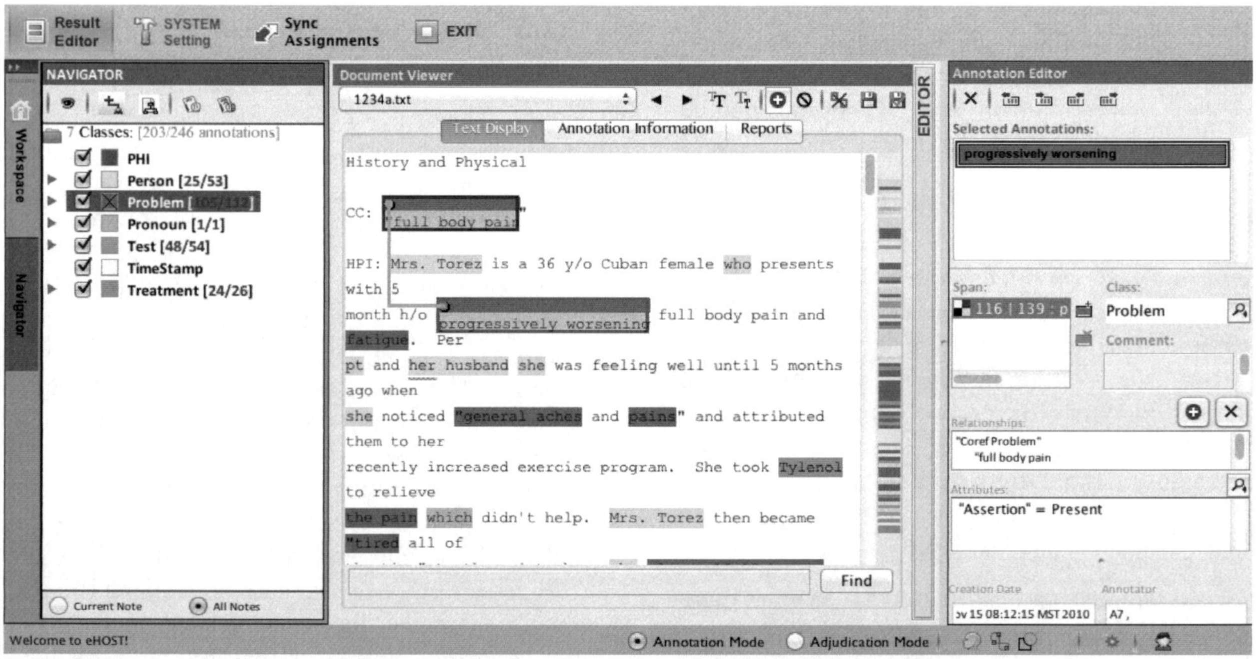

Figure 2. Example annotations using the eHOST interface

2.3.3 Server Integration

Annotation projects of any scale benefit from an automated means of building and distributing batches of texts to annotators, managing standoff XML files generated from annotation tasks or written directly to a database and getting and submitting assignments with minimal user input. Coupling eHOST with server components that comply with the web services API defined for eHOST allows these functionalities. The CASPR module under development by VINCI provides a means to automate the administration of annotation efforts that could include crowdsourced annotation projects.

Clicking on the sync assignments tab in the eHOST client (Figure 2) brings up a GUI that allows annotators to sync with a server location, enter credentials, see documents assigned, and designate documents as on hold, in process, or completed. When a user syncs and gets assignments from CASPR, a project folder is created that contains the annotation schema, text documents, and annotations sent from the server. The CASPR module allows an annotator to open the project and complete their task without needing to manage files or folders. Once completed, annotations can be synced to the server, and the next assignment will be loaded. The CASPR module allows iterative distribution of annotation batches without sending large sets of documents to annotators that may contain sensitive data, decreasing the risk of breaches in privacy and data security.

2.3.4 Additional Features

The document viewer panel employs visual cues to display relationships between annotations using color coding representing a parent and child node and line indicator between them showing the relationship. An "annotation profiler" to the right of the scroll bar shows the density of annotations color-coded to their categories, as well as relative to their positions in the document. This type of data visualization is useful to see the relative location of annotations within a single document or across an en tire document corpus.

An adjudication mode is also included in the stand-alone eHOST client that allows difference matching and side-by-side comparison of annotations for efficient adjudication of discrepancies between annotations. Standard reporting metrics can be calculated including Inter-Annotator Agreement (IAA), Recall, Precision and F_1-Measure.

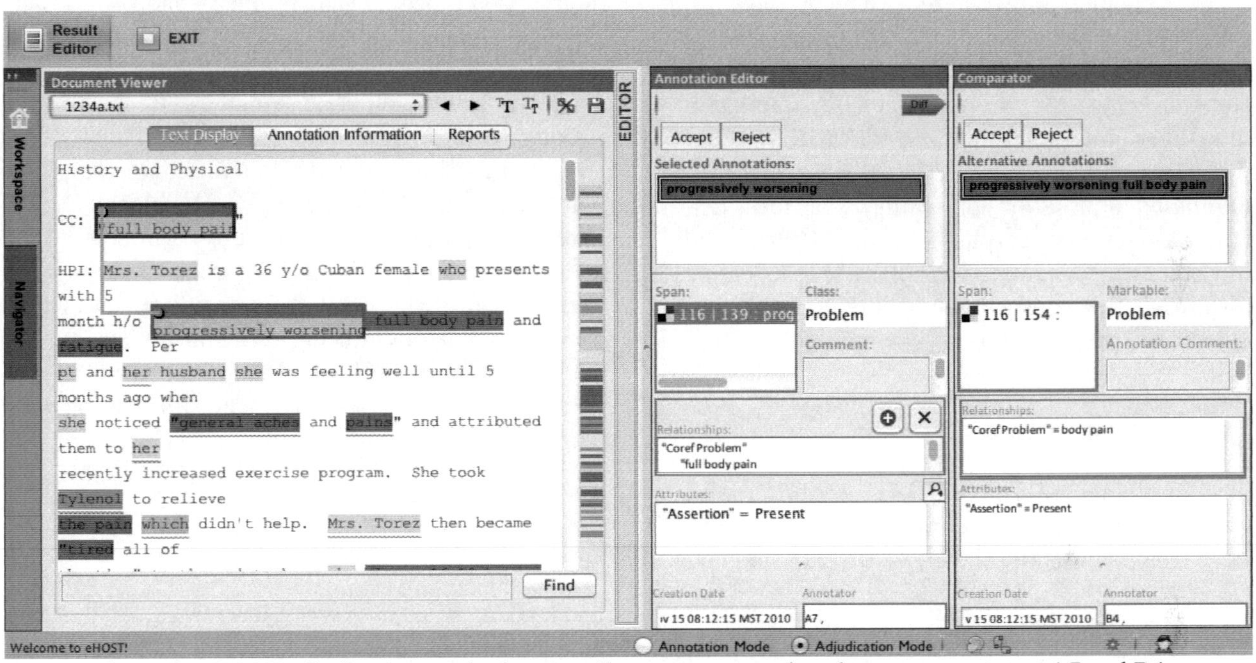

Figure 3. eHOST adjudication mode showing discrepant annotations between annotators A7 and B4

In Adjudication mode discrepant annotations are shown using a wavy red underline in the editor window and by a red bolded outline in a side by side two panel view between the annotation editor and comparator (Figure 3). These metrics and comparison tables between annotator results on the same documents can be output as HTML formatted reports that can be used by an adjudicator to quickly identify discrepancies between annotators (Figure 4). These reports and the editor window display can also be used to quickly train annotators on new clinical domains using a reference standard created by domain experts for training purposes. Using these features error analysis can also be done by importing outputs from an NLP system that have been converted into the XML format used by eHOST.

File: 1234a.txt

h 5 month h/o progressively worsening full body pain and fatigue. Per pt and her husband she was feeling well until 5 months ago when she noticed "general aches and pains" and attributed them to her recently increased exercise program. She took Tylenol to relieve the pain which didn't help. Mrs. Torez the

	Annotator:[A7 ,]	Annotator:[B4 ,]
Annotation Text	"general aches	aches
Span	(251,265)	(260,265)
Class	Problem	Problem
Relationship	linked to "progressively worsening" with relationship: [Coref Problem]	linked to "progressively worsening full body pain" with relationship:[Coref Problem]
Attributes	Assertion = Conditional;	Assertion = Conditional;

Figure 4. HTML Formatted report showing discrepant annotations between annotators A7 and B4

3 Advanced eHOST Features

There are also other more advanced features that have been integrated with eHOST. These include an "Oracle" mode that allows semi-automated annotation of similar spans of text across a document corpus, a means to easily and quickly curate annotated spans of text to create custom dictionaries, and machine-assisted pre-annotation integrated with the annotation tool itself.

3.1 Oracle Mode

Also implemented with eHOST is an "Oracle" mode which uses exact string matching allowing the user to annotate all spans of text that are identical to a new annotation. The oracle lists where these candidate annotations are found along with the surrounding context. The annotator can then accept or reject candidate spans annotated with the same markable class. Oracle mode can run within the current document or across the entire document corpus. This type of functionality is useful for annotation tasks that may involve identifying and marking spans of text that are repetitive or follow the same format For example, the 2011 i2b2/VA annotation task in which annotation of pronominal information was required for co-reference resolution (Figure 5).

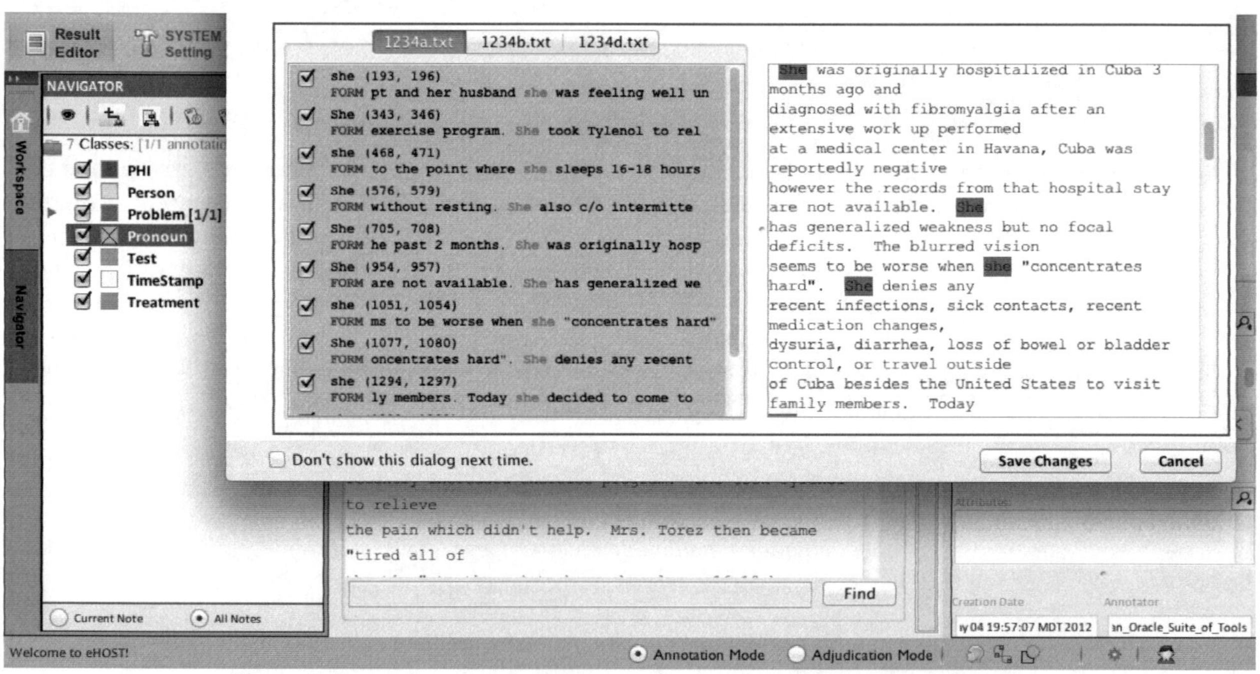

Figure 5. Example annotations generated using the eHOST "Oracle" mode

3.2 Semi-Automated Curation and Dictionary Management

Using the navigator window users can navigate to all annotations in either a single document or across an entire document corpus (Figure 6). The end user can curate annotations directly, create classes on the fly, or add attributes to annotations found from the navigator pane. These functions also allow users to easily identify spurious annotations introduced from machine-assisted approaches correct misclassification errors, and quickly curate all annotations within a single document or across an entire document corpus.

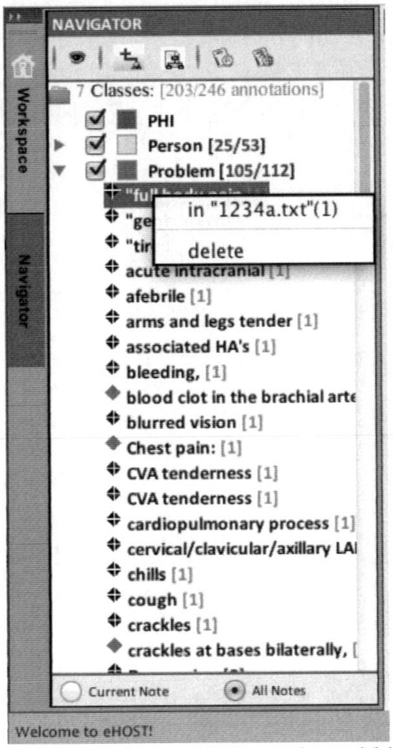

Figure 6. Semi-Automated curation within the document corpus

One task often associated with development of NLP systems involves manually creating or enhancing some existing representation of lexical knowledge that can be used as a domain specific dictionary. Using eHOST users can export annotations to create a dictionary of terms, phrases, or individual tokens that have been identified by human annotators and assigned to markable information classes. Once curated, annotated information can be exported as a new dictionary. User created dictionaries can be integrated with a database or exported and used in the creation of some ontologic representation of information using Protégé. Output from a dictionary manager is in the form of a delimited text file and can therefore be modified to fit any standardized information model or used to pre-annotate subsequent document batches.

3.3 Machine-Assisted Pre-Annotation

An interface is provided in eHOST that can be used for machine-assisted pre-annotation of documents in the active project corpus using either dictionaries or regular expressions based approaches. Users can import libraries of regular expressions or build their own regular expressions using a custom regular expression builder. Users can build and modify dictionaries created as part of annotation tasks that may include semi-automated curation steps. Dictionaries and regular expressions can also be coupled with the ConText algorithm (Chapman et al., 2007) to identify concept attributes such as negation, experiencer, and temporality. Pre-annotations derived from some external third party source such as an NLP system written as Knowtator XML outputs may also be imported into eHOST or passed to eHOST using CASPR.

Computational speed required for pre-annotation can be improved by selecting an option to use an internal statistical dictionary indexing function. This feature is particularly useful in situations where pre-annotation dictionaries are extremely large, such as where a subset of some standard vocabulary may be used to pre-annotate documents. Using the result editor and its associated functions annotators can add missed annotations, modifying existing annotations and delete spurious annotations. Handling pre-annotations in this way allows troubleshooting and error analysis of NLP system outputs imported into eHOST that can be shown to a reviewer in context and also facilitates interactive annotator training.

3.4 Machine-Assisted Verification

One of the more innovative features integrated with eHOST is the ability to verify and produce recommendations that help human annotators comply with syntactic and lexical rules that are specified by annotation task guidelines. Machine-Assisted verification is most useful when used on lexical or syntax rules to ensure that candidate phrases generated by automated systems are similar to those marked by humans. These rules rely more on adherence to patterns than on decision-making, so the strengths of human review with machine approaches to semi-automated verification can be leveraged. When identifying medical concepts, it is common that noun phrases are marked as candidates. The determination of how much of a noun phrase to mark (inclusion of articles, adjectives, noun-modifiers, prepositional phrases) and at what granularity (simple nouns or complex noun phrases) may vary with each project.

The verifier allows portions of an annotation guideline to be programmed into rules that check for consistency. Rules check whether a word appears within a user-defined window before and after an annotation. Each rule can be linked to text that describes why the annotation was flagged. Annotators are then provided suggestions on the correct span based on the rule. Using the surrounding text, the guideline text, and the suggestion, the annotator can determine the final span for an annotation. These machine-assisted verifier functions help support reference standard generation by providing the context of annotations that seem to fail syntactic and lexical rules while allowing human annotators to focus on domain expertise required to identify and classify information found in clinical texts.

Conclusion

Our prototype system provides functionalities that have been created to more efficiently support reference standard generation including machine-assisted annotation approaches. It is our hope that these system features will serve as the basis for the further development efforts that will be part of an enterprise level system. Outputs of such an annotation tool could be used as inputs for pipeline NLP systems or as one component of a common workbench of tools used for clinical NLP development tasks.

We have implemented and tested eHOST for the 2010 and 2011 i2b2/VA challenge annotation tasks and annotation projects for the Consortium for Healthcare Informatics Research (CHIR). The stand-alone eHOST client tool is available from http://code.google.com/p/ehost along with a demonstration project, a users guide, API documentation, and source code. The eHOST/CASPR interfaces will be used to support a large-scale crowd sourced annotation task used for annotation of disorders, temporal expressions, uncertainty, and negation along with data standardization. These efforts will include more rigorous analysis and usability assessment of eHOST/CASPR for crowd sourcing and other small and large-scale annotation projects.

Acknowledgments

Support and funding was provided by the VA Salt Lake City HealthCare System and the VA Consortium for Healthcare Informatics Research (CHIR), VA HSR HIR 08-374, the VA Informatics and Computing Infrastructure (VINCI), VA HIR 08-204, and NIH Grant U54 HL 108460 for integrating Data for Analysis, Anonymization and Sharing (iDASH), NIGMS 7R01GM090187.

References

Beatrice Alex, Claire Grover, Barry Haddow, Mijail Kabadjov, Ewan Klein, Michael Matthews, Stuart Roebuck, Richard Tobin, and Xinglong Wang. 2008. Assisted curation: does text mining really help? In: *Proceedings of the Pacific Symposium on Biocomputing.*

Wendy W. Chapman, David Chu, John N. Dowling. 2007. ConText: An Algorithm for Identifying Contextual Features from Clinical Text. In: *ACL-07 2007.*

Carol Friedman, Hongfang Liu, Lyudmila Shagina, Stephen Johnson, George Hripcsak. 2001. Evaluating the UMLS as a source of lexical knowledge for medical language processing. In: *Proc AMIA Symp, 2001: 189-93.*

Karen Fort and Saggot B. 2010. Influence of Pre-Annotation on POS-tagged Corpus Development. In: *Proceedings of the Fourth Linguistic Annotation Workshop. ACL 2010: 56-63.*

Pei-Yun Hsueh, Prem Melville, Vikas Sindhwani. 2009. Data Quality from Crowdsourcing: A study of Annotation Selection Criteria. In: *Proceedings of the NAACL HLT Workshop on Active Learning for Natural Language Processing. June 2009: 27-35.*

Danielle Mowery, Henk Harkema, Wendy W. Chapman. 2008. Temporal Annotation of Clinical Text. In: *ACL-08 2008.*

Mark A. Musen, John Gennari, Henrik Eriksson, Samson W. Tu, and Angel R. Puerta. 1995. PROTEGE-II: computer support for development of intelligent systems from libraries of components. In: *Medinfo 1995*.

Aurélie Névéol, Rezarta Islamaj-Doğan, Zhiyong Lu. 2011. Semi-automatic semantic annotation of PubMed queries: a study on quality, efficiency, satisfaction. In: *J Biomed Inform. 2011 Apr; 44(2):310-8.*

Stefanie Nowak and Stefan Ruger. 2010. How Reliable are Annotations via Crowdsourcing? A Study about Inter-Annotator Agreement for Multi-label Image Annotation. In: *MIR 10 2010.*

Philip V. Ogren. 2006. Knowtator a protege plug-in for annotated corpus construction. In: *Proceedings of the 2006 Conference of the North American Chapter of the Association for Computational Linguistics on Human Language Technology, 2006: 273-5.*

Philip V. Ogren, Guergana K. Savova, Christopher G. Chute. 2008. Constructing Evaluation Corpora for Automated Clinical Named Entity Recognition. In: *Proceedings of the sixth international conference on Language Resources and Evaluation LREC 2008: 3143-3150.*

Angus Roberts, Robert Gaizauskas, Mark Hepple, Neil Davis, George Demetriou, Yikun Guo, Jay Kola, Ian Roberts, Andrea Setzer, Archana Tapuria, Bill Wheeldin. 2007. The CLEF corpus: semantic annotation of clinical text. In: *AMIA Annu Symp Proc, 625-9.*

Angus Roberts, Robert Gaizauskas, Mark Hepple, George Demetriou, Yikun Guo, Ian Roberts, Andrea Setzer. 2009. Building a semantically annotated corpus of clinical texts. In: *J Biomed Inform, 42(5): 950-66.*

Ellen Riloff. 1993. Automatically constructing a dictionary for information extraction tasks. In: *Proceedings of the Eleventh National Conference on Artificial Intelligence, 811-816.*

Burr Settles, Mark Craven, and Lewis Friedland. 2008. Active Learning with Real Annotation Costs. In: *Proceedings of the NIPS Workshop on Cost-Sensitive Learning*. 2008.

Burr Settles. 2009. Active Learning Literature Survey. In: *Computer Sciences Technical Report 1648. University of Wisconsin-Madison. 2009.*

Brett R. South, Shuying Shen, F. Jeff Friedlin, Matthew H. Samore, and Stephane M. Meystre. 2010. Enhancing Annotation of Clinical Text using Pre-Annotation of Common PHI. In: *AMIA Annu Symp Proc. 2010.*

Brett R. South, Shuying Shen, Robyn Barrus, Scott L. DuVall, Ozlem Uzuner, and Charlene Weir. 2011. Qualitative analysis of workflow modifications used to generate the reference standard for the 2010 i2b2/VA challenge. In: *AMIA Annu Symp Proc. 2011.*

Özlem Uzuner, Imre Solti, Fei Xia, and Eithon Cadag. 2010. Community annotation experiment for ground truth generation for the i2b2 medication challenge. In: *J Am Med Inform Assoc, 2010. 17(5):519-23.*

Özlem Uzuner, Brett R. South, Shuying Shen, and Scott L. DuVall. 2011. 2010 i2b2/VA challenge on concepts, assertions, and relations in clinical text. In: *JAMIA 18(5): 552-556.*

Özlem Uzuner, Andreea Bodnari, Shuying Shen, Tyler B. Forbush, John Pestian, and Brett R. South. 2012. Evaluating the state of the art in coreference resolution for electronic medical records. In: *JAMIA doi: 10.1136/amiajnl-2011-000784.*

Aobo Wang, Cong Duy Vu Hoang, and Min-Yen Kan. 2010. Perspectives on Crowdsourcing Annotations for Natural Language Processing. In: *CSIDM Project No. CSIDM-200805.*

Alexander S. Yeh, Lynette Hirschman, and Alexander A. Morgan. 2003. Evaluation of text data mining for database curation: Lessons learned from the KDD challenge cup. In: *Bioinformatics, 19(Suppl 1): i331–339, 2003.*

MedLingMap: A growing resource mapping the Bio-Medical NLP field

Marie Meteer, Bensiin Borukhov, Michael Crivaro,
Michael Shafir, Attapol Thamrongrattanarit
{mmeteer, bborukhov, mcrivaro, mshafir, tet}@brandeis.edu
Department of Computer Science
Brandeis University
Waltham, MA 02453, USA

Abstract

The application of natural language processing (NLP) in the biology and medical domain crosses many fields from Healthcare Information to Bioinformatics to NLP itself. In order to make sense of how these fields relate and intersect, we have created "MedLingMap" (www.medlingmap.org) which is a compilation of references with a multi-faceted index. The initial focus has been creating the infrastructure and populating it with references annotated with facets such as topic, resources used (ontologies, tools, corpora), and organizations. Simultaneously we are applying NLP techniques to the text to find clusters, key terms and other relationships. The goal for this paper is to introduce MedLingMap to the community and show how it can be a powerful tool for research and exploration in the field.

1 Introduction

In any field, understanding the scope of the field as well as finding materials relevant to a particular project paradoxically gets more difficult as the field gets larger. This is even more difficult in a field such as Bio-Medical NLP, since it is at the crossroads of multiple disciplines. The drawbacks of keyword search, even using a specific engine such as Google Scholar, are well documented (Stoica et. al 2007) and recent trends in content aggregation and content curation have emerged to attempt to address the problem. Uses of curation range from those in library science to ensure material remain accessible as format and electronic readers change and to make that information more findable (e.g. Peer and Green 2012) to those in marketing to increase revenue by providing more relevant content (Beaulaurier 2012).

However, these approaches still have challenges. Automatic aggregation over a large body of content still provides too many results without additional filtering mechanism. Content curation, which filters content by value and annotates it to ensure higher quality returns, is expensive since annotating large collections of content with the metadata needed to support faceted search and navigation is a huge challenge.

The goal of the work described in this paper is to provide a framework for creating a useful resource tool bounded by the interests of a specific community which can take advantage of automated clustering and keyword extraction techniques and the use of community based annotation through crowd sourcing and social tatting to provide valuable curation. What is an impossible task for a single team because doable when we successfully harness and empower the community.

The MedLingMap site is available at www.medlingmap.org. Currently MedLingMap has over 300 references many of which are annotated according to a set of "contextual" facets (described below). We first provide some use cases for the system and then go into more detail on the content, infrastructure and origins of the system.

We welcome members of the field to join MedLingMap as a curator to help extend the resource. Just email info@medlingmap.org to get an account. Please include your affiliation.

2 Use Cases

MedLingMap was started as a class project in a Brandeis graduate course on NLP in the Medical Domain to provide a means of finding and organizing the publications in the field and as a data source for exploring trends in topics and relationships among researchers. While there are many use cases for such a resource, three stand out.

The first is simply the ability to find material that meets very specific criteria. for example, to find papers using "MetaMap" for named entity extraction over clinical data. MedLingMap's growing collection of references and the necessary meta-data to make it useful is well suited to this task.

The second is to support the exploration of an area. If I'm interested in clinical coding, I can select that subject area and am presented with a number of papers. I notice that Phil Resnik is on a number of papers and may want to follow up on his work. I also see many of the papers are tagged with AHIMA, including an entire proceedings that is worth exploring. I select a paper and see the abstract mentions a particular challenge that is also worth following up on. We are in the process of developing a personal "workspace" that will let researchers record searches, annotate findings, and keep a queue of the "next directions" that might be worth following up.

The third use case gets back to one of the original premises of the work, which is that a "map" of a field goes beyond a collection of materials, it also provides context and can be used to see "hot spots" and trends. In order to provide this information and visualization, we have developed a set of tools applying a variety of NLP techniques, such as clustering, topic identification and tf-idf to the content of the papers. This work is described in more detail in (Thamrongrattanarit, et al, 2012).

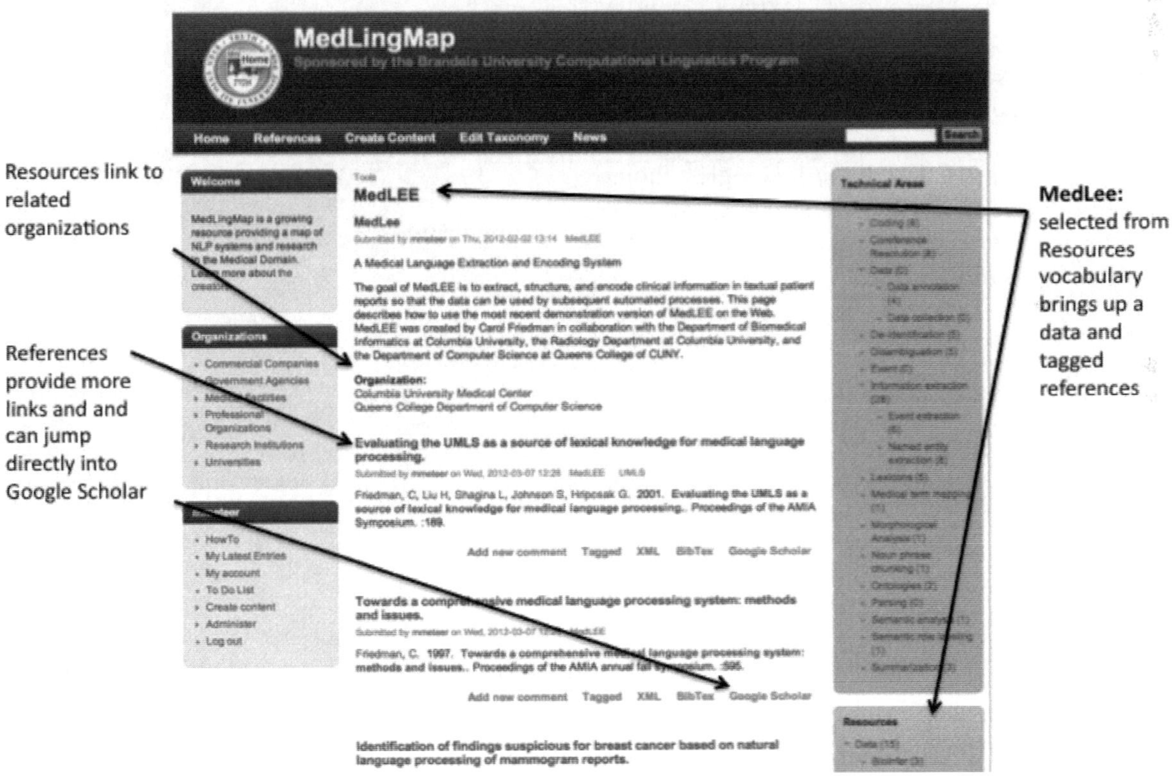

Figure 1: MedLingMap site: www.medlingmap.org

3 Content and Context

The core content in MedLingMap are the references themselves. The underlying representation is based on bibtex and references can be added by either pasting in a single bibtex item or uploading an entire file. A reference can be added through a form interface as well.

We have added BioNLP and related workshops dating back to 2002 as well as many other documents. In addition to the references, there are entries for a variety of organizations and resources with a short description and links for each. These elements are entered by hand. The assumption is that there are a limited number of them and editorial control is more important than speed of entry.

3.1 Examples of the interface

The MedLingMap interface is shown in Figure 1. All references, resources and organizations are linked through a set of "taxonomies" (described below), which have been developed bottom up based on the material tagged to date. Selecting any item from the taxonomies will select content annotated by that tag. So selecting a "Technical area" from the box on the right brings in all the papers annotated by that topic. A similar box of "resources" allows the user to select all papers that have been annotated as using a particular resource.

For example, in Figure 1 the user has selected MedLEE from the "Resources" taxonomy and is shown the information on MedLee as well as references that have been annotated as discussing MedLEE. In addition to the basic bibliographic information, the user can export the reference in bibtex or xml or jump directly to it Google scholar, which can provide multiple ways of accessing the resource. Alternative views show all of the references by year, author or title.

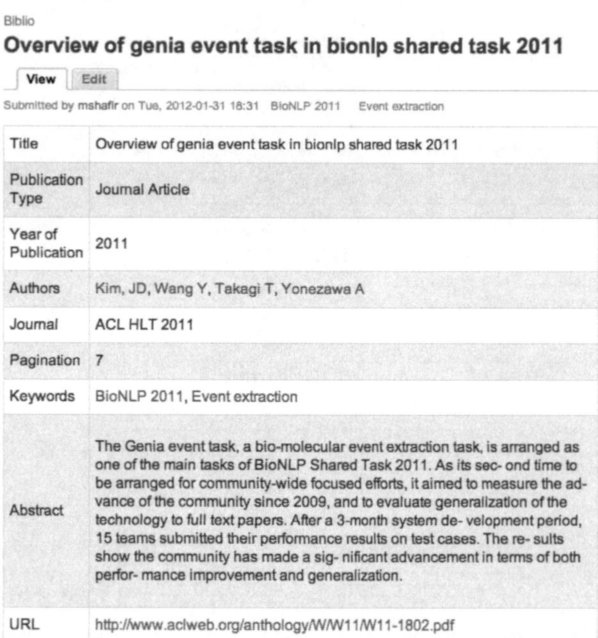
Biblio

Overview of genia event task in bionlp shared task 2011

View | Edit

Submitted by mshafir on Tue, 2012-01-31 18:31 BioNLP 2011 Event extraction

Title	Overview of genia event task in bionlp shared task 2011
Publication Type	Journal Article
Year of Publication	2011
Authors	Kim, JD, Wang Y, Takagi T, Yonezawa A
Journal	ACL HLT 2011
Pagination	7
Keywords	BioNLP 2011, Event extraction
Abstract	The Genia event task, a bio-molecular event extraction task, is arranged as one of the main tasks of BioNLP Shared Task 2011. As its sec- ond time to be arranged for community-wide focused efforts, it aimed to measure the ad- vance of the community since 2009, and to evaluate generalization of the technology to full text papers. After a 3-month system de- velopment period, 15 teams submitted their performance results on test cases. The re- sults show the community has made a sig- nificant advancement in terms of both perfor- mance improvement and generalization.
URL	http://www.aclweb.org/anthology/W/W11/W11-1802.pdf

Figure 2: Information on a particular reference

By selecting a reference in MedLingMap, additional information is available, as shown in Figure 2. By selecting any of the key terms from the taxonomy at the top of the "view", the user can go to more papers tagged with that term. By selecting any of the authors, the user is shown other papers by that author. Those with a "curator" account (described below) can select "edit" and make changes or provide additional tags.

In addition, there is a standard search mechanism, as shown in Figure 3. We are in the process of implementing true faceted search, similar to "advanced search" for recipes, where you can select one or more item from each taxonomy to constrain the search.

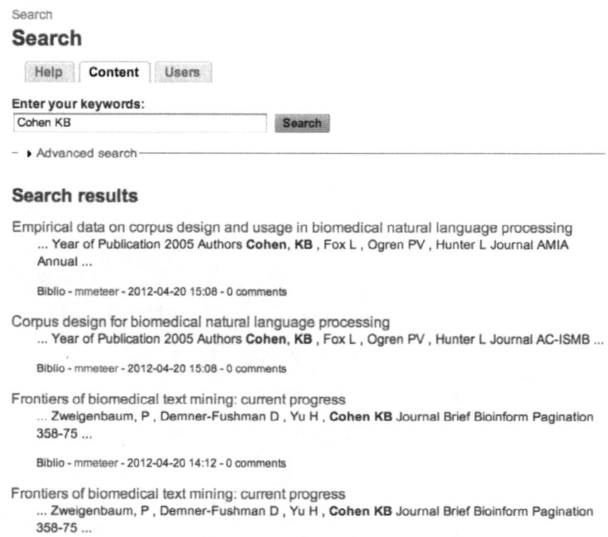

Figure 3: Open search

3.2 Faceted indexing

Indexing content along multiple dimensions or "facets" is not new to search (Alan 1995) and significant work has gone into creating effective interfaces for faceted search (Hearst 2006). When searching for research materials, the context the work was done can be a significant contributor to being able to find related materials. "Necessity is the mother of invention" implies that if you want to find similar solutions, look for similar needs.

To try to capture this kind of information, MedLingMap has facets organized into taxonomies:

- Technical area or topic of the work (shown in the screen shot above)
- Resources used:

- Data: Corpora such as Genia, CRAFT, i2B2, BioInfer
- Lexical Resources, which are organized into dictionaries, and ontologies and include UMLS, PubMEd, MedLine, MeSH, and Medical Wordnet
- Tools, such as parsers, taggers, annotation toolkits and more complete systems, such as MedLee, GATE, and MIST
- Shared tasks, such as the BioNLP 2009 and 2011 shared tasks, BioCreative, and i2b2

- Institution the work was done in or is associated with in some way (e.g. funding, providing resources, etc)

As the project continues, these facets will grow and new ones will be added. Additional facets under consideration include the program (e.g. across multiple institutions, generally associated with a single funding source), target data (e.g. medical literature or clinical records).

4 Origins of MedLingMap

As mentioned above, MedLingMap was started as a class project in a graduate course NLP in the Medical Domain and the creation of the taxonomies and population of the material was done as part of the class. However, the underlying architecture itself is based on a system that has been under development for speech recognition for the past two years (www.stcspeechmap.org) by author Marie Meteer as part of the Speech Technology Consortium's effort to improve prior art research in non-patent literature.

The driving principle is that the "art" in any field (the papers, documentation, product descriptions, etc) can only be understood in terms of the context in which they were produced, contexts which show relations between them that is usually not available in the individual documents. For example, much of the early work in speech recognition addressed the challenges of multimodal interfaces well before we had sophisticated mobile devices. Solutions are being reinvented and patents applied for that would not considered novel if the original research were more readily available. Similar issues arise in multidisciplinary fields such as Bio-Medical NLP where different groups come together who do not have the same historical context and may not know about previous research.

5 Infrastructure

MedLingMap and SpeechMap are built on Drupal[1] an open architecture Content Management System (CMS), which underlies many web sites ranging from www.whitehouse.gov to BestBuy.

Using Drupal ensures that MedLingMap can be a living, growing resource. Drupal provides the following functionality:

- A database to store, retrieve, and maintain large documents sets and web pages, providing multiple views into the contents.
- Specific content types for resources, organizations, authors, and references, all linked though a set of taxonomies.
- The capability to load in references in bibtex format either in a group or individually and annotate them with terms from the taxonomies.
- Maintenance facilities, such as suggesting when multiple authors may be the same person and merging them.
- User profiles with different permission levels to accommodate viewers, contributors, social tagging, and private workspaces with the appropriate levels of security.
- The ability to integrate powerful search components, such as SOLR[2], as well as specific modules, such as the Bibliography module which provides automatic links to Google Scholar to retrieve those documents.
- Web-based to allow easy outside access and be more compatible with other systems.
- Extensibility both for more content, more content types, and more functionality. For example while there is a module that produces a warning if a possible duplicate reference, we are still looking for one that would search out potential duplicates and propose merges. If none exists, such a module can be written and easily integrated.

[1] http://drupal.org/

[2] SOLR is an open-source search server based on the Lucene Java search library. http://lucene.apache.org/solr/

6 Value for Stakeholders

The value of MedLingMap varies with the audience. We first talk about the value to the current community and contrast MedLingMap to similar resources already available. We then look at stakeholders outside or entering the community and the value MedLingMap brings to them.

6.1 BioMedical NLP community

For members of the community, a central repository for papers in the field is a "nice to have". There is information that is surfaced by seeing the organization of the information and links to resources in one place, but if you have been attending conferences and workshops regularly, this is not new information. You know the players and already follow the work you are interested in.

In addition, similar information is available elsewhere, though in a more distributed form. ACL has made all of the proceedings to conferences and workshops available[3]. Similarly ACM and IEEE Xplore provide access to all of the papers they control. The significant difference is that in these collections even the advance search is relying on standard bibliographic elements, such as author and title, and keyword search and there is no segmentation of the material by field, which introduces significant ambiguity as the same term can mean different things in different fields. Similarly PubMed and GoPubMed offer documents and advanced search on a huge body of literature, but focused on biology and medicine, not the application of NLP techniques to those fields. MedLingMap is designed to be focused on a smaller community with more like interests.

It is also important to note that MedLingMap is providing links to papers, not the actual papers, which are controlled by the publishers. While many papers are readily available using the links provided or can be found through the Google Scholar link for each reference, if you need a subscription to see the entire paper such as for IEEE, you still need to go through your standard method to get those papers.

LREC's Resource Map is more similar in that it provides more in depth information that the aggregations described above, however the focus is on mapping the resources themselves, not necessarily all of the publications that have taken advantage of those resources, though some of that information may be available by following the links. LREC is also using a crowd sourcing method for growing the resource by asking those who submit papers also submit the information about the resources they used. This is an interesting model in that it assures that those contributing have a stake in the result since they are members of the community by virtue of submitting a paper.

Organizations such as BioNLP.org and SigBioMed are also important resource aggregators for the community. Neither are focused on publications and we hope that MedLingMap willl become one more resource they would point to.

6.2 From the outside

For students or those who come to the field from a neighboring field, the aggregation of the material in MedLingMap can save considerable time and provide overview or "map" of the field. Queries that are ambiguous in Google Scholar are more precise when the domain is limited. <example>

This increase in the ability of newcomers in the field to find what they are looking for actual turns into benefits for those in the field in two ways: First, one's own papers become more findable, increasing citations and potential collaborations. Second, for those who teach, MedLingMap provides a great environment for the students to do targeted research. Letting them loose in a constrained search environment increases the likelihood they will find a rich body of material to learn from and build on without having to always hand select the papers.

7 Growing the resource

The real challenge for a community resource such as MedLingMap is how to grow it to be comprehensive and keep it up to date, specifically how to:

- grow the number of references and resources
- increase high quality annotations that go beyond what can be extracted automatically.
- provide visualizations that bring to light the connections in the material.
- maintain the quality of the data, for example by fining and merging duplicate entries and

[3] http://aclweb.org/anthology-new/

ensuring information about resources and organizations is up to data.

The two choices for growing are automatic techniques and human annotation. We discuss the former in a related paper (Thamrongrattanarit 2012). Here we describe how manual annotation can be feasible.

7.1 Distributed Power

The key to high quality documents and tagging is community involvement. There are two complementary approaches that are key to the MedLingMap project: crowd sourcing and social tagging. Crowd sourcing involves the community in finding relevant resources, particularly those that are fairly obscure and predate the internet. The second is social tagging which lets individuals check on their own materials or materials in areas related to their own work and adding or adjusting the tags to make the content more searchable.

The key to making these tactics work is setting up the right support in the underlying system. Fortunately, the MedLingMap infrastructure allows for easy signup for those volunteering to contribute. These technique have been used successfully in patent prior art search by Article One, Inc.[4] which puts out a call to researchers to find art on a particular patent. If the client selects that art to support their case, the contributor is paid. The patent office itself attempted something similar in the Peer to Patent program[5], which depended on people's desire to improve the quality of patents by letting them contribute art. It was moderately successful, but without the kind of specific reward the Article One provides, they did not get nearly as much material as they would have liked.

MedLingMap, SpeechMap and other efforts of its kind have the same problem: no one has enough time. So how do we address it? How do we create a convincing value proposition? Here are a couple suggestions:

Teaching: MedLingMap is a great teaching tool. Not only can students use it to do research on the material that's in it, we as educators can enlist them to both tag material and go out on the web to find additional material to tag and add. In just one semester we have made considerable progress. If everyone teaching a similar course enlisted their students, the students would gain and the resource would grow.

Research support: With the implementation of the personal workspace described above, the system will provide a unique service not available from other aggregators or content owners.

Funded project: Being able to hire student annotators would accelerate the process. For the SpeechMap project we have a proposal into the US Patent Office for funding. We are open to suggestions about funding sources for MedLingMap.

[4] http://www.articleonepartners.com/

[5] http://peertopatent.org/

Conclusion

With MedLingMap's infrastructure in place and enough content to provide an exemplar of how it can grow, the challenge now is engaging the community in what we see as an exiting experiment in harnessing the resources of the internet through crowd sourcing and social tagging to create a living resource that will benefit both the current and future members of the field. MedLingMap also provides a resource for exploring automated ways of annotating and organizing research materials. We also hope that this can be a map itself, to build similar "maps" in other subfields.

References

Allen, RB. 1995. Retrieval from Facet Spaces, Electronic Publishing Chichester, Vol. 8(2 & 3), 247–257.

Beaulaurier, Joe. 2012. Content Curating for Fun and Profit, http://whatcommarketing.com/content-curating-for-fun-and-profit/.

Hearst, Marti. 2006. Design Recommendations for Hierarchical Faceted Search Interfaces. ACM SIGIR Workshop on Faceted Search, August, 2006

Peer, L. Green, A. 2012. Building an Open Data Repository for a Specialized Research Community: Process, Challenges and Lessons, International Journal of Digital Curation, Vol 7, No 1.

E. Stoica, M.A. Hearst, and M. Richardson. 2007. Automating Creation of Hierarchical Faceted Metadata Structures. In Human Language Technologies: The Annual Conference of the North American Chapter of the Association for Computational Linguistics (NAACL-HLT 2007), pages 244–251.

Thamrongrattanarit, A., Shafir, M., Crivaro, M., Borukhov, B. , Meteer, M. What can NLP tell us about Bio NLP? BioNLP Workshop, Montreal, CA, 2012

Exploring Label Dependency in Active Learning for Phenotype Mapping

Shefali Sharma[1], **Leslie Lange**[2], **Jose Luis Ambite**[1], **Yigal Arens**[1], **Chun-Nan Hsu**[1,3]
[1]Information Sciences Institute, University of Southern California, Marina del Rey, CA 90292, USA
[2]Department of Genetics, University of North Carolina, Chaple Hills, NC 27599, USA
[3]Institute of Information Sciences, Academia Sinica, Taipei 115, Taiwan
chunnan@isi.edu

Abstract

Many genetic epidemiological studies of human diseases have multiple variables related to any given phenotype, resulting from different definitions and multiple measurements or subsets of data. Manually mapping and harmonizing these phenotypes is a time-consuming process that may still miss the most appropriate variables. Previously, a supervised learning algorithm was proposed for this problem. That algorithm learns to determine whether a pair of phenotypes is in the same class. Though that algorithm accomplished satisfying F-scores, the need to manually label training examples becomes a bottleneck to improve its coverage. Herein we present a novel active learning solution to solve this challenging phenotype-mapping problem. Active learning will make phenotype mapping more efficient and improve its accuracy.

1 Introduction

Phenotypes are observable traits of an individual organism resulting from the presence and interaction of its genotype with the environment. Phenotypes potentially related to human health are of interest in genetics and epidemiology, including common clinical conditions, inheritance disorders, as well as various risk factors such as diet. Substantial amounts of genomic data, including genome-wide genotyping from GWAS (Genome-Wide Association Studies) (Hardy and Singleton, 2009; Consortium, 2007) and sequencing, are being produced in conjunction with the collection of carefully defined and measured phenotypes to study the role of genetic variations in a wide variety of inherited traits and disorders for many decades.

Recently, there is an emerging need to re-use these valuable phenotype-genotype association data to boost the statistical power and improve sensitivity and specificity of the search of associations between various disorders and genetic variations. New paradigms of genomic studies may be fostered once a map of related phenotypes is easily accessible. In fact, one of such new paradigms, PheWAS (Phenome Wide Association Studies), has been developed and producing interesting findings (Denny et al., 2010; Pendergrass et al., 2011) with the help of phenotype mapping and harmonization. Unlike GWAS, which focus on calculating the association between the variation of hundreds of thousands of genotyped single nucleotide polymorphisms (SNPs) and a single or small number of phenotypes, PheWAS uses an extensive range of detailed phenotypic measurements for comprehensively exploring the association between genetic variations and phenotypes. The investigation of a broad range of phenotypes has the potential to identify pleiotropy, reveal novel mechanistic insights, generate new hypotheses, and define a more complete picture of genetic variations and their impact on human diseases.

To facilitate integration of genomic data sets, the research community needs to categorize comparable phenotype measurements and match them across multiple genomic studies to identify data sets of interest as well as potential future collaborations. While the naming strategy for genetic variants is largely standardized across studies (e.g. *rs* numbers for single nucleotide polymorphisms or SNPs), this is often not the case for phenotype variables. Due to the lack of a standardized terminologies or other controlled vocabularies, it becomes increasingly difficult to find studies with comparable phenotypes as the genomic data accumulate. A researcher searching for the availability of comparable phenotypes across multiple studies is confronted with a veritable mountain of variables to sift through. Even within a study, there are often numerous versions of semantically equivalent phenotypic variables. Manually mapping and harmonizing these phenotypes is a time-consuming process that may still miss the most appropriate variables.

Previously, (Hsu et al., 2011) have developed a supervised learning algorithm that learns to determine whether a pair of phenotypes is semantically related from their descriptors. Though that algorithm accomplished satisfying F-scores, the need to manually label training examples becomes a bottleneck to improve its coverage. Moreover, the algorithm treats each pair independently, but pairs that consist of common phenotypes are not independent. Exploring this dependency may potentially improve its performance. In this paper, we investigate how to apply active learning to solve this challenging phenotype-mapping problem. Application of effective active learning techniques will make phenotype mapping more efficient and improve its accuracy and, along with intuitive phenotype query tools, would provide a major resource for researchers utilizing these genomic data.

Active learning queries a user for labels of unlabeled phenotypes that may improve the learning of phenotype mapping the most and thereby reduce the need of labeling efforts. To select the most useful training examples to query, different selection strategies have been proposed in the past (Settles, 2010):

- **Uncertainty Sampling** In this strategy, an active learner chooses an instance that is the most uncertain for the current model to label (Lewis and Catlett, 1994).

- **Query-By-committee** This strategy (Seung et al., 1992) is also known as **maximum disagreement** (Ayache and Quénot, 2007; Di and Crawford, 2011) because the idea is to choose an instance for which a committee of models disagrees the most among its members about its label.

- **Expected Model Change** The general principle of this strategy is to choose an instance to query when if its label is available, the model will be changed the most (Settles and Craven, 2008).

- **Expected Error Reduction** Active learning is useful when the selected instance reduce the error the most and this strategy looks for an instance that can achieve this ultimate goal directly.

- **Variance Reduction** Inspired by the bias-variance analysis of the generalization performance, the variance reduction principle seeks to query for instances that reduce the variance of the model the most. A similar approach is applied in the *optimal experimental design* in statistics (Federov, 1972). However, usually this also requires to solve expensive optimization problems.

- **Density-Weighted Methods** By considering the distribution of the instances, this strategy addresses an issue of uncertainty sampling and query-by-committee where outliers are likely to be selected but contribute limitedly to improving the learning (Fujii et al., 1998; Dasgupta and Hsu, 2008).

The method reported here basically follows the maximum disagreement principle of query-by-committee to select unlabeled pairs of phenotypes to query. A committee must be formed in order for this strategy to be applied, but it has been shown that even a small committee works well in practice. Various approaches can be applied to create committees. For example, co-testing (Muslea et al., 2006) applies this principle by combining forward and backward parsing models for information extraction. A key to the success of this strategy is that member models in the committee complement strengths and weaknesses.

The idea of our method is to compare the match-or-not assignments by the model trained by supervised learning and the class assignments derived from exploring linkages of the labeled and unlabeled phenotypes. The most useful pairs to query are those whose assignments from the two different sources disagree with the highest confidence.

Exploring linkages may improve classifier learning when the classes of instances depend on each other. This idea has been studied in the context of classification of network data, such as pages on the Web, co-reference resolution, word sense disambiguation, and statistical relational learning (see e.g., (Macskassy, 2007; McCallum and Wellner, 2005; Popescul et al., 2003)).

In this paper, we present an algorithm that implement our idea. This algorithm can be divided into two major steps. The first step of the algorithm explores the linkages and the second step prioritizes pairs of phenotypes to query. By identifying maximum disagreement pair instances between the model classification results and exploring linkages between labeled and unlabeled phenotype variables, our active learner queries users for labels of unlabeled phenotypes that may improve the mapping the most and therefore will reduce the need of labeling efforts. Our experimental results show that exploring linkages can perfectly infer the match-or-not labels for a large number of pairs, and that active learning from maximum disagreement pairs improves the performance faster than from randomly selected pairs, suggesting that active learning by exploring linkages is a promising approach to the problem of phenotype mapping.

2 Phenotype Mapping

2.1 Problem Definition

Phenotype mapping is a task of searching for all databases of participating studies to find a set of phenotype variables that match a requested variable that the researcher is interested in. This is similar to the definition given in (Hsu et al., 2011) where the task is defined as the assignment of every phenotype variable from each participating study to one of a set categories, or classes, which corresponds to the "requested variable."

Table 1 shows a fragment of the phenotype mapping results of the phenotype variables that we matched manually from a consortium of cohort studies for a set of 70 requested variables. In this fragment, we show the phenotype variables assigned to one of the requested variables, the phenotype class `'hypertension'`. The real ID of a phenotype in a **Cohort** is given in column **Variable**. In this example, seven cohort studies have a total of 13 phenotype measurements related to hypertension.

Column **Description** is the main clue for automatic matching. The variable descriptions usually contain less than 10 words. As we can see in Table 1, the description contains abbreviations (e.g., `'HTN'`, `'HBP'`,`dx`), aliases (e.g., `'High Blood Pressure'` vs. `Hypertension`), measurement criteria (e.g., `DBP>90 MMHG`, `sys GE 140`, `per JNC7`, `JNC VI`), and tokens irrelevant to our task. As a result, word-by-word string similarity or sophisticated edit-distance based metrics can only match a small number of them. These examples are phenotypes that share similar semantics and are manually mapped to the same classes but their descriptions contain few or no common words. It is impossible for a model solely using the given descriptions to figure out that they refer to related phenotypes without bringing to bear additional information.

Other challenges of the phenotype problem include: not knowing in advance how many classes there are, unavailability of comprehensive categorization of phenotypes, and that the solution should scale well for a large number of phenotypes.

2.2 Supervised Learning for Phenotype Mapping

Here, we review the supervised learning method described in (Hsu et al., 2011), where phenotype mapping was casted as a pair matching problem and applied supervised learning to learn to tag a pair as a match or not. A pair of phenotypes are considered as a match if they are assigned to the same class, otherwise it is not. 13 phenotype variables in Table 1 will yield 78 pairs of positive examples of matched pairs. A maximum entropy classifier (MaxEnt) (Hastie et al., 2009) was used as the model to estimate the probability that a pair is a match. Two types of features were considered. The first type is based on string similarity metrics to combine the strength of a variety of string similarity metrics to measure the edit distance between the descriptions of a pair of phenotypes and use the result to determine if they

Cohort	Requested Variables	Variable	Description
ARIC	Hypertension	HYPERT06	HYPERTENSION, DEFINITION 6
CARDIA	Hypertension	Y01DBP	HYPERTENSION BASED ON DBP> 90 MMHG
CARDIA	Hypertension	Y01HTN	HIGH BLOOD PRESSURE
CARDIA	Hypertension	Y01HTNTP	TYPE OF HYPERTENSION
CFS	Hypertension	htn	HTN: abnormal bp (sys GE 140 or dia GE 90) or meds
CFS	Hypertension	htndx	HTN: self report of MD dx of HTN
CHS	Hypertension	HYPER	CALCULATED HTN STATUS
FHS	Hypertension	A70	HISTORY OF HYPERTENSION
FHS	Hypertension	B373	HYPERTENSION-ON TREAT OR ELEVATED BP
FHS	Hypertension	C332	HBP status
JHS	Hypertension	HTN017	Hypertension Status Per JNC7
MESA	Hypertension	HIGHBP1	HYPERTENSION: SELF-REPORT
MESA	Hypertension	HTN1C	Hypertension by JNC VI (1997) criteria

Table 1: Example variables of phenotype class 'hypertension'

match each other. The other type is the *weighted Jaccard* where appearence of tokens and bi-grams in both or one of the descriptions of a given phenotype pair is used as the features. The training algorithm for MaxEnt will virtually assign to each token or bi-gram a weight when it appears in the descriptions of an input phenotype pair. Weighted Jaccard is superior to string similarity features because string similarity metrics treat all tokens equally and the information provided by these metrics is limited. Therefore weighted jaccard was shown to outperform string similarity features by a large margin in the experimental evaluation.

Before the feature extraction step, descriptions will be augmented with the definitions given in the Merriam-Webster Medical Dictionary (2006)[1]. For example, 'hypertension' will be augmented with its definition in the dictionary 'abnormally high arterial blood pressure' and converted into 'hypertension abnormally high arterial blood pressure'. Augmented 'hypertension' will have many shared tokens with 'high blood pressure'. This augmentation step was proven to be effective in boosting recall, as semantically equivalent pairs described by totally different sets of tokens can be matched.

(Hsu et al., 2011) also reported a transitive inference method to take advantage of the transitive relationship of matched phenotype pairs. The idea is that if v_1 and v_2 are a match, so are v_2 and v_3, then v_1 and v_3 must be a match, too. Applying transitive inference did improve the performance, but when all possible transitive relations are explored, the performance degraded because false positives accumulated. The transitive inference method does not fully explore the dependency between pairs that share common phenotype variables. A more sophisticated approach is required.

[1] www.m-w.com/browse/medical/a.htm

3 Methods

Figure 1 illustrates our active learning idea. The idea is that, given a training set of phenotype variables X manually matched with class labels and a test set of unlabeled phenotype variables, the first step is to infer the class of each unlabeled variable by exploring the pairwise match scores assigned by the model trained by the training set. When we obtain a plausible class assignment to each unlabeled variable, we can classify each pair of unlabeled variables v_1 and v_2 by the trained model again to determine if they are a match or not and compare the result with their plausible class assignments.

If it turns out that the results agree with each other, we will move the pair to a set called *sure pairs*, otherwise, we will move the pair to a queue which will be sorted in descreasing order by how much the results disagree. Then we can query for true labels of the pairs in the queue to add to the training set the most useful examples and thus accomplish the active learning.

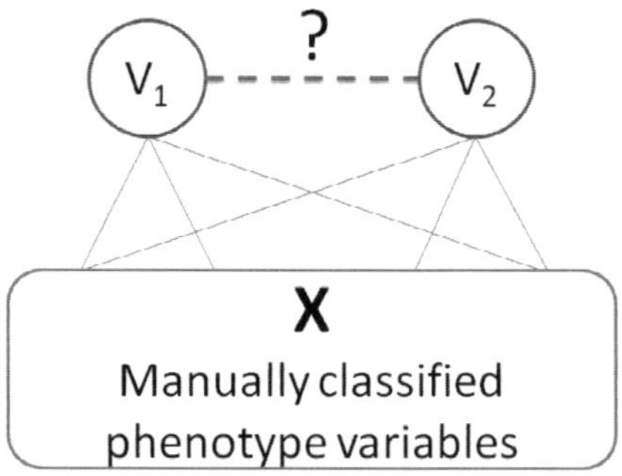

Figure 1: Inferernce of match between unlabeled phenotype variables by exploring their linkages to labeled pairs

Algorithm 1 Procedure LabelA

1: **Initialization**
- Training variables X with their class annotated $class(x) = c \in C, \forall x \in X$
- Test variables V with unknown class $class(v), \forall v \in V$

2: $H \leftarrow Train(\{(x_1, x_2, m) | x_1, x_2 \in X, m = I(class(x_1) = class(x_2))\})$
3: $A \leftarrow \{(v, x) | v \in V \wedge x \in X\}$
4: **procedure** LABELA(A, H)
5: Output $\leftarrow \emptyset$
6: **for** $v \in V$ **do**
7: $\forall x \in X, P^H_{vx} \leftarrow H(v, x)$
8: $c \leftarrow \arg\max_c(P^H_{vx})$
9: $L^H_{vx} \leftarrow \max_C(logP^H_{vx})$
10: **if** $L^H_{vx} < -2$ **then**
11: $c \leftarrow$ null,
12: $s \leftarrow \log(1 - 2^{L^H_{vx}})$
13: **else**
14: $s \leftarrow L^H_{vx}$
15: **end if**
16: Add (v, c, s) to Output
17: **end for**
18: Return Output
19: **end procedure**

3.1 Assigning Phenotype Categories

Procedure LabelA is to assign a class label to each unlabeled test variable by matching them to labeled training variables. Let A denote the set of all pairs between a test variable and a training variable. For each variable, the output contains an element of the variable, its assigned class label (may be null) and a score (log-likelihood). Function $I(.)$ in line 2 is the indicator function that returns 1 if its parameter is true and 0 otherwise. H is the model learned by calling the supervised training procedure. In line 7, P^H_{vx} is the probability that variables v and x are a match estimated by H. In line 8, LabelA assigns v to a class c, which is the class of the training variable x that maximizes P^H_{vx}. That is to assign the class of x as that of v if P^H_{vx} is the largest. Other selection can be used. For example, for each class c, we can estimate P^H_{vx} for all training variables x in c, and select c as the class of v if $\frac{1}{n}\sum \log P^H_{vx}$, the geometric mean of the probabilities, is the largest. These selection criteria are based on different assumptions and we will empirically compare which one is a better choice. In fact, any type of average can potentially be considered here.

3.2 Prioritizing Unlabeled Pairs

Procedure LabelB orders pairs of test variables to query for match-or-not and class labels. Let B be the set of all pairs of test variables. LabelB also generates a set called SurePairs. For each pair in B, LabelB checks if the model H considers the pair as a match ($P^H_{vx} \geq 0.5$) or not, and then checks if the pair is assigned by LabelA to the same class or not. If it is a match and assigned to the same class, or not a match and assigned to different classes, that is, if H and LabelA agree, then the pair will be moved to SurePairs, otherwise, the pair will be moved to Queue. For a disagreed pair, LabelB also estimate the degree of disagreement by the sum of the log-probabilities of the class assignments (L^H_{c1} and L^H_{c2}) and the match-or-not by the model ($P^H_{v_1 v_2}$). SurePairs can then be used for training.

We can then query for true labels of pairs in Queue. We can either query whether a pair is a match or not or query for their class label. After a certain number of queries, we can repeat the procedure to compute a new set of SurePairs and Queue, until all phenotypes are correctly assigned to a class.

Algorithm 2 Procedure LabelB

Initialization
H, A as in LabelA
$B \leftarrow \{(v_1, v_2) | v_1, v_2 \in V\}$
SurePairs $\leftarrow \emptyset$; Queue $\leftarrow \emptyset$
$\forall v_1, v_2 \in V, P^H_{v_1 v_2} \leftarrow H(v_1, v_2)$
$(v, class(v), L^H_c), \forall v \in V \leftarrow$ LabelA(A,H)
procedure LABELB(B, A, H)
    for $(v_1, v_2) \in B$ **do**
        if $P^H_{v_1, v_2} \geq 0.5$ **then**
            if $c_1 = c_2$ **then**
                Add $(v_1, v_2, 1)$ to SurePairs
            else
                $s \leftarrow L^H_{c_1} + L^H_{c_2} + \log(1 - P^H_{v_1 v_2})$
                Add (v_1, v_2, s) to Queue
            end if
        else
            if $c_1 = c_2$ **then**
                $s \leftarrow L^H_{c_1} + L^H_{c_2} + \log P^H_{v_1 v_2}$
                Add (v_1, v_2, s) to Queue
            else
                Add $(v_1, v_2, 0)$ to SurePairs
            end if
        end if
    end for
    Sort (v_1, v_2, m) in Queue by m
    Return Queue and SurePairs
end procedure

4 Results

4.1 Data

We manually selected 1,177 phenotype variables from a total of 35,041 in the databases of seven cohort studies as shown in Table 1 and assigned them to one of 70 requested variables that are common trait classes related to a large consortium study of cardiovascular disorders. These seven cohorts include ARIC (the Atherosclerosis Risk In Communities study `www.cscc.unc.edu/aric/`), CARDIA (the Coronary Artery Risk In Young Adults study `www.cardia.dopm.uab.edu`), CFS (the Cleveland Family study `dceweb1.case.edu/serc/collab/project_family.shtml`), CHS (the Cardiovascular Heart Study `www.chs-nhlbi.org/`), FHS (Framingham Heart Study `www.framinghamheartstudy.org/`), JHS (Jackson Heart Study `jhs.jsums.edu/jhsinfo/`), and MEC (the Multi-Ethnic Cohort `www.crch.org/multiethniccohort/`, `www.uscnorris.com/mecgenetics/`).

From these 1,177 phenotypes, 21,886 pairs are considered matches, that is, they are positive pairs with both phenotype variables in the same class. 670,190 pairs are negatives.

Method / Model	Precision	Recall	F-score
String similarity			
MaxEnt	0.5557	0.0660	0.1179
Weighted Jaccard			
MaxEnt	0.8791	0.4848	0.6250
w/ dictionary	0.9200	0.6104	0.7339
w/ transitive infer.	0.7735	0.6612	0.7129
w/ both	0.7728	0.8402	0.8051

Table 2: Performance results of supervised learning

4.2 Result of Supervised Learning

We divided all pairs in our data set by half into training and test sets and evaluate different options of the supervised learning algorithm with different options as described in (Hsu et al., 2011). The results as shown in Table 2 are consistent with the conclusions given in (Hsu et al., 2011). That is, weighted Jaccard features with dictionary augmentation plus transitive inference yields the best performance.

We also performed a split-by-variable test, where the set of all variables is divided into three equal parts. Two of them are used for training and the other for testing. This is closer to the realistic application scenario and provides a better estimation of the generalization performance of a trained model. The results are given as the first two rows in Table 3.

4.3 Result of Active Learning

We implemented the two algorithms and evaluate the performance. We still applied split-by-variable to divide the data with $\frac{1}{3}$ for testing and $\frac{2}{3}$ for training. We measured the performance when `SurePairs` produced by procedure `LabelB` was added to the training set, and then increasingly add more pairs in `Queue`, also produced by `LabelB`, to the training set, and measured the performance of the trained models to simulate an active learning

Method/Model	Precision	Recall	F-score
w/o dictionary	0.8344	0.4106	0.5504
w/ dictionary	0.6310	0.5287	0.5753
Test on A	0.7956	0.5243	0.6321
GM SurePairs (62622)	0.8772	0.5909	0.7061
Model (62622)	0.9577	0.2936	0.4494
MP SurePairs (74229)	0.8845	0.6196	0.7287
Model (74229)	0.9660	0.2875	0.4431

Table 3: Performance results of splitting by variables. Numbers in the parentheses show the number of pairs in `SurePairs`.

query sequence.

To ensure a fair comparison, we always use the set A, the pairs between a labeled and unlabeled phenotype variables, as the hold-out set for testing in all performance evaluations. Note that pairs in the set A never appear in either `SurePairs` or `Queue`, because pairs in `SurePairs` or `Queue` are selected from the set B, which contains the pairs between unlabeled phenotype variables. The third row of Table 3 shows the performance of the model tested only on A.

We implemented two versions of procedure `LabelA` that are different in the methods they used to assign a class to an unlabeled variable. The first, `MP`, is to use the maximum probability and the other, `GM`, is to use the maximum geometric mean of the probabilities (see Section 3.1).

We start by evaluating the quality of `SurePairs`. `GM` produced 62,622 pairs (1,642 positives) while `MP` had 74,229 pairs (1,816 positives). The match-or-not labels assigned by `LabelB` for both methods turn out to be perfectly correct, suggesting that combining model training and linkage exploration can effectively infer the match-or-not labels.

Adding `SurePairs` to the training set boosts F-scores, as shown in Table 3, which also shows that, in contrast, if we add the same number of pairs to the training set, but assign them match-or-not labels with the trained model, they will degrade F-scores.

Next, we added pairs in `Queue` to the training set, 280 pairs at a time, and measured the F-scores achieved by the resulting model. Figure 2 shows the learning curves of three different ways to order `Queue` produced with `GM`: descreasing, increasing, and random scores. The decreasing-score one performed the best by improving F-scores the fastest, confirming that higher-scored pairs are more useful. The end points of the three curves do not meet because we have not exhausted all training examples.

Similarly, we evaluated decreasing and random ordering of `Queue` produced by applying `MP`. We note that `MP` already produced a large set of `SurePairs`. As a result, less pairs are in `Queue` compared to that by `GM`. Therefore, after 9 passes, all pairs are exhausted and no obvious difference can be observed between decreasing and random ordering in the end.

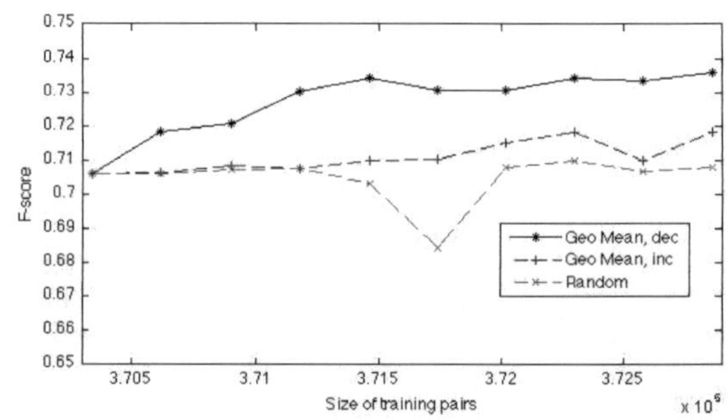

Figure 2: Learning curves of active learning: class assignment by maximum geometric mean of probabilities

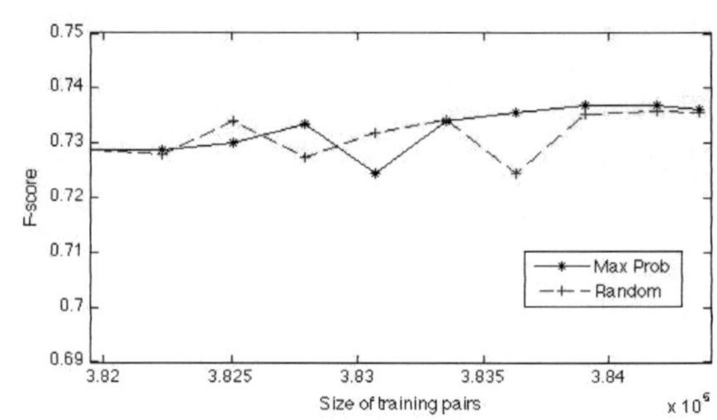

Figure 3: Learning curves of active learning: class assignment by maximum probabilities

5 Conclusions and Future Works

Despite the vast amounts of genomic data available in repositories, identification of relevant datasets can be challenging for researchers interested in specific phenotypic measures. This paper presents our active learning approach that will be implemented as a component of new informatics tools for the research community to categorize phenotype measurements from genomic studies.

We show that comparing class assignment by exploring linkages and by the model can be effective in both improving the match-or-not assignments and ordering unlabeled pairs as queries for active learning. It is interesting that when two sources of class assignment agree, the pairs' match-or-not assignments are perfectly correct. How generalizable for this result deserves further investigation. We note that in order to perform a fair comparison, no pair between labeled and unlabeled phenotype variables are used for training. In a real application, they can be added to either `SurePairs` or `Queue` by extending procedure `LabelB` to include them.

Acknowledgments

We thank Cheng-Ju Kuo and Congxing Cai for their help in producing the results reported in Section 4.2. This research is supported by NHLBI-NIH grant 1UH2HL108780-01.

References

Stéphane Ayache and Georges Quénot. 2007. Evaluation of active learning strategies for video indexing. *Signal Processing: Image Communication*, 22(78):692 – 704. Special Issue on Content-Based Multimedia Indexing and Retrieval.

The Wellcome Trust Case Control Consortium. 2007. Genome-wide association study of 14,000 cases of seven common diseases and 3,000 shared controls. *Nature*, 447(7145):661–678, June.

Sanjoy Dasgupta and Daniel Hsu. 2008. Hierarchical sampling for active learning. In *Machine Learning, Proceedings of the Twenty-Fifth International Conference on Machine Learning (ICML-08)*, pages 208–215.

Joshua C. Denny, Marylyn D. Ritchie, Melissa A. Basford, Jill M. Pulley, Lisa Bastarache, Kristin Brown-Gentry, Deede Wang, Dan R. Masys, Dan M. Roden, and Dana C. Crawford. 2010. Phewas: demonstrating the feasibility of a phenome-wide scan to discover gene-disease associations. *Bioinformatics*, 26(9):1205–1210.

Wei Di and Melba M. Crawford. 2011. View generation for multiview maximum disagreement based active learning for hyperspectral image classification. *Geoscience and Remote Sensing, IEEE Transactions on*, PP(99):1 –13.

Valeriĭ V. Federov. 1972. *Theory of Optimal Experiments*. Academic Press.

Atsushi Fujii, Takenobu Tokunaga, Kentaro Inui, and Hozumi Tanaka. 1998. Selective sampling for example-based word sense disambiguation. *Computational Linguistics*, 24(4):573–597.

John Hardy and Andrew Singleton. 2009. Genomewide association studies and human disease. *New England Journal of Medicine*, 360(17):1759–1768.

Trevor Hastie, Robert Tibshirani, and Jerome H. Friedmann. 2009. *The Elements of Statistical Learning (2nd Edition)*. Springer-Verlag, New York, NY, USA.

Chun-Nan Hsu, Cheng-Ju Kuo, Congxing Cai, Sarah A. Pendergrass, Marylyn D. Ritchie, and Jose Luis Ambite. 2011. Learning phenotype mapping for integrating large genetic data. In *Proceedings of BioNLP 2011 Workshop*, BioNLP '11, pages 19–27, Portland, OR, USA. Association for Computational Linguistics.

David D. Lewis and Jason Catlett. 1994. Heterogeneous uncertainty sampling for supervised-learning. In *Proceedings of the International Conference on Machine Learning (ICML-94)*, pages 148–156.

Sofus A. Macskassy. 2007. Improving learning in networked data by combinng explicit and mined links. In *Proceedings of the National Conference on Artificial Intelligence (AAAI-07)*, page 590.

Andrew McCallum and Ben Wellner. 2005. Conditional models of identity uncertainty with application to noun coreference. In Lawrence K. Saul, Yair Weiss, and Léon Bottou, editors, *Advances in Neural Information Processing Systems 17*, pages 905–912. MIT Press, Cambridge, MA.

Merriam-Webster. 2006. *Medical Dictionary*. Merriam-Webster, Springfield, MA, USA.

Ion Muslea, Steve Minton, and Craig A. Knoblock. 2006. Active learning with multiple views. *Journal of Artificial Intelligence Research*, 27:203–233.

S. A. Pendergrass, K. Brown-Gentry, S. M. Dudek, E. S. Torstenson, J. L. Ambite, C. L. Avery, S. Buyske, C. Cai, M. D. Fesinmeyer, C. Haiman, G. Heiss, L. A. Hindorff, C. N. Hsu, R. D. Jackson, C. Kooperberg, L. Le Marchand, Y. Lin, T. C. Matise, L. Moreland, K. Monroe, A. P. Reiner, R. Wallace, L. R. Wilkens, D. C. Crawford, and M. D. Ritchie. 2011. The use of phenome-wide association studies (phewas) for exploration of novel genotype-phenotype relationships and pleiotropy discovery. *Genetic Epidemiology*, 35(5):410–422.

Alexandrin Popescul, Rin Popescul, and Lyle H. Ungar. 2003. Statistical relational learning for link prediction.

In *In Workshop on Learning Statistical Models from Relational Data at the International Joint Conference on Articial Intelligence (IJCAI-2003)*.

Burr Settles and Mark Craven. 2008. An analysis of active learning strategies for sequence labeling tasks. In *Proceedings of the Conference on Empirical Methods in Natural Language Processing (EMNLP)*, pages 1069–1078. ACL Press.

Burr Settles. 2010. Active learning literature survey. Computer Science Technical Report 1648, University of Wisconsin-Madison, January.

H. Sebastian Seung, Manfred Opper, and Haim Sompolinsky. 1992. Query by committee. In *Proceedings of the Fifth Annual Workshop on Computational Learning Theory (COLT'92)*, pages 278–294.

Evaluating Joint Modeling of Yeast Biology Literature and Protein-Protein Interaction Networks

Ramnath Balasubramanyan and **Kathryn Rivard** and **William W. Cohen**
School of Computer Science
Carnegie Mellon University
rbalasub,krivard,wcohen@cs.cmu.edu

Jelena Jakovljevic and **John Woolford**
Deparment of Biological Sciences
Carnegie Mellon University
jelena,jw17@andrew.cmu.edu

Abstract

Block-LDA is a topic modeling approach to perform data fusion between entity-annotated text documents and graphs with entity-entity links. We evaluate Block-LDA in the yeast biology domain by jointly modeling PubMed® articles and yeast protein-protein interaction networks. The topic coherence of the emergent topics and the ability of the model to retrieve relevant scientific articles and proteins related to the topic are compared to that of a text-only approach that does not make use of the protein-protein interaction matrix. Evaluation of the results by biologists show that the joint modeling results in better topic coherence and improves retrieval performance in the task of identifying top related papers and proteins.

1 Introduction

The prodigious rate at which scientific literature is produced makes it virtually impossible for researchers to manually read every article to identify interesting and relevant papers. It is therefore critical to have automatic methods to analyze the literature to identify topical structure in it. The latent structure that is identified can be used for different applications such as enabling browsing, retrieval of papers related to a particular sub-topic etc. Such applications assist in common scenarios such as helping a researcher identify a set of articles to read (perhaps a set of well-regarded surveys) to familiarize herself with a new sub-field; helping a researcher to stay abreast with the latest advances in his field by identifying relevant articles etc.

In this paper, we focus on the task of organizing a large collection of literature about yeast biology to enable topic oriented browsing and retrieval from the literature. The analysis is performed using topic modeling(Blei et al., 2003) which has, in the last decade, emerged as a versatile tool to uncover latent structure in document corpora by identifying broad topics that are discussed in it. This approach complements traditional information retrieval tasks where the objective is to fulfill very specific information needs.

In addition to literature, there often exist other sources of domain information related to it. In the case of yeast biology, an example of such a resource is a database of known protein-protein interactions (PPI) which have been identified using wetlab experiments. We perform data fusion by combining text information from articles and the database of yeast protein-protein interactions, by using a latent variable model — Block-LDA (Balasubramanyan and Cohen, 2011) that jointly models the literature and PPI networks.

We evaluate the ability of the topic models to return meaningful topics by inspecting the top papers and proteins that pertain to them. We compare the performance of the joint model i.e. Block-LDA with a model that only considers the text corpora by asking a yeast biologist to evaluate the coherence of topics and the relevance of the retrieved articles and proteins. This evaluation serves to test the utility of Block-LDA on a real task as opposed to an internal evaluation (such as by using perplexity metrics for example). Our evaluaton shows that the joint model outperforms the text-only approach both in topic coherence and in top paper and protein retrieval as measured by precision@10 values.

The rest of the paper is organized as follows. Section 2 describes the topic modeling approach used in the paper. Section 3 describes the datasets used followed by Section 4 which details the setup of the experiments. The results of the evaluation are presented in Section 5 which is followed by the conclusion.

2 Block-LDA

The Block-LDA model (plate diagram in Figure 1) enables sharing of information between the component on the left that models links between pairs of entities represented as edges in a graph with latent block structure, and the component on the right that models text documents, through shared latent topics. More specifically, the distribution over the entities of the type that are linked is shared between the block model and the text model.

The component on the right, which is an extension of the LDA models documents as sets of "bags of entities", each bag corresponding to a particular type of entity. Every entity type has a topic wise multinomial distribution over the set of entities that can occur as an instance of the entity type. This model is termed as Link-LDA(Nallapati et al., 2008) in the literature.

The component on the left in the figure is a generative model for graphs representing entity-entity links with an underlying block structure, derived from the sparse block model introduced by Parkkinen et al. (2009). Linked entities are generated from topic specific entity distributions conditioned on the topic pairs sampled for the edges. Topic pairs for edges (links) are drawn from a multinomial defined over the Cartesian product of the topic set with itself. Vertices in the graph representing entities therefore have mixed memberships in topics. In contrast to Mixed-membership Stochastic Blockmodel (MMSB) introduced by Airoldi et al. (2008), only observed links are sampled, making this model suitable for sparse graphs.

Let K be the number of latent topics (clusters) we wish to recover. Assuming documents consist of T different types of entities (i.e. each document contains T bags of entities), and that links in the graph are between entities of type t_l, the generative process is as follows.

1. Generate topics: For each type $t \in 1, \ldots, T$, and topic $z \in 1, \ldots, K$, sample $\beta_{t,z} \sim$ Dirichlet(γ), the topic specific entity distribution.

2. Generate documents. For every document $d \in \{1 \ldots D\}$:

- Sample $\theta_d \sim$ Dirichlet(α_D) where θ_d is the topic mixing distribution for the document.
- For each type t and its associated set of entity mentions $e_{t,i}$, $i \in \{1, \cdots, N_{d,t}\}$:
  - Sample a topic $z_{t,i} \sim$ Multinomial(θ_d)
  - Sample an entity $e_{t,i} \sim$ Multinomial($\beta_{t,z_{t,i}}$)

3. Generate the link matrix of entities of type t_l:

- Sample $\pi_L \sim$ Dirichlet(α_L) where π_L describes a distribution over the Cartesian product of the set of topics with itself, for links in the dataset.
- For every link $e_{i1} \rightarrow e_{i2}$, $i \in \{1 \cdots N_L\}$:
  - Sample a topic pair $\langle z_{i1}, z_{i2} \rangle \sim$ Multinomial(π_L)
  - Sample $e_{i1} \sim$ Multinomial($\beta_{t_l, z_{i1}}$)
  - Sample $e_{i2} \sim$ Multinomial($\beta_{t_l, z_{i2}}$)

Note that unlike the MMSB model, this model generates only realized links between entities.

Given the hyperparameters α_D, α_L and γ, the joint distribution over the documents, links, their topic distributions and topic assignments is given by

$$p(\pi_L, \boldsymbol{\theta}, \boldsymbol{\beta}, \mathbf{z}, \mathbf{e}, \langle \mathbf{z_1}, \mathbf{z_2} \rangle, \langle \mathbf{e_1}, \mathbf{e_2} \rangle | \alpha_D, \alpha_L, \gamma) \propto \tag{1}$$

$$\prod_{z=1}^{K} \prod_{t=1}^{T} \mathrm{Dir}(\beta_{t,z} | \gamma_t) \times$$

$$\prod_{d=1}^{D} \mathrm{Dir}(\theta_d | \alpha_D) \prod_{t=1}^{T} \prod_{i=1}^{N_{d,t}} \theta_d^{z_{t,i}^{(d)}} \beta_{t, z_{t,i}^{(d)}}^{e_{t,i}} \times$$

$$\mathrm{Dir}(\pi_L | \alpha_L) \prod_{i=1}^{N_L} \pi_L^{\langle z_{i1}, z_{i2} \rangle} \beta_{t_l, z1}^{e_{i1}} \beta_{t_l, z2}^{e_{i2}}$$

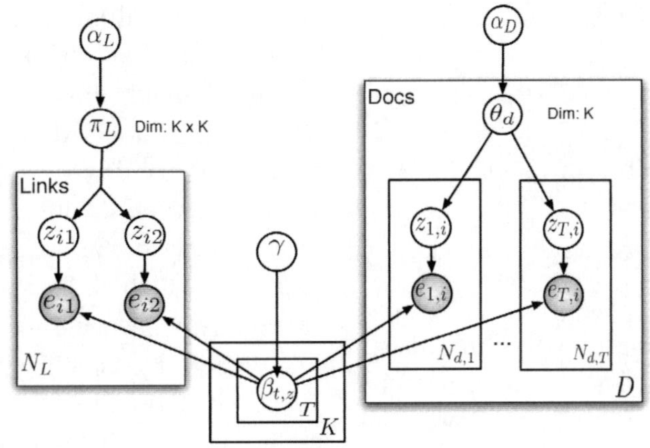

α_L - Dirichlet prior for the topic pair distribution for links
α_D - Dirichlet prior for document specific topic distributions
γ - Dirichlet prior for topic multinomials
π_L - multinomial distribution over topic pairs for links
θ_d - multinomial distribution over topics for document d
$\beta_{t,z}$ - multinomial over entities of type t for topic z
$z_{t,i}$ - topic chosen for the i-th entity of type t in a document
$e_{t,i}$ - the i-th entity of type t occurring in a document
z_{i1} and z_{i2} - topics chosen for the two nodes participating in the i-th link
e_{i1} and e_{i2} - the two nodes participating in the i-th link

Figure 1: Block-LDA

A commonly required operation when using models like Block-LDA is to perform inference on the model to query the topic distributions and the topic assignments of documents and links. Due to the intractability of exact inference in the Block-LDA model, a collapsed Gibbs sampler is used to perform approximate inference. It samples a latent topic for an entity mention of type t in the text corpus conditioned on the assignments to all other entity mentions using the following expression (after collapsing θ_D):

$$p(z_{t,i} = z | e_{t,i}, \mathbf{z}^{\neg \mathbf{i}}, \mathbf{e}^{\neg \mathbf{i}}, \alpha_D, \gamma) \propto (n_{dz}^{\neg i} + \alpha_D) \frac{n_{zte_{t,i}}^{\neg i} + \gamma}{\sum_{e'} n_{zte'}^{\neg i} + |E_t|\gamma} \tag{2}$$

Similarly, we sample a topic pair for every link conditional on topic pair assignments to all other links after collapsing π_L using the expression:

$$p(\mathbf{z_i} = \langle z_1, z_2 \rangle | \langle e_{i1}, e_{i2} \rangle, \mathbf{z}^{\neg \mathbf{i}}, \langle \mathbf{e_1}, \mathbf{e_2} \rangle^{\neg \mathbf{i}}, \alpha_L, \gamma) \propto \left(n_{\langle z_1, z_2 \rangle}^{L \neg i} + \alpha_L \right) \times \frac{\left(n_{z_1 t_l e_{i1}}^{\neg i} + \gamma \right) \left(n_{z_2 t_l e_{i2}}^{\neg i} + \gamma \right)}{\left(\sum_e n_{z_1 t_l e}^{\neg i} + |E_{t_l}|\gamma \right) \left(\sum_e n_{z_2 t_l e}^{\neg i} + |E_{t_l}|\gamma \right)} \tag{3}$$

E_t refers to the set of all entities of type t. The n's are counts of observations in the training set.

- n_{zte} - the number of times an entity e of type t is observed under topic z
- n_{zd} - the number of entities (of any type) with topic z in document d
- $n_{\langle z_1, z_2 \rangle}^L$ - count of links assigned to topic pair $\langle z_1, z_2 \rangle$

The topic multinomial parameters and the topic distributions of links and documents are easily recovered using their MAP estimates after inference using the counts of observations.

$$\beta_{t,z}^{(e)} = \frac{n_{zte} + \gamma}{\sum_{e'} n_{zte'} + |E_t|\gamma}, \quad (4)$$

$$\theta_d^{(z)} = \frac{n_{dz} + \alpha_D}{\sum_{z'} n_{dz'} + K\alpha_D} \text{and} \quad (5)$$

$$\pi_L^{\langle z_1,z_2\rangle} = \frac{n_{\langle z_1,z_2\rangle} + \alpha_L}{\sum_{z_1',z_2'} n_{\langle z_1',z_2'\rangle} + K^2\alpha_L} \quad (6)$$

A de-noised form of the entity-entity link matrix can also be recovered from the estimated parameters of the model. Let B be a matrix of dimensions $K \times |E_{t_l}|$ where row $k = \beta_{t_l,k}$, $k \in \{1, \cdots, K\}$. Let Z be a matrix of dimensions $K \times K$ s.t $Z_{p,q} = \sum_{i=1}^{N_L} \mathbf{I}(z_{i1} = p, z_{i2} = q)$. The de-noised matrix M of the strength of association between the entities in E_{t_l} is given by $M = B^T ZB$.

In the context of this paper, de-noising the protein-protein interaction networks studied is an important application. The joint model permits information from the large text corpus of yeast publications to be used to de-noise the PPI network and to identify potential interactions that are missing in the observed network. While this task is important and interesting, it is outside the scope of this paper and is a direction for future work.

3 Data

We use a collection of publications about yeast biology that is derived from the repository of scientific publications at PubMed®. PubMed® is a free, open-access on-line archive of over 18 million biological abstracts and bibliographies, including citation lists, for papers published since 1948. The subset we work with consists of approximately 40,000 publications about the yeast organism that have been curated in the Saccharomyces Genome Database (SGD) (Dwight et al., 2004) with annotations of proteins that are discussed in the publication. We further restrict the dataset to only those documents that are annotated with at least one protein from the protein-protein interactions databases described below. This results in a protein annotated document collection of 15,776 publications. The publications in this set were written by a total of 47,215 authors. We tokenize the titles and abstracts based on white space, lowercase all tokens and eliminate stopwords. Low frequency (< 5 occurrences) terms are also eliminated. The vocabulary that is obtained consists of 45,648 words.

The Munich Institute for Protein Sequencing (MIPS) database (Mewes et al., 2004) includes a hand-crafted collection of protein interactions covering 8000 protein complex associations in yeast. We use a subset of this collection containing 844 proteins, for which all interactions were hand-curated.

Finally, we use another dataset of protein-protein interactions in yeast that were observed as a result of wetlab experiments by collaborators of the authors of the paper. This dataset consists of 635 interactions that deal primarily with ribosomal proteins and assembly factors in yeast.

4 Setup

We conduct three different evaluations of the emergent topics. Firstly, we obtain topics from only the text corpus using a model that comprises of the right half of Figure 1 which is equivalent to using the Link-LDA model. For the second evaluation, we use the Block-LDA model that is trained on the text corpus and the MIPS protein-protein interaction database. Finally, for the third evaluation, we replace the MIPS database with the interaction obtained from the wetlab experiments. In all the cases, we set K, the number of topics to be 15. In each variant, we represent documents as 3 sets of entities i.e. the words in the abstracts of the article, the set of proteins associated with the article as indicated in the SGD database and finally the authors who wrote the article. Each topic therefore consists of 3 different multinomial distributions over the sets of the 3 kinds of entities described.

Topics that emerge from the different variants can possibly be assigned different indices even when they discuss the same semantic concept. To compare topics across variants, we need a method to determine which topic indices from the different variants correspond to the same semantic concept. To obtain the mapping between topics from each variant, we utilize the Hungarian algorithm (Kuhn, 1955) to solve the assignment problem where the cost of aligning topics together is determined using the Jensen-Shannon divergence measure.

Once the topics are obtained, we firstly obtain the proteins associated with the topic by retrieving the

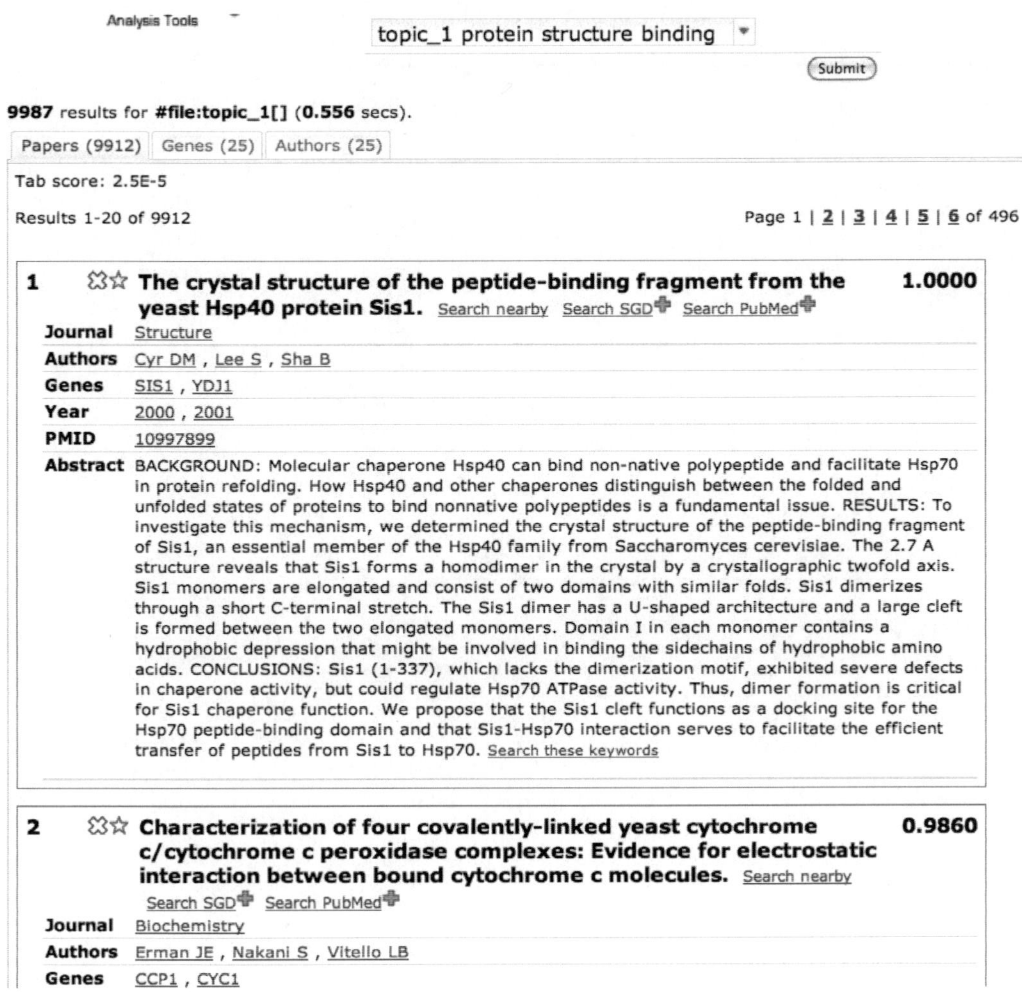

Figure 2: Screenshot of the Article Relevance Annotation Tool

Variant	Num. Coherent Topics
Only Text	12 / 15
Text + MIPS	13 / 15
Text + Wetlab	15 / 15

Table 1: Topic Coherence Evaluation

top proteins from the multinomial distribution corresponding to proteins. Then, the top articles corresponding to each topic is obtained using a ranked list of documents with the highest mass of their topic proportion distributions (θ) residing in the topic being considered.

4.1 Manual Evaluation

To evaluate the topics, a yeast biologist who is an expert in the field was asked to mark each topic with a binary flag indicating if the top words of the distribution represented a coherent sub-topic in yeast biology. This process was repeated for the 3 different variants of the model. The variant used to obtain results is concealed from the evaluator to remove the possibility of bias. In the next step of the evaluation, the top articles and proteins assigned to each topic were presented in a ranked list and a similar judgement was requested to indicate if the article/protein was relevant to the topic in question. Similar to the topic coherence judgements, the process was repeated for each variant of the model. Screenshots of the tool used for obtaining the judgments can be seen in Figure 2. It should be noted that since the nature of the topics in the literature considered was highly technical and specialized, it was impractical to get judgements from multiple annotators.

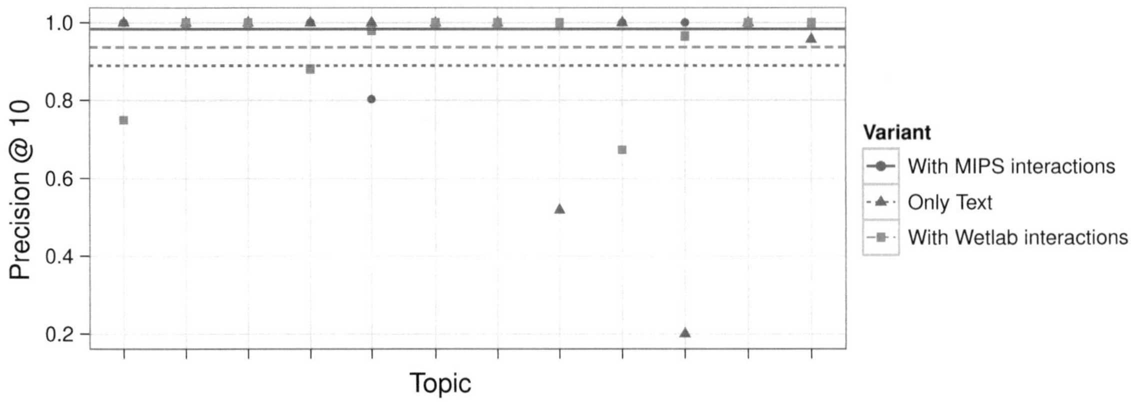

(a) Article Retrieval

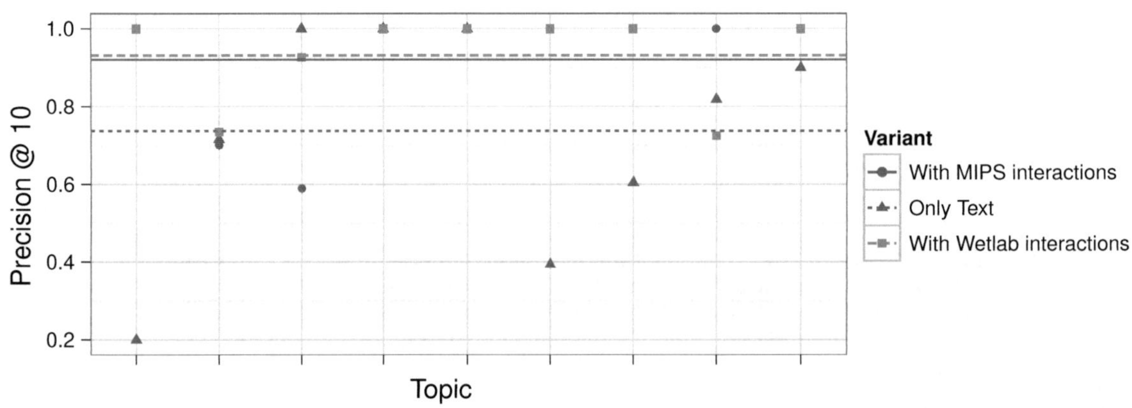

(b) Protein Retrieval

Figure 3: Retrieval Performance Evaluation (Horizontal lines indicate mean across all topics)

To evaluate the retrieval of the top articles and proteins, we measure the quality of the results by computing its precision@10 score.

5 Results

First we evaluate the coherence of the topics obtained from the 3 variants described above. Table 1 shows that out of the 15 topics that were obtained, 12 topics were deemed coherent from the text-only model and 13 and 15 topics were deemed coherent from the Block-LDA models using the MIPS and wetlab PPI datasets respectively.

Next, we study the precision@10 values for each topic and variant for the article retrieval and protein retrieval tasks, which is shown in Figure 3. The plots also show horizontal lines representing the mean of the precision@10 across all topics. It can be seen from the plots that for both the article and protein retrieval tasks, the joint models work better than the text-only model on average. For the article retrieval task, the model trained with the text + MIPS resulted in the higher mean precision@10 whereas for the protein retrieval task, the text + Wetlab PPI dataset returned a higher mean precision@10 value. For both the protein retrieval and paper retrieval tasks, the improvements shown by the joint models using either of the PPI datasets over the text-only model (i.e. the Link LDA model) were statistically significant at the 0.05 level using the paired Wilcoxon sign test. The difference in performance between the

Topic: Protein Structure & Interactions	
Top articles using Publications Only	**Top articles using Block-LDA with Wetlab PPI**
* X-ray fiber diffraction of amyloid fibrils.	* X-ray fiber diffraction of amyloid fibrils.
* Molecular surface area and hydrophobic effect.	* Scalar couplings across hydrogen bonds.
* Counterdiffusion methods for macromolecular crystallization.	* Dipolar couplings in macromolecular structure determination.
* Navigating the ClpB channel to solution.	* Structure of alpha-keratin.
* Two Rippled-Sheet Configurations of Polypeptide Chains, and a Note about the Pleated Sheets.	* Stable configurations of polypeptide chains.
* Molecular chaperones. Unfolding protein folding.	* The glucamylase and debrancher of S. diastaticus.
* The molten globule state as a clue for understanding the folding and cooperativity of globular-protein structure.	* A study of 150 cases of pneumonia.
* Unfolding and hydrogen exchange of proteins: the three-dimensional ising lattice as a model.	* Glycobiology.
* Packing of alpha-helices: geometrical constraints and contact areas.	* The conformation of thermolysin.
Topic: DNA Repair	
Top articles using Publications Only	**Top articles using Block-LDA with Wetlab PPI**
* Passing the baton in base excision repair.	* Telomeres and telomerase.
* The bypass of DNA lesions by DNA and RNA polymerases.	* Enzymatic photoreactivation: overview.
* The glucamylase and debrancher of S. diastaticus.	* High-efficiency transformation of plasmid DNA into yeast.
* DNA replication fidelity.	* The effect of ultraviolet light on recombination in yeast.
* Base excision repair.	* T-loops and the origin of telomeres.
* Nucleotide excision repair.	* Directed mutation: between unicorns and goats.
* The replication of DNA in Escherichia Coli.	* Functions of DNA polymerases.
* DNA topoisomerases: why so many?	* Immortal strands? Give me a break.

Table 2: Sample of Improvements in Article Retrieval

two joint models that used the two different PPI networks were however insignificant which indicates that there is no observable advantage in using one PPI dataset over the other in conjunction with the text corpus.

Table 2 shows examples of poor results of article retrieval obtained using the publications-only model and the improved set of results obtained using the joint model.

5.1 Topics

Table 3 shows 3 sample topics that were retrieved from each variant described earlier. The table shows the top words and proteins associated with the topics. The topic label on the left column was assigned manually during the evaluation by the expert annotator.

Conclusion

We evaluated topics obtained from the joint modeling of yeast biology literature and protein-protein interactions in yeast and compared them to topics that were obtained from using only the literature. The topics were evaluated for coherence and by measuring the mean precision@10 score of the top articles and proteins that were retrieved for each topic. Evaluation by a domain expert showed that

Topic	Top Words & Proteins
Protein Structure & Interactions (Publications Only)	**Words:** protein structure binding residues domain structural beta complex atp proteins alpha interactions folding structures form terminal peptide helix model interaction bound domains molecular changes conformational **Proteins:** CYC1 SSA1 HSP82 SUP35 HSP104 HSC82 SSA2 YDJ1 URE2 KAR2 SSB1 SSA4 GCN4 SSA3 SSB2 PGK1 PDI1 SSC1 HSP60 STI1 SIS1 RNQ1 SEC61 SSE1 CCP1
DNA Repair (Using MIPS PPI)	**Words:**dna recombination repair replication strand single double cells mutations stranded induced base uv mutants mutation homologous virus telomere human type yeast activity telomerase mutant dna_polymerase **Proteins:** RAD52 RAD51 RAD50 MRE11 RAD1 RAD54 SGS1 MSH2 RAD6 YKU70 REV3 POL30 RAD3 XRS2 RAD18 RAD2 POL3 RAD27 YKU80 RAD9 RFA1 TLC1 TEL1 EST2 HO
Vesicular Transport (Using Wetlab PPI)	**Words:**membrane protein transport proteins atp golgi er atpase membranes plasma_membrane vesicles cells endoplasmic_reticulum complex fusion ca2 dependent translocation vacuolar intracellular yeast lipid channel hsp90 vesicle **Proteins:** SSA1 HSP82 KAR2 PMA1 HSC82 SEC18 SSA2 YDJ1 SEC61 PEP4 HSP104 SEC23 VAM3 IRE1 SEC4 SSA4 SEC1 PMR1 PEP12 VMA3 VPH1 SSB1 VMA1 SAR1 HAC1

Table 3: Sample Topics

the joint modeling produced more coherent topics and showed better precision@10 scores in the article and protein retrieval tasks indicating that the model enabled information sharing between the literature and the PPI networks.

References

Edoardo M. Airoldi, David Blei, Stephen E. Fienberg, and Eric P. Xing. 2008. Mixed membership stochastic blockmodels. *Journal of Machine Learning Research*, 9:1981–2014, September.

Ramnath Balasubramanyan and William W. Cohen. 2011. Block-LDA: Jointly modeling entity-annotated text and entity-entity links. In *SDM*, pages 450–461. SIAM / Omnipress.

David. M Blei, Andrew Y. Ng, and Michael I. Jordan. 2003. Latent dirichlet allocation. *The Journal of Machine Learning Research*, 3:993–1022.

Selina S. Dwight, Rama Balakrishnan, Karen R. Christie, Maria C. Costanzo, Kara Dolinski, Stacia R. Engel, Becket Feierbach, Dianna G. Fisk, Jodi Hirschman, Eurie L. Hong, Laurie Issel-Tarver, Robert S. Nash, Anand Sethuraman, Barry Starr, Chandra L. Theesfeld, Rey Andrada, Gail Binkley, Qing Dong, Christopher Lane, Mark Schroeder, Shuai Weng, David Botstein, and Michael Cherry J. 2004. Saccharomyces genome database: Underlying principles and organisation. *Briefings in bioinformatics*, 5(1):9.

Harold W. Kuhn. 1955. The Hungarian method for the assignment problem. *Naval Research Logistics Quarterly*, 2(1-2):83–97.

Hans-Werner Mewes, C. Amid, Roland Arnold, Dmitrij Frishman, Ulrich Gldener, Gertrud Mannhaupt, Martin Mnsterktter, Philipp Pagel, Normann Strack, Volker Stmpflen, Jens Warfsmann, and Andreas Ruepp. 2004. MIPS: Analysis and annotation of proteins from whole genomes. *Nucleic Acids Res*, 32:41–44.

Ramesh M. Nallapati, Amr Ahmed, Eric P. Xing, and William W. Cohen. 2008. Joint latent topic models for text and citations. In *Proceeding of the 14th ACM SIGKDD international conference on Knowledge discovery and data mining*, pages 542–550, Las Vegas, Nevada, USA. ACM.

Juuso Parkkinen, Janne Sinkkonen, Adam Gyenge, and Samuel Kaski. 2009. A block model suitable for sparse graphs. In *Proceedings of the 7th International Workshop on Mining and Learning with Graphs (MLG 2009)*, Leuven. Poster.

$Rank_{\text{Pref}}$: Ranking Sentences Describing Relations between Biomedical Entities with an Application

Catalina O Tudor **K Vijay-Shanker**
Department of Computer and Information Sciences
University of Delaware, Newark, DE, USA
tudor@cis.udel.edu vijay@cis.udel.edu

Abstract

This paper presents a machine learning approach that selects and, more generally, ranks sentences containing clear relations between genes and terms that are related to them. This is treated as a binary classification task, where preference judgments are used to learn how to choose a sentence from a pair of sentences. Features to capture how the relationship is described textually, as well as how central the relationship is in the sentence, are used in the learning process. Simplification of complex sentences into simple structures is also applied for the extraction of the features. We show that such simplification improves the results by up to 13%. We conducted three different evaluations and we found that the system significantly outperforms the baselines.

1 Introduction

Life scientists, doctors and clinicians often search for information relating biological concepts. For example, a doctor might be interested to know the impact of a drug on some disease. One source of information is the knowledge bases and ontologies that are manually curated with facts from scientific articles. However, the curation process is slow and cannot keep up with ongoing publications. Moreover, not all associations between biological concepts can be found in these databases.

Another source of information is the scientific literature itself. However, searching for biological facts and how they might be related is often cumbersome. The work presented in this paper tries to automate the process of finding sentences that clearly describe relationships between biological concepts. We rank all sentences mentioning two concepts and pick the top one to show to the user. In this paper, we focused on certain specific types of concepts (i.e., genes[1] and terms believed to be related to them), although our approach can be generalized.

Systems to facilitate knowledge exploration of genes are being built for the biomedical domain. One of them, eGIFT (Tudor et al., 2010), tries to identify iTerms (informative terms) for a gene based on frequency of co-occurrence (see Figure 1 for top 15 terms selected for gene *Groucho*). iTerms are unigrams, bigrams, and exact matches of biomedical terms gathered from various controlled vocabularies. Thus, iTerms can be of any type (e.g., processes, domains, drugs, other genes, etc.), the types being determined by what is being described about the gene in the literature. The iTerms for a gene are ranked based on a score that compares their frequencies of occurrence in publications mentioning the gene in question with their frequencies in a background set of articles about a wide variety of genes.

Previous evaluation of eGIFT by life scientists suggested that there is almost always some kind of relationship between a gene and its iTerms. These relationships can be many and varied from one gene-term pair to another. Sometimes a user might make an erroneous assumption about a gene-term association if sentences supporting the association are not immediately inspected. For example, upon seeing "co-repressor" in connection to gene *Groucho*, eGIFT users might correctly assume that *Groucho* is

[1]Throughout the paper, the word "gene" will be used for both the gene and its products.

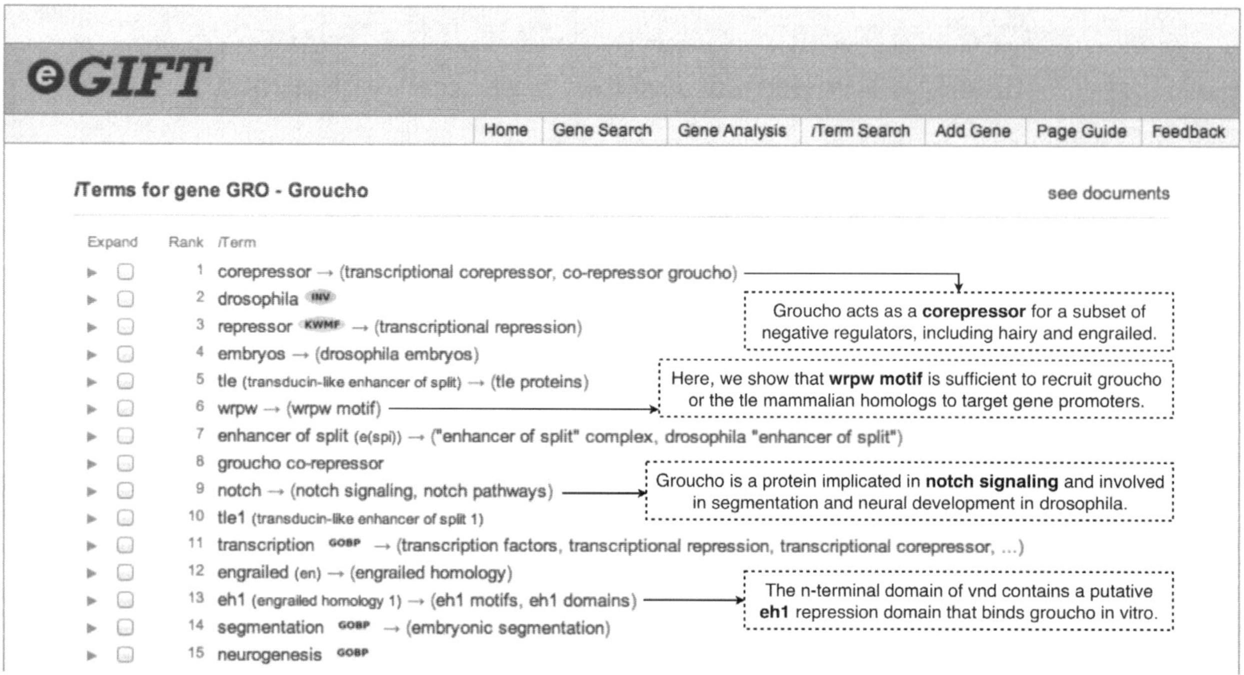

Figure 1: Top iTerms for gene *Groucho*, and sentences picked by $Rank_{\text{Pref}}$ for various iTerms.

a co-repressor (i.e., a protein that binds to transcription factors). However, upon seeing "wrpw motif", a user might assume that this is a motif contained within gene *Groucho*, as this is typically the association that we make between genes and information annotated for them in knowledge bases. But this would be a wrong assumption, since in actuality the wrpw motif is contained within other genes that interact with *Groucho*, fact which is evident from reading sentences containing the gene and the motif. To get a quick overall understanding of a gene's functionalities, users of eGIFT could be presented with terms extracted for the gene, as well as sentences clearly describing how they are related.

Our method selects sentences using a model that is trained on preference judgments provided by biologists. Example sentences chosen by our method are shown in Figure 1. While we evaluate our approach on sentences from eGIFT, this work could have equally applied on other similar systems (Smalheiser et al., 2008; Gladki et al., 2008; Kim et al., 2008; Kaczanowski et al., 2009). These systems also identify "important terms" from a set of documents retrieved for a given search (either a gene name or other biomedical concept).

The main contributions of this work are: (1) a method for ranking sentences by employing machine learning; (2) the use of preference judgments; (3) features to capture whether two terms are clearly related and in focus in a sentence; (4) another application of sentence simplification, showing a significant gain in performance when utilized.

We continue with a description of our approach, which includes the use of preference judgments to learn the models, how the features are extracted, and how the sentence simplifier is used for this task. The evaluation of the trained model and the system's results are presented in the following section. Related work, conclusions, and future directions are provided at the end of the manuscript.

2 Methods

Rather than pre-judging what is important for this task and manually determining a weighting schema to automatically score sentences for a gene-term pair, we approached this task using machine learning. We asked a group of annotators to rank sentences relating genes and iTerms, and we used their annotations, together with features described in Section 2.3, to learn how to rank sentences.

2.1 Preference Judgments

For the annotation task, we presented biologists with sentences containing a gene-term pair and asked them to specify which sentence they prefer. One way to do this is by employing the pointwise approach, which requires absolute judgments (i.e. the annotator scores each sentence in a list or ranks the sentences based on their relevance to the given task). A second approach is the pairwise approach, which requires the iteration of preference judgments (i.e., the annotator is presented with two sentences at a time, and is asked to chose one as more relevant to the task than the other).

In order to simplify the annotator's task, as well as construct a more reliable training set, we used the pairwise approach. Our decision was influenced by Carterette et al. (2008), who showed that preference judgments are faster and easier to make than absolute judgments. Thus, we can obtain many annotated instances in a relatively short amount of time. Moreover, since there are only two possible outcomes in choosing one sentence, we need at most three judges for a majority vote. This will also ensure consistency in the annotations. We discuss the model trained on preference judgments in Section 2.2.

2.2 Learned Models: $Pref_{\text{SVM}}$ and $Rank_{\text{Pref}}$

We used the preference judgments to learn a model, $Pref_{\text{SVM}}$, that picks one sentence from a pair of sentences. This model was built using SVM^{Light} with a linear kernel. The examples used in the learning process correspond to pairs of sentences. For each pair, we constructed a vector of feature values, by subtracting the feature values corresponding to the first sentence from the feature values corresponding to the second sentence. We assigned a positive value to a pair vector if the first sentence was preferred and a negative value if the second one was preferred.

We can also use $Pref_{\text{SVM}}$ to design a system that can rank all the sentences containing a gene and an iTerm, by performing comparisons between sentences in the list. We call $Rank_{\text{Pref}}$ the system that picks one sentence from a group of sentences, and which also ranks the entire set of sentences. This method recursively applies $Pref_{\text{SVM}}$ in the following manner: Two sentences are randomly picked from a given list of sentences. $Pref_{\text{SVM}}$ chooses one sentence and discards the other. A third sentence is then randomly picked from the list, and $Pref_{\text{SVM}}$ makes its choice by comparing it to the sentence kept in the previous step. This process of picking, comparing and discarding sentences is continued until there is only one sentence left. We keep track of comparison results and apply transitivity, in order to speed up the process of ranking all the sentences.

2.3 Features

Each sentence is first chunked into base phrases. We used Genia Tagger (Tsuruoka et al., 2005), which provides part-of-speech tags for every word in the sentence. We trained a chunker (i.e., shallow parser that identifies base NPs) using the Genia corpus.

We considered typical features that are used in machine learning approaches, such as distance between gene and iTerm, length of sentence, etc. Moreover, we included additional groups of features that we felt might be important for this task: one group to capture how the relationship is described textually, another group to capture how central the relationship is in terms of what is being described in the sentence, and the last to capture whether the relation is stated as a conjecture or a fact. The weights for these features will be determined automatically during the learning process and they will be dependent on whether or not the features were effective, given the annotation set.

The first type of features is to capture how the relationship is described textually. As an example, consider the sentence "*Bmp2* stimulates **osteoblastic differentiation**"[2], where the gene and the iTerm are in subject and object (direct object or otherwise) positions, and the verb is a common biological verb. Thus, we constructed a set of *lexico-syntactic patterns* to capture the different kinds of argument relations served by the two concepts. We grouped 25 lexico-syntactic patterns into 8 groups, corresponding to different relational constructions that can exist between a gene and an iTerm. Example patterns are shown in Table 1 for each group, and the symbols used in these patterns are explained in Table 2. When a sentence matches a pattern group, the corresponding value is set to 1 for that feature.

[2]In our examples, the gene will be marked in *italics* and the iTerm will be marked in **bold**.

Group	Example Pattern
G1	**G** VG+ **I**
G2	**G/I** via/by/through **I/G**
G3	**G** VG+ (NP/PP)* by/in VBG **I**
G4	**G/I** by/in VBG **I/G**
G5	**G/I** VB **I/G**
G6	**G/I** of **I/G**
G7	**G/I** other_preposition **I/G**
G8	including/such as/etc. **G/I** and **I/G**

Table 1: Examples of lexico-syntactic patterns

Symb	Definition
NP	a base noun phrase
PP	a preposition followed by a base noun phrase
VG+	a series of one or more verb groups
VBG	a verb group in which the head is a gerund verb
VBN	a verb group in which the head is a participle verb
VB	a verb group in which the head is a base verb
G, I	base noun phrases, with 0 or more prepositional phrases, containing the gene/iTerm

Table 2: Symbols used in the pattern notation

For example, the following sentence, in which the gene is *Lmo2* and the iTerm is "erythropoiesis", matches the pattern in G1: [**G** VG+ **I**].

> While Tal1 has been shown to induce erythroid differentiation , *Lmo2* appears to suppress fetal **erythropoiesis**.

where "Lmo2" matches **G**, "appears to suppress" matches VG+, and "fetal erythropoiesis" matches **I**.

Notice how the verb plays an important role in the patterns of groups G1, G3, G4, and G5. We also have a *verb type* feature which differentiates groups of verbs having the gene and the iTerm as arguments (e.g., "activates", "is involved in", "plays a role", etc. are treated as different types).

The second type of features captures how central the relationship is in terms of what is being described in the sentence. The *subject feature* records whether the gene and iTerm appear in the subject position, as this will tell us if they are in focus in the sentence. While we do not parse the sentence, we take a simplified sentence (see Section 2.4) and see if the gene/term appear in a noun phrase preceding the first tensed verb. Another feature, the *gene-iTerm position*, measures how close the gene and the term are to each other and to the beginning of the sentence, as this makes it easier for a reader to grasp the relation between them. For this, we add the number of words occurring to the left of the segment spanning the gene and iTerm, and half of the number of words occurring between them. Finally, we included a *headedness feature*. The idea here is that if the gene/term are not the head of the noun group, but rather embedded inside, then this potentially makes the relation less straightforward. These groups are denoted by **G** and **I** in the patterns shown in Table 1.

The third type of features captures information about the sentence itself. The *sentence complexity* feature is measured in terms of the number of verbs, conjunctions, commas, and parentheticals that occur in the sentence. We use a *conjecture* feature for detecting whether the sentence involves a hypothesis. We have a simple rule for this feature, by seeing if words such as "may", "could", "probably", "potentially", etc., appear in proximity of the gene and iTerm. Additionally, we have a *negation* feature to detect whether the relationship is mentioned in a negative way. We look for words such as "not", "neither", etc., within proximity of the gene and iTerm.

Although the features and lexico-syntactic patterns were determined by analyzing a development set of sentences containing genes and their iTerms, we believe that these features and patterns can be used to rank sentences involving other biomedical entities, not just genes.

2.4 Sentence Simplification

Notice that the lexico-syntactic patterns are written as sequences of chunks and lexical tags. If a sentence matches a pattern, then the sentence expresses a relation between the gene and the iTerm. However, sometimes it is not possible to match a pattern if the sentence is complex.

For example, consider sentence A in Table 3, for gene *Cd63*. Let us assume that the iTerm is "protasomes". Clearly, there is a relationship between the gene and the iTerm, namely that *Cd63* was found in pc-3 cell-derived protasomes. However, none of the lexico-syntactic patterns is able to capture this relation, because of all the extra information between

A	*Cd63*, an integral membrane protein found in multivesicular lysosomes and **secretory granules**, was also found in pc-3 cell-derived **protasomes**.
S1	*Cd63* was found in pc-3 cell-derived **protasomes**.
S2	*Cd63* is an integral membrane protein.
CS1	*Cd63* is found in multivesicular lysosomes.
CS2	*Cd63* is found in **secretory granules**.

Table 3: Simplified sentences for gene *Cd63*. Example *i*Terms: "protasomes" and "secretory granules".

the gene and the term. While we may have multiple patterns in each group, we cannot necessarily account for each lexical variation at this level of granularity.

We are using a sentence simplifier, built in-house, to ensure a match where applicable. The simplifier identifies appositions, relative clauses, and conjunctions/lists of different types, using regular expressions to match chunked tags. In the sentence of Table 3, the simplifier recognizes the apposition "an integral membrane protein", the reduced relative clause "found in multivesicular bodies/lysosomes and secretory granules" and the noun conjunction "multivesicular bodies/lysosome and secretory granules". It then produces several simplified sentences containing the gene. S1 and S2, shown in Table 3, are simplified sentences obtained from the simplifier. CS1 and CS2 are additional simplified sentences, which required the combination of multiple simplifications: the appositive, the relative clause, and the noun conjunction.

Notice how each of the simplified sentences shown in Table 3 is now matching a pattern group. If we are interested in the relationship between *Cd63* and "protasomes", we can look at S1. Likewise, if we are interested in the relationship between *Cd63* and "secretory granules", we can look at CS2.

We have a *matching* feature that tells whether the pattern was matched in the original sentence, a simplified sentence, or a combined sentence, and this feature is taken into account in the learning process.

3 Results and Discussion

We evaluated both $\mathit{Pref}_{\mathrm{SVM}}$ and $\mathit{Rank}_{\mathrm{Pref}}$. Each required a different set of annotated data. For the evaluation of $\mathit{Pref}_{\mathrm{SVM}}$, we used the preference judgments and leave-one-out cross validation. And for the evaluation of $\mathit{Rank}_{\mathrm{Pref}}$, we asked the annotators to order a group of sentences mentioning gene-*i*Term pairs. Six life science researchers, with graduate degrees, annotated both sets.

3.1 Evaluation of $\mathit{Pref}_{\mathrm{SVM}}$

First, we evaluated the performance of $\mathit{Pref}_{\mathrm{SVM}}$ using leave-one-out cross validation.

3.1.1 Annotation of Preference Judgements

We started by selecting a group of pairs of sentences. We randomly picked gene-*i*Term combinations, and for each combination, we randomly picked two sentences containing both the gene and the term. To alleviate bias, the order of the sentences was chosen randomly before displaying them to the annotators. In our guidelines, we asked the annotators to choose sentences that clearly state the relationship between the gene and the *i*Term. Because the focus here is on the relationship between the two terms, we also asked them to refrain from choosing sentences that describe additional information or other aspects. It is conceivable that, for other applications, extra information might be an important determining factor, but for our task we wanted to focus on the relationship only.

For each pair of sentences, we wanted to have three opinions so that we can have a majority vote. To alleviate the burden on the annotators, we started by giving each pair of sentences to two annotators, and asked for an extra opinion only when they did not agree. Each biologist was given an initial set of 75 pairs of sentences to annotate, and shared the same amount of annotations (15) with each of the other biologists. 225 unique pairs of sentences were thus annotated, but six were discarded after the annotators informed us that they did not contain the gene in question.

In 34 out of 219 pairs of sentences, the two biologists disagreed on their annotations. These cases included pairs of similar sentences, or pairs of sentences that did not describe any relationship between

System	Performance	Correct
Baseline 1	65.75%	144
Baseline 2	71.69%	157
$Pref_{SVM}$ without Simp	72.14%	158
$Pref_{SVM}$ with Simp	83.10%	182

Table 4: Results for $Pref_{SVM}$

the gene and the iTerm. An example of sentences for which the annotators could not agree is:

1. The tle proteins are the mammalian homologues of *gro*, a member of the drosophila **notch signaling** pathway.
2. In drosophila, *gro* is one of the neurogenic genes that participates in the **notch signalling** pathway .

For these 34 pairs, we randomly selected another annotator and considered the majority vote.

3.1.2 Baselines

We chose two baselines against which to compare $Pref_{SVM}$. The first baseline always chooses the shortest sentence. For the second baseline, we looked at the proximity of the gene/term to the beginning of the sentence, as well as the proximity of the two to each other, and chose the sentence that had the lowest accumulated proximity. The reason for this second baseline is because the proximity of the gene/term to the beginning of the sentence could mean that the sentence focuses on the gene/term and their relation. Furthermore, the proximity of the gene to the iTerm could mean a clearer relation between them.

3.1.3 Results

We evaluated $Pref_{SVM}$ by performing leave-one-out cross validation on the set of 219 pairs of sentences. Each pair of sentences was tested by using the model trained on the remaining 218 pairs. The results are shown in Table 4.

The first baseline performed at 65.75%, correctly choosing 144 of 219 sentences. The second baseline performed slightly better, at 71.69%. $Pref_{SVM}$ outperformed both baselines, especially when the sentence simplifier was used, as this facilitated the match of the lexico-syntactic patterns used as features. $Pref_{SVM}$ performed at 83.10%, which is 17.35% better than the first baseline, and 11.41% better than the second baseline.

3.2 Evaluation of $Rank_{Pref}$

The previous evaluation showed how $Pref_{SVM}$ performs at picking a sentence from a pair of sentences. But ultimately, for the intended eGIFT application, the system needs to choose one sentence from many. We evaluated $Rank_{Pref}$ for this task.

3.2.1 Annotating Data for Sentence Selection

For this evaluation, we needed to create a different set of annotated data that reflects the selection of one sentence from a group of sentences.

Since a gene and an iTerm can appear in many sentences, it is too onerous a task for a human annotator to choose one out of tens or hundreds of sentences. For this reason, we limited the set of sentences mentioning a gene and an iTerm to only 10. We randomly picked 100 gene-term pairs and for the pairs that contained more than ten sentences, we randomly chose ten of them. On average, there were 9.4 sentences per set.

We asked the same annotators as in the previous evaluation to participate in this annotation task. Because the task is very time consuming, and because it is hard to decide how to combine the results from multiple annotators, we assigned each set of sentences to only one annotator. We showed the sentences in a random order so that biasing them would not be an issue.

We initially asked the annotators to order the sentences in the set. However, this task proved to be impossible, since many sentences were alike. Instead, we asked the annotators to assign them one of three categories:

(Cat.1) Any sentence in this category could be considered the "best" among the choices provided;

(Cat.2) These sentences are good, but there are other sentences that are slightly better;

(Cat.3) These sentences are not good or at least there are other sentences in this set that are much better.

Classifying the sentences into these categories was less cumbersome, fact which was confirmed by our evaluators after a trial annotation.

Out of the total of 936 sentences, 322 (34.4%) were placed in the first category, 332 (35.5%) were

System	Cat.1	Cat.2	Cat.3
Baseline 1	58	30	12
Baseline 2	61	24	15
$Rank_{\text{Pref}}$ without Simp	67	21	12
$Rank_{\text{Pref}}$ with Simp	80	17	3

Table 5: Results for $Rank_{\text{Pref}}$

placed in the second category, and 282 (30.1%) were placed in the third category. On average, it took about 15 minutes for an annotator to group a set's sentences into these three categories. So each annotator volunteered approximately 5 hours of annotation time.

3.2.2 Results

Table 5 shows how the top sentences picked for the 100 gene-term pairs by the four systems matched with the annotations. 80 of 100 sentences that $Rank_{\text{Pref}}$ picked were placed in Cat.1 by the annotators, 17 were placed in Cat.2, and 3 sentences were placed in Cat.3. These results compare favorably with results obtained for the two baselines and $Rank_{\text{Pref}}$ without the use of the simplifier.

Furthermore, instead of just focussing on the top choice sentence, we also considered the ranking of the entire set of sentences. We looked at how the ranked lists agree with the categories assigned by the annotators. We used the normalized discounted cumulative gain (nDCG) (Jarvelin and Kekalainen, 2002), a standard metric used in information retrieval to evaluate the quality of the ranked lists. DCG at rank p is defined as:

$$DCG_p = rel_1 + \sum_{i=2}^{p} \frac{rel_i}{log_2 i}$$

where rel_i is the relevance of the item at position i. We normalize DCG by dividing it by an ideal gain (i.e., DCG of same list, when ordered from highest to lowest relevance).

For our task, we took the relevance score to be 1 for a sentence placed in Cat.1, a relevance score of 0.5 for a sentence placed in Cat.2, and a relevance score of 0 for a sentence placed in Cat.3. We report a normalized discounted cumulative gain of 77.19%.

This result compares favorably with results reported for the two baselines (68.36% for B1 and 68.32% for B2) as well as for when the sentence simplifier was removed (69.45%).

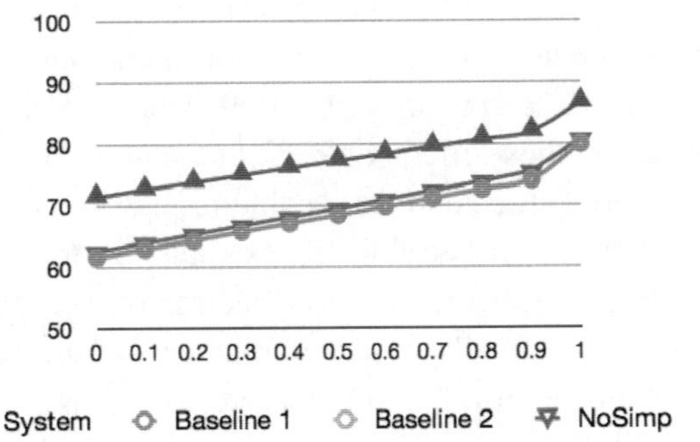

Figure 2: Distribution of nDCG for different relevance scores assigned to sentences placed in category Cat.2.

Figure 2 shows different results for nDCG when the relevance score for Cat.2 is varied between 0 (same as sentences placed in Cat.1) and 1 (same as sentences placed in Cat.3).

4 Related Work

To the best of our knowledge, no one has attempted to rank sentences from the biomedical literature, using machine learning on a set of data marked with preference judgments. However, different approaches have been described in the literature that use preference judgments to learn ranked lists. For example, Radlinski and Joachims (2005) used preference judgments to learn ranked retrieval functions for web search results. These judgments were generated automatically from search engine logs. Their learned rankings outperformed a static ranking function. Similar approaches in IR are those of Cohen et al. (1999) and Freund et al. (2003).

Ranking of text passages and documents has been done previously in BioNLP for other purposes. Suomela and Andrade (2005) proposed a way to rank the entire PubMed database, given a large training set for a specific topic. Goldberg et al. (2008) and Lu et al. (2009) describe in detail how they identified and ranked passages for the 2006 Trec Genomics Track (Hersh et al., 2006). Yeganova et al. (2011) present a method for ranking positively labeled data within large sets of data, and this method was applied by Neveol et al. (2011) to rank sentences containing deposition relationships between biological data and public repositories.

Extraction of sentences describing gene functions has also been applied for creating gene summaries (Ling et al., 2007; Jin et al., 2009; Yang et al., 2009). However, these methods differ in that their goal is not to look for sentences containing specific terms and their relations with genes, but rather for sentences that fall into some predefined categories of sentences typically observed in gene summaries.

Sentence simplification has been used to aid parsing (Chandrasekar et al., 1996; Jonnalagadda et al., 2009). Devlin and Tait (1998) and Carroll et al. (1998) use it to help people with aphasia. Siddharthan (2004) was concerned with cohesion and suggested some applications.

The idea of using lexico-syntactic patterns to identify relation candidates has also been applied in the work of Banko et al. (2007), although their patterns are not used in the learning process.

5 Conclusion and Future Directions

We have developed a system which aims to identify sentences that clearly and succinctly describe the relation between two entities. We used a set of preference judgements, as provided by biologists, to learn an SVM model that could make a choice between any two sentences mentioning these entities.

The model compares favorably with baselines on both the task of choosing between two sentences, as well as ranking a set of sentences. The performance for choosing between two sentences was 83.10%, as compared to 65.75% and 71.69% for the two baselines, respectively. For choosing one sentence from a list of sentences, the performance was 80%, as compared to 58% and 61%. Furthermore, when the entire list of ranked sentences was evaluated, the system reported a nDCG of 77.19%, compared to 68.36% and 68.32% for the two baselines.

The model's performance was also shown to be significantly better when sentence simplification was used. We were able to match relation patterns on complex sentences, and observed an increase of 10.96%, 13%, and 7.74% for the three evaluations afore-mentioned, respectively. It is noteworthy that, without the simplification, the performance is only slightly better than the second baseline. This is because the second baseline uses information that is also used by our system, although this does not include the lexico-syntactic patterns that identify the type of relation between the gene and the term.

Given that the full system's performance is much better than both baselines, and that the system's performance without simplification is only slightly better than the second baseline, we believe that: (1) the pattern and type of relation determination are important, and (2) sentence simplification is crucial for the determination of the relationship type.

We are currently pursuing summaries for genes. Since iTerms have been shown in previous evaluations to represent important aspects of a gene's functionality and behavior, we are investigating whether they are represented in gene summaries found in EntrezGene and UniProtKB. If so, an *extractive* summary can be produced by choosing sentences for the gene and its iTerms. We are also considering developing *abstractive* summaries. Our use of lexico-syntactic patterns can be extended to pick the exact relation between a gene and the iTerm. For example, by using the lexico-syntactic patterns, coupled with simplification, we can extract the following exact relations from the four sentences shown in Figure 1: "Groucho is a corepressor", "The wrpw motif recruits groucho", "Groucho is implicated in notch signaling", and "The eh1 repression domain binds groucho". With these relations extracted, using text generation algorithms for textual realization and cohesion, we can produce *abstractive* summaries.

We would also like to investigate how to generalize this work to other pairs of entities, as well as how to generalize this work for other applications which may or may not require the same features as the ones we used.

Acknowledgments

This work has been supported in part by USDA Grant 2008-35205-18734 and the Agriculture and Food Research Initiative Competitive USDA Grant 2011-67015-3032. We thank Cecilia Arighi, Kevin Bullaughey, Teresia Buza, Fiona McCarthy, Lakshmi Pillai, Carl Schmidt, Liang Sun, Hui Wang, and Qinghua Wang for participating in the annotation task and/or for various discussions. We also thank the anonymous reviewers for their comments and suggestions, which helped us improve the manuscript.

References

Michele Banko, Michael J Cafarella, Stephen Soderland, Matt Broadhead, and Oren Etzioni. 2007. Open Information Extraction from the Web. In *Proceedings of IJCAI.*

John Carroll, Guido Minnen, Yvonne Canning, Siobhan Devlin, and John Tait. 1998. Practical simplification of English newspaper text to assist aphasic readers. *Proceedings of the AAAI98 Workshop on Integrating AI and Assistive Technology*, pages 7–10.

Ben Carterette, Paul N Bennett, David Maxwell Chickering, and Susan T Dumais. 2008. Here or there: Preference judgments for relevance. In *Proceedings of the IR research, 30th European conference on Adv. in IR.*

R Chandrasekar, Christine Doran, and B Srinivas. 1996. Motivations and methods for text simplification. In *Proceedings of the 16th conference on Computational linguistics*, volume 2, pages 1041–1044. Association for Computational Linguistics.

Wiliam W Cohen, Robert E Schapire, and Yoram Singer. 1999. Learning to order things. *Journal of Artificial Intelligence Research*, 10:243–270.

Siobhan Devlin and John Tait. 1998. The use of a psycholinguistic database in the simplification of text for aphasic readers. *Linguistic Databases*, pages 161–173.

Yoav Freund, Raj Iyer, Robert E Schapire, and Yoram Singer. 2003. An efficient boosting algorithm for combining preferences. *Journal of Machine Learning Research*, 4:933–969.

Arek Gladki, Pawel Siedlecki, Szymon Kaczanowski, and Piotr Zielenkewicz. 2008. e-LiSe–an online tool for finding needles in the 'Medline haystack'. *Bioinformatics*, 24(8):1115–1117.

Andrew B Goldberg, David Andrzejewski, Jurgen Van Gael, Burr Settles, Xiaojin Zhu, and Mark Craven. 2008. Ranking biomedical passages for relevance and diversity. In *Proceedings of TREC.*

William Hersh, Aaron M Cohen, Phoebe Roberts, and Hari Krishna Rekapalli. 2006. TREC 2006 Genomics Track Overview.

Kalervo Jarvelin and Jaana Kekalainen. 2002. Cumulated gain-based evaluation of IR techniques. *ACM Transactions on Information Systems*, 20(4):422–446.

Feng Jin, Minlie Huang, Zhiyong Lu, and Xiaoyan Zhu. 2009. Towards automatic generation of gene summary. In *Proceedings of the BioNLP 2009 Workshop*, pages 97–105. Association for Computational Linguistics, June.

Siddhartha Jonnalagadda, Luis Tari, Jorg Hakenberg, Chitta Baral, and Graciela Gonzalez. 2009. Towards effective sentence simplification for automatic processing of biomedical text. In *Proceedings of NAACL HLT 2009: Short Papers*, pages 177–180.

Szymon Kaczanowski, Pawel Siedlecki, and Piotr Zielenkewicz. 2009. The high throughput sequence annotation service (HT-SAS) - the shortcut from sequence to true medline words. *BMC Bioinformatics*, 10:148–154, May.

Jung-Jae Kim, Piotr Pezik, and Dietrich Rebholz-Schuhmann. 2008. MedEvi: Retrieving textual evidence of relations between biomedical concepts from medline. *Bioinformatics*, 24(11):1410–1412.

Xu Ling, Jing Jiang, Xin He, Qiaozhu Mei, Chengxiang Zhai, and Bruce Schatz. 2007. Generating gene summaries from biomedical literature: A study of semi-structured summarization. *Information Processing and Management*, 43:1777–1791, March.

Yue Lu, Hui Fang, and Chengxiang Zhai. 2009. An empirical study of gene synonym query expansion in biomedical information retrieval. *Information Retrieval*, 12:51–68, February.

Aurélie Névéol, W John Wilbur, and Zhiyong Lu. 2011. Extraction of data deposition statements from the literature: a method for automatically tracking research results. *Bioinformatics*, 27(23):3306–3312.

Filip Radlinski and Thorsten Joachims. 2005. Query chains: Learning to rank from implicit feedback. In *Proceedings of KDD'05.*

Advaith Siddharthan. 2004. *Syntactic Simplification and Text Cohesion.* Ph.D. thesis, University of Cambridge.

Neil R Smalheiser, Wei Zhou, and Vetle I Torvik. 2008. Anne O'Tate: A tool to support user-driven summarization, drill-down and browsing of PubMed search results. *Journal of Biomedical Discovery and Collaboration*, 3(1):2–11.

Brian P Suomela and Miguel A Andrade. 2005. Ranking the whole MEDLINE database according to a large training set using text indexing. *BMC Bioinformatics*, 6(75), March.

Yoshimasa Tsuruoka, Yuka Tateishi, Jing-Dong Kim, Tomoko Ohta, John McNaught, Sophia Ananiadou, and Jun'ichi Tsujii. 2005. Developing a robust part-of-speech tagger for biomedical text. In *Advances in Informatics – 10th Panhellenic Conference on Informatics, LNCS 3746*, pages 382–392.

Catalina O Tudor, Carl J Schmidt, and K Vijay-Shanker. 2010. eGIFT: Mining Gene Information from the Literature. *BMC Bioinformatics*, 11:418.

Jianji Yang, Aaron Cohen, and William Hersh. 2009. Evaluation of a gene information summarization system by users during the analysis process of microarray datasets. *BMC Bioinformatics*, 10(Suppl 2):S5.

Lana Yeganova, Donald C Comeau, Won Kim, and W John Wilbur. 2011. Text Mining Techniques for Leveraging Positively Labeled Data. In *Proceedings of ACL Workshop BioNLP*, pages 155–163.

Finding small molecule and protein pairs in scientific literature using a bootstrapping method

Ying Yan, Jee-Hyub Kim, Samuel Croset, Dietrich Rebholz-Schuhmann
European Bioinformatics Institute
Wellcome Trust Genome Campus
Hinxton
Cambridge
UK
{yan, jhkim, croset, rebholz}@ebi.ac.uk

Abstract

The relationship between small molecules and proteins has attracted attention from the biomedical research community. In this paper a text mining method of extracting small-molecule and protein pairs from natural text is presented, based on a semi-supervised machine learning approach. The technique has been applied to the complete collection of MEDLINE abstracts and pairs were extracted and evaluated. The results show the feasibility of the bootstrapping system, which will subsequently be further investigated and improved.

1 Introduction

Information extraction has become a major task in text-mining. A large number of studies have been carried out with the objective of developing techniques to overcome the highly ambiguous and variable nature of natural language for the extraction of information from scientific text (Song et al., 2006). Natural language processing (NLP) of biomedical text has been initiated and used for different knowledge discovery tasks such as the extraction of relationships between different types of biological objects.

Relationships between proteins and small molecules are of particular concern in the biomedical research domain. The importance of target specific small molecule research is vital in the scientific community's understanding of numerous biological processes with potential discoveries yielding various translational benefits and outcomes to public health and industry. While there has been a great number of traditional studies already completed in this field, the underlying difficulty with this type of research has been trying to understand how one molecule interacts with a target protein. Given the biological background, many researchers in Cheminformatics and Metabolomics are attempting to find the connections between small molecules and other biological entities in order to bridge the chemical and biological domains.

Of the few reported text mining approaches to this problem, Temkin and Gilder (2003) was concerned with the extraction of protein and small molecule interaction, and used a rule-based approach utilising a lexical analyser and context free grammar. Jiao and Wild (2009) presented a technique for detecting protein and small molecule interaction using a maximum entropy based learning method; this work also uses corpus-based machine learning. The main drawback of both of these studies is that they require a fully annotated corpus which is difficult to generate.

1.1 The bootstrapping method

At present a gold standard annotated corpus is not available, and constructing a reasonable annotated corpus would require an infeasible amount of manual work. Our proposed solution to this problem is to develop a semi-supervised machine learning method. In this paper a bootstrapping algorithm is presented which requires only unannotated training texts and a handful of protein small molecule pairs, known as seeds. The basic work of a bootstrapping system can be presented as an expansion engine which uses the initial seed pairs fed into the system to generate patterns that are used, in turn, to find more pairs. The operation of the algorithm is controlled by certain criteria that are delivered from a measurement of the quality or selectivity of patterns and discovered pairs.

Bootstrapping systems have been maturely used for information extraction purposes in other research domains, and it has been empirically shown to be a powerful method in learning lexico-syntactic patterns for extracting specific relations (Riloff and Jones, 1999). Bootstrapping systems can operate with a greatly reduced number of training examples. A bootstrapping system seems promising for the purpose of relation extraction, making it a suitable candidate method for protein and small molecule pair extraction.

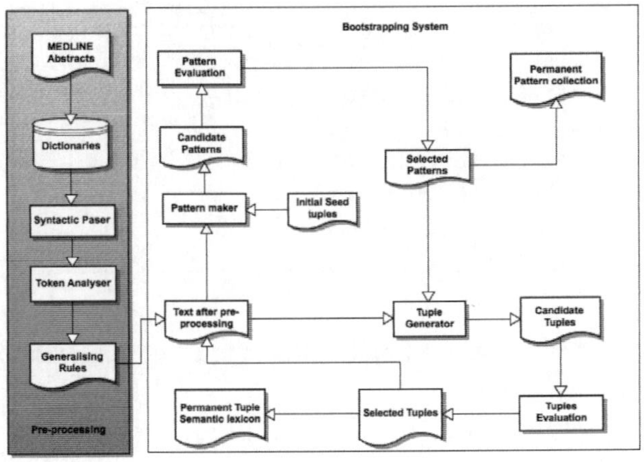

Figure 1: Extraction system architecture

2 Implementation

The typical bootstrapping method was tailored in order to improve its suitability for our extraction task, operating in the biomedical literature resource MEDLINE. The bootstrapping architecture is presented in Figure 1. The whole collection of MEDLINE was filtered using a co-occurrence approach and a named entity recogniser. In this way the sentences which contained both a protein and a small molecule were selected. The structure of patterns which are suitable to extract protein and small molecule pairs from MEDLINE was defined. Each sentence is tokenized and then normalised based on the results of syntactic parsing in order to obtain a more generalised view of the pattern. In the following sections, we describe in more detail these aspects.

2.1 Protein and small molecule recognition

Two dictionary-based named entity recognisers were used to detect the names of proteins and small molecules in the full collection of MEDLINE abstracts, with the two source dictionaries constructed using the resources UniProt (Apweiler et al., 2004) and ChEBI (De Matos et al., 2006) respectively. The following example shows the two recognisers identify a chemical object and a protein object in a sentence from a MEDLINE extract:

<chebi>Paracetamol</chebi>, 100 mg/kg, inhibited <uniprot>COX-1</uniprot> in stomach mucosa ex vivo much less effectively than in other tissues.

2.2 Sentence analysis for normalisation

It was anticipated that variations in tense and other language characteristics would cause problems in pattern generation. We therefore applied a list of normalisation steps for pattern generation. The surrounding context in the biomedical text is not normally useful and makes it difficult to identify the text and observe a clear sentence structure. The parsing result normalises patterns by eliminating non-useful components in a sentence. The step of normalisation hence increases the quality of the pattern.

The complete list of normalisation steps is as follows:

1. Replaced the representation of measurement units, such as mg/L and ml/day.
2. Employed the part-of-speech (POS) tagger GENIA (Tsuruoka et al., 2005) to analyse each token, and the tokens which are weakly related to the sentence structure were removed. So that, the only remaining tokens are the head noun of a noun phrase (NP), the verb phrase, and prepositional phrase chunks.
3. Finally a simple rule to identify the head noun was defined. In a general case, for a NP sequence, the last token is considered as the head noun. When the last token is a single character, the second last token is considered as the head noun.

Table 1: An example of a generated pattern

Seed tuple: Paracetamol, COX-1
Found string: *"CHEBI, UNIT, inhibit UNIPROT in mucosa than in tissue."*
Pattern: NP_List1, UNIT, inhibit NP_List2
Constraints: NP_List1="CHEBI*" NP_List2="UNIPROT*"
Keywords: ",UNIT,inhibit"

The above example after these normalisation steps becomes:

CHEBI, UNIT, inhibit UNIPROT* in mucosa than in tissue.*

where *CHEBI** and *UNIPROT** are the seeds in context.

2.3 Bootstrapping

The bootstrapping system is applied to the normalised sentences. The process starts with 100 high precision protein small molecule pairs collected from the ChEBI ontology. These pairs were retrieved by querying the ChEBI sub-ontology for the relation "has role". From the resulting data we extracted small molecules that are enzyme inhibitors together with the name of the enzyme.

2.3.1 Pattern generation and pair extraction

The concept of a bootstrapping system is that using a high precision seed pair to start the extraction engine, the system can effectively learn the pattern construction rule and the pattern constraints. Searching for the seed pairs in the corpus returns strings which are candidate extraction patterns for other pairs. The candidate patterns are made up of 'slots' and 'context strings', where the slots are either of type small-molecule or protein, and context is the text connecting the slots and the words immediately before and after the pair. By analysing the surrounding context of the slots new elements of the pattern are discovered, which can subsequently be used to search for new small-molecule protein pairs. The process of deriving a pattern from the above example is shown in Table 1.

The generated pattern can then be used to search the corpus and find other matching contexts. New pairs are retrieved from the matching context by simply locating the protein and small molecule names from the same positions as they are in the pattern.

For instance, the pattern produced in Table 1 is matched against a normalised sentence *"data suggest CHEBI, UNIT, inhibit UNIPROT"*, extracting the new pair *<trifluoperazine, CaMKII>*.

2.3.2 Evaluating seeds and patterns

The quality of the pattern is critical since patterns that generate a bad pair can introduce more false positive seeds. Therefore, within a bootstrapping system it is necessary to have a stage of pattern evaluation. Estimations of the confidence score of a pattern can be used as one of the stopping criteria. We implemented an evaluation step for both patterns and pairs based on an evaluation method developed by Agichtein and Gravano (2000). Adapting the approach to this work, if $pattern_i$ predicts tuple $t = <chemical, protein>$, and there is already a tuple $t' = <chemical, protein'>$ with high confidence, and $chemical$ from t is same as $chemical$ from t', then we could define this as a positive match of pattern ($P_{positive}$), otherwise the pattern is considered as a negative match ($P_{negative}$). So that the confidence score of pattern (P) is estimated as:

$$Conf(P) = \frac{P_{positive}}{P_{positive} + P_{negative}} \quad (1)$$

To evaluate the pairs we again employ the method described by Agichtein and Gravano (2000). The confidence of a particular pair is a function of the number of patterns that generate it. Equation 2 shows how to calculate a confidence score for tuple T, where P is the set of patterns that derive T. C_i is the context that also contains T, $Match(C_i, P_i)$ is the degree of match of C_i and P_i.

$$Conf(T) = 1 - \prod_{I=0}^{|P|} \left(1 - \left(Conf\left(P_i\right) \cdot Match\left(C_i, P_i\right)\right)\right) \quad (2)$$

3 Results and discussion

Table 2 shows the top 10 generated patterns ranked by the frequency that they appear in MEDLINE. As can be seen the patterns all have very simple structures. Simple patterns are more likely to be productive, i.e the simpler the structure of the pattern, the more pairs it generates. However, simple structures are also likely to generate more false negative pairs.

The pairs produced by these top 10 patterns were collected, and the confidence score then calculated using equation 1. The result implies that the confidence score of a pattern, and in turn the selectivity and productivity of the pattern, are strongly associated with the pattern's structure.

Table 2: The top 10 comment patterns

Frequency	Pattern	Confidence
68	UNIPROT* CHEBI* CHEBI	0.16
61	CHEBI* UNIPROT* UNIPROT	0.15
51	CHEBI* UNIPROT* be	0.10
49	CHEBI* UNIPROT* CHEBI	0.10
41	UNIPROT* CHEBI* be	0.21
40	CHEBI* UNIPROT*	0.08
38	UNIPROT* CHEBI* UNIPROT	0.16
37	UNIPROT* CHEBI*	0.30
26	be CHEBI* UNIPROT*	0.26
24	UNIPROT* CHEBI CHEBI* CHEBI	0.17

3.1 Quality of the extracted pairs

One hundred pairs extracted by first and second generation patterns were randomly selected for manual inspection by a domain expert curator. It was found that over 60% were valid pairs. From further examination of the cases together with their extraction patterns, it can be seen that the patterns have a high confidence score, ensuring the quality of the extracted pair. For instance, from the original text *Paracetamol, 100 mg/kg, inhibited COX-1 in stomach mucosa ex vivo much less effectively than in other tissues*, the pattern "CHEBI*, UNIT, inhibit UNIPROT*" with 0.62 confidence score derives a correct pair <*Paracetamol, COX-1*>.

Generally speaking, simple patterns are more likely to have lower confidence scores. However it was also found that the pattern quality heavily depends on the quality and reliability of the name entity recognition (NE) system.

4 Conclusions and future work

We have presented a method of detecting small molecule and protein pairs in MEDLINE abstracts. It employs semi-supervised machine learning methods to enable patterns to be automatically generated, rather than requiring human input. The approach can be used for high throughput text mining applications where manual curation is unrealistic.

The first and second iteration of results are promising and show that the approach enables many useful small molecule protein pairs to be extracted from MEDLINE using just a small number of seed pairs as input. The approach makes use of a rigorous method of evaluating the quality of generated patterns and extracted pairs. Manual inspection has been used to validate these preliminary results and has shown that approximately half of the discovered pairs represent valid small molecule protein relationships, and we expect to improve this significantly.

In future we will develop the method further and analyse the results after further algorithm iterations, enabling discovery of new patterns and consequently new pairs of proteins and small molecules that are currently undetected.

References

E. Agichtein and L. Gravano. 2000. Snowball: Extracting relations from large plain-text collections. In *Proceedings of the fifth ACM conference on Digital libraries*, pages 85–94. ACM.

R. Apweiler, A. Bairoch, C.H. Wu, W.C. Barker, B. Boeckmann, S. Ferro, E. Gasteiger, H. Huang, R. Lopez, M. Magrane, et al. 2004. UniProt: the universal protein knowledgebase. *Nucleic acids research*, 32(suppl 1):D115–D119.

P. De Matos, M. Ennis, M. Darsow, M. Guedj, K. Degtyarenko, and R. Apweiler. 2006. ChEBI-chemical entities of biological interest. *Nucleic Acids Research*, Database Summary: 646.

D. Jiao and D.J. Wild. 2009. Extraction of CYP chemical interactions from biomedical literature using natural language processing methods. *Journal of chemical information and modeling*, 49(2):263–269.

E. Riloff and R. Jones. 1999. Learning dictionaries for information extraction by multi-level bootstrapping. In *Proceedings of the National Conference on Artificial Intelligence*, pages 474–479. John Wiley & Sons Ltd.

M. Song, I.Y. Song, X. Hu, and H. Han. 2006. Information extraction in biomedical literature. In J. Wang, editor, *Encyclopedia of Data Warehousing and Data Mining*, pages 615–620. Information Science Reference.

J.M. Temkin and M.R. Gilder. 2003. Extraction of protein interaction information from unstructured text using a context-free grammar. *Bioinformatics*, 19(16):2046–2053.

Y. Tsuruoka, Y. Tateishi, J.D. Kim, T. Ohta, J. McNaught, S. Ananiadou, and J. Tsujii. 2005. Developing a robust part-of-speech tagger for biomedical text. *Advances in informatics*, pages 382–392.

Grading the Quality of Medical Evidence

Binod Gyawali, Thamar Solorio
CoRAL Lab
Department of Computer and Information Sciences
University of Alabama at Birmingham, AL, USA
{bgyawali,solorio}@cis.uab.edu

Yassine Benajiba
Clinical Decision Support Solutions Department
Philips Research North America, Briarcliff Manor, NY, USA
yassine.benajiba@philips.com

Abstract

Evidence Based Medicine (EBM) is the practice of using the knowledge gained from the best medical evidence to make decisions in the effective care of patients. This medical evidence is extracted from medical documents such as research papers. The increasing number of available medical documents has imposed a challenge to identify the appropriate evidence and to access the quality of the evidence. In this paper, we present an approach for the automatic grading of evidence using the dataset provided by the 2011 Australian Language Technology Association (ALTA) shared task competition. With the feature sets extracted from publication types, Medical Subject Headings (MeSH), title, and body of the abstracts, we obtain a 73.77% grading accuracy with a stacking based approach, a considerable improvement over previous work.

1 Introduction

"Evidence Based Medicine (EBM) is the conscientious, explicit, and judicious use of current best evidence in making decisions about the care of individual patients" (Sackett et al., 1996). EBM requires to identify the best evidence, understand the methodology and strength of the approaches reported in the evidence, and bring relevant findings into clinical practice. Davidoff et al. (1995) express EBM in terms of five related ideas. Their ideas imply that the conclusions should be derived based on the best evidence available, the clinical decisions should be made based on the conclusions derived, and the performance of the clinical decisions should be evaluated constantly. Thus, physicians practicing EBM should be constantly aware of the new ideas and the best methodologies available based on the most recent literature. But the amount of clinical documents available is increasing everyday. For example, Pubmed, a service of the US National Library of Medicine contains more than 21 million citations for biomedical literature from MEDLINE, life science journals, and online books (last updated on December 7, 2011) [1]. The abundance of digital information makes difficult the task of evaluating the quality of results presented and the significance of the conclusions drawn. Thus, it has become an important task to grade the quality of evidence so that the most significant evidence is incorporated into the clinical practices.

There are several scale systems available to grade medical evidence. Some of them are: hierarchy of evidence proposed by Evans (2003), Grading of Recommendations Assessment, Development, and Evaluation (GRADE) scale by GRADE (2004), and Strength of Recommendation Taxonomy (SORT) scale by Ebell et al. (2004). The SORT scale addresses the quality, quantity, and consistency of evidence and proposes three levels of ratings: A, B, and C. Grade A is recommended based on the consistent, good-quality patient-oriented evidence, grade B is based on the inconsistent or limited-quality patient-oriented evidence, and grade C is based on consensus, disease-oriented evidence, usual practice, expert opinion or case studies.

[1] http://www.ncbi.nlm.nih.gov/books/NBK3827/

The Australasian Language Technology Association (ALTA) 2011 organized the shared task competition[2] to build an automatic evidence grading system for EBM based on the SORT grading scale. We carry out our experiments using the data set provided for the competition and compare the accuracy of grading the evidence by applying basic approaches and an ensemble (stacking) based approach of classification. We show that the later approach can achieve 73.77% of grading accuracy, a significant improvement over the basic approaches. We further extend our experiments to show that, using feature sets generated from the method and conclusion sections of the abstracts helps to obtain higher accuracy in evidence grading than using a feature set generated from the entire body of the abstracts.

2 Related Work

To the best of our knowledge, automatic evidence grading based on a grading scale was initiated by Sarker et al. (2011). Their work was based on the SORT scale to grade the evidence using the corpus developed by Molla-Aliod (2010). They showed that using only publication types as features could yield an accuracy of 68% while other information like publication types, journal names, publication years, and article titles could not significantly help to improve the accuracy of the grading. Molla-Aliod and Sarker (2011) worked on the evidence grading problem of 2011 ALTA shared task and achieved an accuracy of 62.84% using three sequential classifiers, each trained by one of the following feature sets: word n-grams from the abstract, publication types, and word n-grams from the title. They applied a three way classification approach where the instances classified as A or C were removed from the test set and labeled as such, while instances classified as B were passed to the next classifier in the pipeline. They repeated this process until they reached the end of three sequential classifiers.

Most of the EBM related work is focused on either the identification of important statements from the medical abstracts or the classification of medical abstracts to facilitate the retrieval of important documents. Work by Demner-Fushman et al. (2006), Dawes et al. (2007), Kim et al. (2011) automatically identify the key statements in the medical abstracts and classify them into different levels that are considered important for EBM practitioners in making decisions. Kilicoglu et al. (2009) worked on recognizing the clinically important medical abstracts using an ensemble learning method (stacking). They used different combinations of feature vectors extracted from documents to classify the evidence into relevant or non relevant classes. They approached the problem as a binary classification problem without using any grading scales.

Systematic Reviews (SRs) are very important to support EBM. Creating and updating SRs is highly inefficient and needs to identify the best evidence. Cohen et al. (2010) used a binary classification system to identify the documents that are most likely to be included in creating and updating SRs.

In this work, we grade the quality of evidence based on the SORT scale, that is different from most of the existing works related to classification of abstracts and identification of key statements of abstracts. We work on the same problem as by Molla-Aliod and Sarker (2011) but, we undertake the problem with a different approach and use different sets of features.

3 Dataset

We use the data of 2011 ALTA shared task competition that contains three different sets: training, development and test set. The number of evidence instances present in each set is shown in Table 1. Each data set consists of instances with grades A, B, or C based on the SORT scale. The distribution of evidence grades is shown in Table 2.

Data Set	No. of Evidence Instances
Training Set	677
Development Set	178
Test Set	183

Table 1: Evidence per data set

The evidence instances were obtained from the corpus developed by Molla-Aliod and Santiago-Martinez (2011). The corpus was generated based on the question and the evidence based answer for the question along with SOR grade obtained from the *"Clinical Inquiries"* section of the Journal of

[2]http://www.alta.asn.au/events/sharedtask2011

Grades	Training set (%)	Development set (%)	Test set (%)
A	31.3	27.0	30.6
B	45.9	44.9	48.6
C	22.7	28.1	20.8

Table 2: Evidence distribution per grade

Family Practice (JFP). A sample question from the JFP Clinical Inquiries section is *"How does smoking in the home affect children with asthma?"*. Each evidence contains at least one or more publications depending upon from which publications the evidence was generated. Each publication is an XML file containing information such as abstract title, abstract body, publication types, and MeSH terms. Each publication is assigned at least one publication type and zero or more MeSH terms. The MeSH terms vocabulary [3] is developed and maintained by the National Library of Medicine and is used in representation, indexing and retrieval of medical documents. Some of the medical document retrieval work emphasizes the use of MeSH terms in the efficient retrieval of documents (Trieschnigg et al., 2009; Huang et al., 2011). MeSH terms are also used in document summarization (Bhattacharya et al., 2011).

```
28092 B 10593430
28094 C 14712967 12269676 12165283 12618157
18163 A 8381089 7972972
18164 C 7972972 8621845
18166 A 8386917
16192 B 10920726
18162 A 8386917
52371 C 11642617 2328431 10532723
16193 B 9569395 12069675
```

Figure 1: Sample data file

Each data set contains an additional grade file with the information related to the evidence instances, their grades, and the publications. A sample of the file is shown in Figure 1. The first column contains the evidence id, the second column contains the grades A, B, or C of the evidence based on the SORT scale, and the remaining columns show the publication id of each publication in the evidence.

[3]http://www.nlm.nih.gov/mesh

The problem in this task is to analyze the publications in each evidence provided and classify them into A, B or C.

The dataset available for our research has abstracts in two different formats. One of them contains abstracts divided into sections: background, objective, method, result, and conclusion. The other format contains abstracts with all the information in a single block without any sections. A sample of an abstract having only four sections in the given data is shown below:

Objectives: To determine the effectiveness of a muscle strengthening program compared to a stretching program in women with fibromyalgia (FM).

Methods: Sixty-eight women with FM were randomly assigned to a 12 week, twice weekly exercise program consisting of either muscle strengthening or stretching. Outcome measures included muscle strength (main outcome variable), flexibility, weight, body fat, tender point count, and disease and symptom severity scales.

Results: No statistically significant differences between groups were found on independent t tests. Paired t tests revealed twice the number of significant improvements in the strengthening group compared to the stretching group. Effect size scores indicated that the magnitude of change was generally greater in the strengthening group than the stretching group.

Conclusions: Patients with FM can engage in a specially tailored muscle strengthening program and experience an improvement in overall disease activity, without a significant exercise induced flare in pain. Flexibility training alone also results in overall improvements, albeit of a lesser degree.

In the abstract above, we see that the approaches applied for the study are described in the method section, and the outcome and its effectiveness are described in the conclusion section.

4 Proposed Methodology

In this paper we propose a system to identify the correct grade of an evidence given publications in the evidence. We deal with the problem of evidence grading as a classification problem. In evidence grading, basic approaches have been shown to have poor performance. Molla-Aliod and Sarker (2011) showed that a basic approach of using simple bag-of-word features and a Naive Bayes classifier achieved 45% accuracy and proposed a sequential approach to improve the accuracy at each step. Our preliminary studies of applying the simple classification approach also showed similar results. Here, we propose a stacking based approach (Wolpert,

1992) of evidence grading. Stacking based approach builds a final classifier by combining the predictions made by multiple classifiers to improve the prediction accuracy. It involves two steps. In the first step, multiple base-level classifiers are trained with different feature sets extracted from a dataset and the classifiers are used to predict the classes of a second dataset. Then, a higher level classifier is trained using the predictions made by the base-level classifiers on the second dataset and used to predict the classes of the actual test data. In this approach, base-level classifiers are trained independent of each other and allowed to predict the classes. Based on the predictions made by these base-level classifiers, the higher level classifier learns from those predictions and makes a new prediction that is the final class.

Our stacking based approach of classification uses five feature sets. In the first step of classification, we train five classifiers using different feature sets per classifier and use the classifiers to predict the grades of the development dataset. Thus, at the end of the first step, five different predictions on the development dataset are obtained. In the second step, a new classifier is trained using the grades predicted by the five classifiers as features. This new classifier is then used to predict the grades of the test dataset.

5 Features

We extracted six sets of features from the publications to perform our experiments. They are as follows:

1. Publication types
2. MeSH terms
3. Abstract title
4. Abstract body
5. Abstract method section
6. Abstract conclusion section

For feature set 1, we extracted 30 distinct publication types from the training data. For the MeSH terms feature set, we selected 452 unique MeSH terms extracted from the training data. The publications contained the descriptor name of the MeSH terms having an attribute "majortopicyn" with value 'Y' or 'N'. As MeSH terms feature set, we selected only those MeSH term descriptor names having majortopicyn='Y'.

We extracted the last four sets of features from the title, body, method, and conclusion sections of the abstracts. Here, the body of an abstract means the whole content of the abstract, that includes background, objective, method, result, and conclusion sections. We applied some preprocessing steps to generate these feature sets. We also applied a feature selection technique to reduce the number of features and include only the high informative features from these feature sets. The details about preprocessing and feature selection techniques are described in Section 6.

We performed all the experiments on the basis of evidence, i.e. we created a single feature vector per evidence. If an evidence contained more than one publication, we generate its features as the union of the features extracted from all its publications.

The grades of the evidence in the SORT scale are based on the quality of evidence, basis of experiments, the methodologies used, and the types of analysis done. Grades also depend upon the effectiveness of the approach used in the experiments. The method section of an abstract contains the information related to the basis of the experiments, such as randomized controlled trails, systematic review, cohort studies, and the methods used in their research. The conclusion section of the abstract usually contains the assertion statements about how strongly the experiment supports the claims. Analysis of the contents of abstracts shows that the information needed for grading on SORT scale is typically available in the method and conclusion sections, more than in the other sections of the abstracts. Thus, we used the method and conclusion sections of the abstracts to generate two different feature sets so that only the features more likely to be important in grading using the SORT rating would be included.

Separating method and conclusion sections of the abstracts

In order to extract features from the method and conclusion sections, we should separate them from the body of abstracts, which is a challenging task for those abstracts without section headers. Of the total number of abstracts, more than one-third of the abstracts do not contain the section headers. In order to separate these sections, we used a very simple approach based on the number of sentences present in the method and conclusion sections, and the body of the abstracts. We used the following information to separate the method and conclusion sections from these abstracts: i) Order of sections in the abstracts, ii) Average number of sentences in the method and conclusion sections of the abstracts having sections, and iii) Average number of sentences in the entire body of the abstracts not having sections. All the abstracts having section headers contained the sections in the same order: background, objective, method, result and conclusion. From the available training dataset, we calculated:

i. The average number of sentences in the method (4.14) and conclusion (2.11) sections of the abstracts divided into sections

ii. The average number of sentences (8.78) of the abstracts not having sections

Based on these values, we fragmented the abstracts that do not have the section headers and separated the method and conclusion sections from them. Table 3 shows how the method and conclusion sections of those abstracts were generated. For example, the fourth row of the table says that, if an abstract without section headers has 6, 7 or 8 sentences (let it be n), then the 3rd, 4th and 5th sentences were considered as the method section, and the nth sentence was considered as the conclusion section.

Total sentences in Abstracts(n)	Method	Conclusion
1	None	1
2 or 3	1	n
4 or 5	2 and 3	n
6 or 7 or 8	2, 3 and 4	n
More than 8	3, 4 and 5	n-1 and n

Table 3: Selecting method and conclusion of the abstracts having a single block

6 Experiments and Results

This section describes the two sets of experiments performed to compare the performance of the stacking based approach and the effectiveness of the base-level classifiers used. The first set of experiments was done to provide a baseline comparison against our stacking based approach. The second set consists of five experiments to evaluate different configurations of stack based classifiers. The basic approach of classification implies the use of a single classifier trained by using a single feature vector.

We applied preprocessing steps to generate feature sets from the title, body, method and conclusion sections of the abstracts. The preprocessing steps were: detecting sentences using OpenNLP Sentence Detector[4], stemming words in each sentence using Porter Stemmer (Porter, 1980), changing the sentences into lower-case, and removing punctuation characters from the sentences. After the preprocessing step, we generated features from the unigrams, bigrams and trigrams in each part. We removed those features from the feature sets that contained the stopwords listed by Pubmed[5] or contained any token having a length less than three characters. To remove the less informative features, we calculated the information gain of the features in the training data using Weka (Hall et al., 2009) and selected only the top 500 high informative features for each feature set. We used the Weka SVM classifier for all the experiments. Based on the best result obtained after a series of experiments run with different kernel functions and regularization parameters, we chose the SVM classifier with a linear kernel and regularization parameter equals 1 for all the experiments. We used a binary weight for all the features.

6.1 First set of experiments

In the first set, we performed nine experiments using the basic classification approach and one experiment using the stacking based approach. The details of the experiments and the combinations of the features used in them are as shown in Table 4.

The first six experiments in the table were implemented by applying a basic approach of classification and each using only a single set of features. Experiments 7, 8, and 9 were similar to the first six experiments except, they used more than one set of features to create the feature vector. Each feature in the experiments 7, 8, and 9 encode the section of its origin. For example, if feature $abdomen$ is present in method as well as conclusion sections, it is represented as two distinct features $conc\_abdomen$ and $method\_abdomen$. In experiment 10, we applied

[4]http://incubator.apache.org/opennlp

[5]http://www.ncbi.nlm.nih.gov/books/NBK3827 /table/pubmedhelp.T43/?report=objectonly)

the stacking approach of classification using five base-level classifiers. The base-level classifiers in this experiment are the basic classifiers used in experiments 1 to 5.

Exp. No.	Features used	Exp. type
1.	Publication types	Basic approach
2.	MeSH terms	
3.	Abstract title	
4.	Abstract method	
5.	Abstract conclusion	
6.	Abstract body	
7.	Publication types, MeSH terms	
8.	Publication types, MeSH terms, Abstract title, Abstract body	
9.	Publication types, MeSH terms, Abstract title, Abstract method, Abstract conclusion	
10.	Publication types	Stacking based approach
	MeSH terms	
	Abstract title	
	Abstract method	
	Abstract conclusion	

Table 4: Experiments to compare basic approaches to a stacking based approach

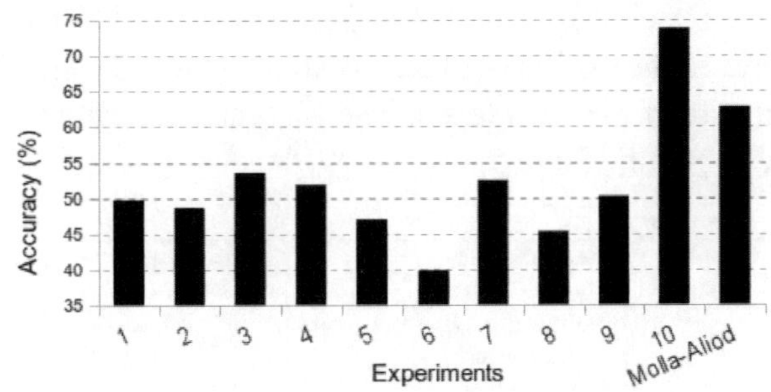

Figure 2: Comparison of accuracy of basic approaches to a stacking based approach. X-axis shows the experiments and Y-axis shows the accuracy of the experiments. The first nine experiments are based on the basic approach and the tenth experiment is based on the stacking based approach.

Figure 2 shows the results of the 10 experiments described in Table 4 in the same order, from 1st to 10th place and the result of the experiment by Molla-Aliod and Sarker (2011). The results show that the stacking based approach gives the highest accuracy (73.77%), outperforming all the basic approaches applying any combination of feature sets. The stacking based approach outperforms the baseline of a single layered classification approach (Exp 9) that uses all the five sets of features. Molla-Aliod and Sarker (2011) showed that a simple approach of using a single classifier and bag-of-words features could not achieve a good accuracy (45.9%) and proposed a new approach of using a sequence of classifiers to achieve a better result. Similar to their simple approach, our basic approaches could not achieve good results, but their performance is comparable to Molla-Aliod and Sarker (2011)'s baseline system. The result of our stacking based approach shows that our approach has a better accuracy than the sequencial classification approach (62.84%) proposed by Molla-Aliod and Sarker (2011).

Our stacking based approach works on two levels. In the first level, the base-level classifiers predict the grades of the evidence. In the next level, these predictions are used to train a new classifier that learns from the predictions to identify the grades correctly. Moreover, the five feature sets used in our experiments were unrelated to each other. For example, the features present in MeSH headings were different from the features used in publication types, and similarly, the features present in the method section of the abstract were different from the features present in the conclusion section. Each base-level classifier trained by one of these feature sets is specialized in that particular feature set. Thus, using the predictions made by these specialized base-level classifiers to train a higher level classifier helps to better predict the grades, this cannot be achieved by a single classifier trained by a set of features (Exp. 1, 2, 3, 4, 5, 6), or a group of different feature sets (Exp. 7, 8, 9).

6.2 Second set of experiments

In the second set of experiments, we compared five experiments performed varying the base-level classifiers used in our stack based approach. Experiments 1 and 2 were performed using a single base-level classifier, that means that the second classifier is trained on only one feature. Experiments 3 and 4 were performed by using four base-level classifiers, and experiment 5 was performed using five base-level classifiers. The 5th experiment in this set is same as the 10th experiment in the first set. The details about the feature sets used in each experiment are shown in Table 5.

Exp. No.	Features used	No. of Base level classifiers
1.	Publication types, MeSH terms, Abstract title, Abstract body	1
2.	Publication types, MeSH terms, Abstract title, Abstract method, Abstract conclusion	1
3.	Publication types MeSH terms Abstract title Abstract body	4
4.	Publication types MeSH terms Abstract title Abstract method, Abstract conclusion	4
5.	Publication types MeSH terms Abstract title Abstract method Abstract conclusion	5

Table 5: Experiments to compare stacking based approach

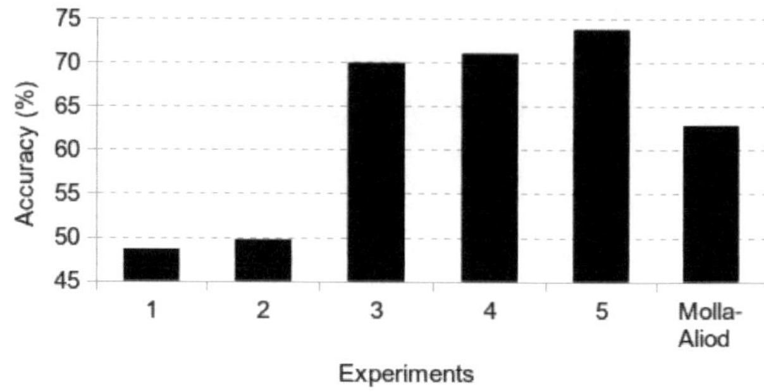

Figure 3: Comparison of accuracy of the stacking based approaches. X-axis shows the experiments and Y-axis shows the accuracy of the experiments. 1st and 2nd experiments use only one base-level classifier, 3rd and 4th experiment are based on four base-level classifiers and 5th one uses five base-level classifiers.

Figure 3 shows the accuracy of the five experiments shown in Table 5 in the same order. It shows that the accuracy of 1st and 2nd experiments is lower than the accuracy of 3rd, 4th, and 5th experiments. In these two experiments, a feature vector generated from the prediction of a single base-level classifier is used to train the higher level classifier, that is not sufficient to make a correct decision.

Experiments 3, 4, and 5 show a considerable improvement in the accuracy of the grading. Comparing the results of experiments 3 and 4, we see that the 4th experiment has higher accuracy than the 3rd one. The difference between these experiments was the use of features from the method and conclusion sections of the abstracts in the 4th experiment, while using features from the entire body of abstracts in the 3rd experiment. The higher accuracy in the 4th experiment shows that the method and conclusion sections of the experiment contain high informative text that is important for evidence grading, while the body of abstracts may contain some information that is not relevant to the task. The same analysis can also be inferred from the results of experiment 8 and 9 in the first set of experiments. The highest accuracy obtained in the 5th experiment of applying 5 base-level classifiers shows that identifying the sections of the abstracts containing high informative features and using a sufficient number of base-level classifiers can help to achieve a good accuracy in evidence grading.

7 Error Analysis

The result obtained by the stacking based approach (5th experiment in Table 5) using five base-level classifiers gave a higher error rate in predicting grades A and C, compared to the error rate in predicting grade B. Most of the error is the misclassification of A to C and vice versa. One of the possible reasons of this might be due to the use of the feature set extracted from the conclusion section. Among the five base-level classifiers used in the experiment, the one trained by the features extracted from the conclusion sections has the lowest accuracy (5th experiment in Figure 2). We evaluated the text contained in the conclusion section of the abstracts in our dataset. The section mostly contains the assertion statements having the words showing strong positive/negative meanings. Conclusion of A grade evidence mostly contains the information that strongly asserts the claim *(e.g. emollient treatment*

***significantly** reduced the high-potency topical corticosteroid consumption in infants with AD)*, while that of C grade evidence is not strong enough to assert the claim *(e.g. PDL therapy should be considered among the better established approaches in the treatment of warts, although data from this trial suggest that this approach is **probably not** superior)*. It seems that the problem might be because of not processing the negations appropriately. So, in order to preserve some negation information present in the conclusion sections, we performed another experiment by merging words *no*, *not*, *nor* with their successor word to create a single token from the two words. This approach still could not reduce the misclassification. Thus, the simple approach of extracting unigram, bigram, and trigram features from the conclusion section might not be sufficient and might need to include higher level analysis related to assertion/certainty of the statements to reduce the misclassification of the evidence.

Other possible reasons of the misclassification of the evidence might be the imbalanced data set. Our dataset (Table 2) contains higher number of instances with grade B than those with grades A and C. Moreover, the number of publications per evidence is not uniform, that ranges from 1 to 8 publications per evidence in the test data. Analyzing the results, we found that misclassification of evidence having only one publication is higher than that of the evidence having more than one publication. If an evidence contains only one publication, the features of the evidence extracted from a single publication might not be sufficient to accurately grade the evidence and might lead to misclassification.

In order to evaluate the appropriateness of our approach in extracting the method and conclusion sections, we performed a manual inspection of abstracts. We could not revise all the abstracts to verify the approach. Thus, we randomly selected 25 abstracts without section headers from the test data and viewed the content in them. We found that the conclusion section was appropriately extracted in almost all abstracts, while the selection of method section was partially effective. Our approach was based on the assumption that all the abstracts having many sentences have all the sections (background, objective, method, result, and conclusion). But we found that the abstracts do not follow the same format, and the start sentence of the method section is not consistent. Even a long abstract might sometimes start with the method section, and sometimes the objective section might not be present in the abstracts. This could lead to increase the error in our grading system.

8 Conclusion

This paper presents an approach of grading the medical evidence applying a stacking based classifier using the features from publication types, MeSH terms, abstract body, and method, and conclusion sections of the abstracts. The results show that this approach achieves an accuracy of 73.77%, that is significantly better than the previously reported work. Here, we present two findings: 1) We show that the stacking based approach helps to obtain a better result in evidence grading than the basic approach of classification. 2) We also show that the method and conclusion sections of the abstracts contain important information necessary for evidence grading. Using the feature sets generated from these two sections helps to achieve a higher accuracy than by using the feature set generated from the entire body of the abstracts.

In this work, all the information available in the method and conclusion sections of the abstracts is treated with equal weight. Evidence grading should not depend upon specific disease names and syndromes, but should be based on how strong the facts are presented. We would like to extend our approach by removing the words describing specific disease names, disease syndromes, and medications, and giving higher weight to the terms that describe the assertion of the statements. In our current work, we apply a simple approach to extract the method and conclusion sections from the abstracts not having sections. Improving the approach by using a machine learning algorithm that can more accurately extract the sections might help to increase the accuracy of grading. Including the information about the strength of assertions made in the conclusion sections could also help in boosting the accuracy. Future work would also include testing the effectiveness of our approach on other diverse data sets having complex structures of the evidence, or on a different grading scale.

References

Sanmitra Bhattacharya, Viet HaThuc, and Padmini Srinivasan. 2011. Mesh: a window into full text for document summarization. *Bioinformatics*, 27(13):i120–i128.

Aaron M. Cohen, Kyle Ambert, and Marian McDonagh. 2010. A Prospective Evaluation of an Automated Classification System to Support Evidence-based Medicine and Systematic Review. *AMIA Annu Symp Proc.*, 2010:121 – 125.

Frank Davidoff, Brian Haynes, Dave Sackett, and Richard Smith. 1995. Evidence based medicine. *BMJ*, 310(6987):1085–1086, 4.

Martin Dawes, Pierre Pluye, Laura Shea, Roland Grad, Arlene Greenberg, and Jian-Yun Nie. 2007. The identification of clinically important elements within medical journal abstracts: Patient-Population-Problem, Exposure-Intervention, Comparison, Outcome, Duration and Results (PECODR). *Informatics in Primary Care*, 15(1):9–16.

Dina Demner-Fushman, Barbara Few, Susan E. Hauser, and George Thoma. 2006. Automatically Identifying Health Outcome Information in MEDLINE Records. *Journal of the American Medical Informatics Association*, 13(1):52 – 60.

M. H. Ebell, J. Siwek, B. D. Weiss, S. H. Woolf, J. Susman, B. Ewigman, and M. Bowman. 2004. Strength of recommendation taxonomy (SORT): a patient-centered approach to grading evidence in the medical literature. *American Family Physician*, 69(3):548–56+.

David Evans. 2003. Hierarchy of evidence: a framework for ranking evidence evaluating healthcare interventions. *Journal of Clinical Nursing*, 12(1):77–84.

GRADE. 2004. Grading quality of evidence and strength of recommendations. *BMJ*, 328(7454):1490, 6.

Mark Hall, Eibe Frank, Geoffrey Holmes, Bernhard Pfahringer, Peter Reutemann, and Ian H. Witten. 2009. The WEKA data mining software: an update. *SIGKDD Explor. Newsl.*, 11(1).

Minlie Huang, Aurlie Nvol, and Zhiyong Lu. 2011. Recommending MeSH terms for annotating biomedical articles. *Journal of the American Medical Informatics Association*, 18(5):660–667.

Halil Kilicoglu, Dina Demner-Fushman, Thomas C Rindflesch, Nancy L Wilczynski, and R Brian Haynes. 2009. Towards Automatic Recognition of Scientifically Rigorous Clinical Research Evidence. *Journal of the American Medical Informatics Association*, 16(1):25–31.

Su Kim, David Martinez, Lawrence Cavedon, and Lars Yencken. 2011. Automatic classification of sentences to support Evidence Based Medicine. *BMC Bioinformatics*, 12(Suppl 2):S5.

Diego Molla-Aliod and Maria Elena Santiago-Martinez. 2011. Development of a Corpus for Evidence Based Medicine Summarisation. In *Proceedings of the Australasian Language Technology Association Workshop*.

Diego Molla-Aliod and Abeed Sarker. 2011. Automatic Grading of Evidence: the 2011 ALTA Shared Task. In *Proceedings of Australasian Language Technology Association Workshop*, pages 4–8.

Diego Molla-Aliod. 2010. A Corpus for Evidence Based Medicine Summarisation. In *Proceedings of the Australasian Language Technology Association Workshop*, volume 8.

MF Porter. 1980. An algorithm for sufx stripping. *Program*, 14(3):130–137.

David L Sackett, William M C Rosenberg, J A Muir Gray, R Brian Haynes, and W Scott Richardson. 1996. Evidence based medicine: what it is and what it isn't. *BMJ*, 312(7023):71–72, 1.

Abeed Sarker, Diego Molla-Aliod, and Cecile Paris. 2011. Towards automatic grading of evidence. In *Proceedings of LOUHI 2011 Third International Workshop on Health Document Text Mining and Information Analysis*, pages 51–58.

Dolf Trieschnigg, Piotr Pezik, Vivian Lee, Franciska de Jong, Wessel Kraaij, and Dietrich Rebholz-Schuhmann. 2009. MeSH Up: effective MeSH text classification for improved document retrieval. *Bioinformatics*, 25(11):1412–1418.

David H. Wolpert. 1992. Stacked generalization. *Neural Networks*, 5(2):241 – 259.

Classifying Gene Sentences in Biomedical Literature by Combining High-Precision Gene Identifiers

Sun Kim, Won Kim, Don Comeau, and W. John Wilbur
National Center for Biotechnology Information
National Library of Medicine, National Institutes of Health
Bethesda, MD 20894, USA
{sun.kim,won.kim,donald.comeau,john.wilbur}@nih.gov

Abstract

Gene name identification is a fundamental step to solve more complicated text mining problems such as gene normalization and protein-protein interactions. However, state-of-the-art name identification methods are not yet sufficient for use in a fully automated system. In this regard, a relaxed task, gene/protein sentence identification, may serve more effectively for manually searching and browsing biomedical literature. In this paper, we set up a new task, gene/protein sentence classification and propose an ensemble approach for addressing this problem. Well-known named entity tools use similar gold-standard sets for training and testing, which results in relatively poor performance for unknown sets. We here explore how to combine diverse high-precision gene identifiers for more robust performance. The experimental results show that the proposed approach outperforms BANNER as a stand-alone classifier for newly annotated sets as well as previous gold-standard sets.

1 Introduction

With the rapidly increasing biomedical literature, text mining has become popular for finding biomedical information in text. Among others, named entity recognition (NER) for bio-entities such as genes and proteins is a fundamental task because extracting biological relationships begins with entity identification. However, NER in biomedical literature is challenging due to the irregularities and ambiguities in bio-entities nomenclature (Yang et al., 2008). In particular, compound entity names make this problem difficult because it also requires deciding word boundaries.

Recent bio-text competitions such as JNLPBA (Kim et al., 2004) and BioCreative (Lu et al., 2011; Smith et al., 2008) have evaluated NER systems for gene mentions. Even though progress has been made in several areas, gene identification methods are not yet sufficient for real-world use without human interaction (Arighi et al., 2011). Thus, at the present, a realistic suggestion is to use these algorithms as an aid to human curation and information retrieval (Altman et al., 2008).

In this paper, we define a new task, gene/protein sentence classification. A gene or protein sentence means a sentence including at least one specific gene or protein name. This new task has advantages over gene mention identification. First, gene name boundaries are not important at the sentence level and human judges will agree more in their judgments. Second, highlighting gene sentences may be more useful in manual search and browsing environments since this can be done more accurately and with less distraction from incorrect annotations.

To classify gene/protein sentences, we here propose an ensemble approach to combine different NER identifiers. Previous NER approaches are mostly developed on a small number of gold-standard sets including GENIA (Kim et al., 2003) and BioCreative (Smith et al., 2008) corpora. These sets help to find regular name patterns in a limited set of articles, but also limit the NER performance for real-world use. In the proposed approach, we use a Semantic Model and a Priority Model along with BANNER (Leaman and Gonzalez, 2008). The Semantic and Priority Models are used to provide more robust performance on gene/protein sentence classification because they utilize larger resources such as SemCat and PubMed® to detect gene names.

For experiments, we created three new gold-standard sets to include cases appearing in the most recent publications. The experimental results show that our approach outperforms machine learning classifiers using unigrams and substring features as well as stand-alone BANNER classification on five gold-standard datasets.

The paper is organized as follows. In Section 2, the ensemble approach for gene/protein sentence classification is described. Section 3 explains the gold-standard sets used for our experiments. Section 4 presents and discusses the experimental results. Conclusions are drawn in Section 5.

2 Methods

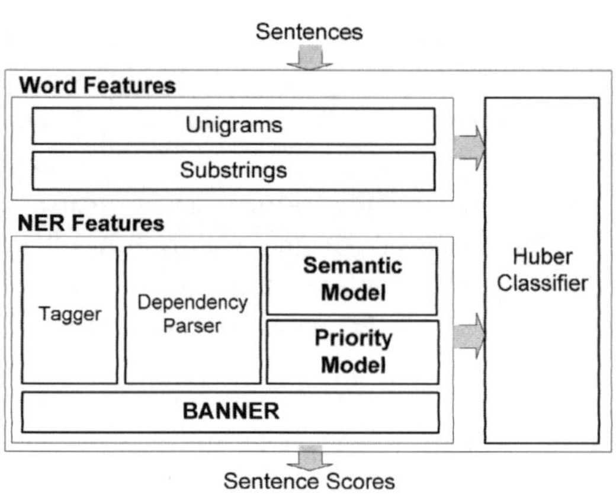

Figure 1. Method Overview.

Figure 1 shows the overall framework for our proposed approach. We basically assume that a main NER module works as a strong predictor, i.e., the majority of outputs obtained from this module are correct. We here use BANNER (Leaman and Gonzalez, 2008) as the main NER method because it adopts features and methods which are generally known to be effective for gene name recognition. While BANNER shows good performance on well-known gold-standard sets, it suffers from relatively poor performance on unknown examples. To overcome this problem, we combine BANNER with two other predictors, a Sematic Model and a Priority Model. First, the Semantic Model and the Priority Model do not use previous gold-standard sets for training. Second, these two models learn name patterns in different ways, i.e., semantic relationships for the Semantic Model and positional and lexical information for the Priority Model. This combination of a strong predictor and two weaker but more general predictors can respond better to unknown name patterns.

As described above, the proposed method mainly relies on outputs from different NER methods, whereas word features can still provide useful evidence for discriminating gene and non-gene sentences. Hence, we alternatively utilize word features such as unigrams and substrings along with NER features. For NER features only, the output is the sum of binary decisions from three NER modules. For word and NER features, the Huber classifier (Kim and Wilbur, 2011) is trained to combine the features. The parameter set in the Huber classifier is optimized to show the best classification performance on test sets. The following subsections describe each feature type used for gene sentence classification.

2.1 Word Features

Unigrams are a set of words obtained from tokenizing sentences on white space. All letters in unigrams are converted to lower case.

Substrings are all contiguous substrings of a sentence, sized three to six characters. This substring feature may help reduce the difference between distributions on training and test sets (Huang et al., 2008). Substrings encode the roots and morphology of words without identifying syllables or stems. They also capture neighboring patterns between words.

2.2 BANNER

BANNER is a freely available tool for identifying gene mentions. Due to its open-source policy and Java implementation, it has become a popular tool.

BANNER uses conditional random fields (CRF) as a discriminative method and utilizes a set of feature types that are known to be good for identifying gene names. The feature sets used are orthographic, morphological and shallow syntax features (Leaman and Gonzalez, 2008):

(1) The part of speech (POS) of a token in a sentence.
(2) The lemma of a word.
(3) 2, 3 and 4-character prefixes and suffixes.
(4) 2 and 3 character n-grams including start-of-token and end-of-token indicators.
(5) Word patterns by converting upper-case letters, lower-case letters and digits to their corresponding representative characters (A, a, 0).
(6) Numeric normalization by converting digits to "0"s.
(7) Roman numerals.
(8) Names of Greek letters.

Even though BANNER covers most popular feature types, it does not apply semantic features or other post-processing procedures such as abbreviation processing. However, these features may not have much impact for reducing performance since our goal is to classify gene sentences, not gene mentions.

2.3 Semantic Model

The distributional approach to semantics (Harris, 1954) has become more useful as computational power has increased, and we have found this approach helpful in the attempt to categorize entities found in text. We use a vector space approach to modeling semantics (Turney and Pantel, 2010) and compute our vectors as described in (Pantel and Lin, 2002) except we ignore the actual mutual information and just include a component of 1 if the dependency relation occurs at all for a word, else the component is set to 0. We constructed our vector space from all single tokens (a token must have an alphabetic character) throughout the titles and abstracts of the records in the whole of the PubMed database based on a snapshot of the database taken in January 2012. We included only tokens that occurred in the data sufficient to accumulate 10 or more dependency relations. There were just over 750 thousand token types that satisfied this condition and are represented in the space. We denote this space by *h*. We then took all the single tokens and all head words from multi-token strings in the categories "chemical", "disease", and "gene/protein" from an updated version of the SemCat database (Tanabe et al., 2006) and placed all the other SemCat categories similarly processed into a category we called "other". We consider only the tokens in these categories that also occur in our semantic vector space *h* and refer to these sets as h_{Chemical}, h_{Disease}, $h_{\text{Gene/Protein}}$, h_{Other}. Table 1 shows the size of overlaps between sets.

	h_{Chemical}	h_{Disease}	$h_{\text{Gene/Protein}}$	h_{Other}
h_{Chemical}	54478	209	4605	5495
h_{Disease}		8801	1139	169
$h_{\text{Gene/Protein}}$			76440	9466
h_{Other}				127337

Table 1. Pairwise overlap between sets representing the different categories.

Class	h'_{Chemical}	h'_{Disease}	$h'_{\text{Gene/Protein}}$	h'_{Other}
Strings	49800	7589	70832	113815
Ave. Prec.	0.8680	0.7060	0.9140	0.9120

Table 2. Row two contains the number of unique strings in the four different semantic classes studied. The last row shows the mean average precisions from a 10-fold cross validation to learn how to distinguish each class from the union of the other three.

In order to remove noise or ambiguity in the training set, we removed the tokens that appeared in more than one semantic class as follows.

$$\begin{aligned} h'_{\text{Chemical}} &= h_{\text{Chemical}} - (h_{\text{Disease}} + h_{\text{Gene/Protein}}) \\ h'_{\text{Disease}} &= h_{\text{Disease}} - (h_{\text{Chemical}} + h_{\text{Gene/Protein}}) \\ h'_{\text{Gene/Protein}} &= h_{\text{Gene/Protein}} - (h_{\text{Chemical}} + h_{\text{Disease}}) \\ h'_{\text{Other}} &= h_{\text{Other}} - (h_{\text{Chemical}} + h_{\text{Disease}} + h_{\text{Gene/Protein}}) \end{aligned} \quad (1)$$

We then applied Support Vector Machine learning to the four resulting disjoint semantic classes in a one-against-all strategy to learn how to classify into the different classes. We used $C = 1.64 \times 10^3$ based upon the size of the training set. As a test of this process we applied this same learning with 10-fold cross validation on the training data and the results are given in the last row of Table 2.

This Semantic Model is an efficient and general way to identify words indicating gene names. Unlike other NER approaches, this model decides a target class solely based on a single word. However, evaluating all tokens from sentences may increase incorrect predictions. A dependency parser analyzes a sentence as a set of head- and dependent-word combinations. Since gene names likely appear in describing a relationship with other entities, a name indicating a gene mention will be mostly placed in a dependent position. Thus, we first apply the C&C CCG parser (Curran et al., 2007), and evaluate words in dependent positions only.

2.4 Priority Model

The Semantic Model detects four different categories for a single word. However, the Priority Model captures gene name patterns by analyzing the order of words and the character strings making up words. Since gene names are noun phrases in general, we parse sentences and identify noun phrases first. These phrases are then evaluated using the Priority Model.

The Priority Model is a statistical language model for named entity recognition (Tanabe and Wilbur, 2006). For named entities, a word to the right is more likely to be the word determining the nature of the entity than a word to the left in general.

Let T_1 be the set of training data for class C_1 and T_2 for class C_2. Let $\{t_\alpha\}_{\alpha \in A}$ denote the set of all tokens used in names contained in $T_1 \cup T_2$. For each token t_α, $\alpha \in A$, it is assumed that there are associated two probabilities p_α and q_α, where p_α is the probability that the appearance of the token t_α in a name indicates that name belongs to class C_1 and q_α is the probability that t_α is a more reliable indicator of the class of a name than any token to its left. Let $n = t_{\alpha(1)} t_{\alpha(2)} \cdots t_{\alpha(k)}$ be composed of the tokens on the right in the given order. Then the probability of n belonging to class C_1 can be computed as follows.

$$p(C_1 \mid n) = p_{\alpha(1)} \prod_{j=2}^{k} \left(1 - q_{\alpha(j)}\right) + \sum_{i=2}^{k} q_{\alpha(i)} p_{\alpha(i)} \prod_{j=i+1}^{k} \left(1 - q_{\alpha(j)}\right) \quad (2)$$

A limited memory BFGS method (Nash and Nocedal, 1991) and a variable order Markov model (Tanabe and Wilbur, 2006) are used to obtain p_α and q_α. An updated version of SemCat (Tanabe and Wilbur, 2006) was used to learn gene names.

2.5 Semantic and Priority Models for High-Precision Scores

The Semantic and Priority Models learn gene names and other necessary information from the SemCat database, where names are semantically categorized based on UMLS® (Unified Medical Language System) Semantic Network. Even though the Semantic and Priority Models show good performance on names in SemCat, they cannot avoid noise obtained from incorrect preprocessing, e.g., parsing errors. The use of a general category for training may also limit performance. To obtain high-precision scores for our ensemble approach, it is important to reduce the number of false positives from predictions. Hence, we apply the Semantic and Priority Models on training sets, and mark false positive cases. These false positives are automatically removed from predictions on test sets. These false positive cases tend to be terms for entities too general to warrant annotation.

Table 3 shows the classification performance with and without false positive corrections on training data. For both Semantic and Priority Models, precision rates are increased by removing false positives. Even though recall drops drastically, this does not cause a big problem in our setup since these models try to detect gene names which are not identified by BANNER.

	SEM	SEM_{FP}	PM	PM_{FP}
Accuracy	0.7907	0.7773	0.7805	0.8390
Precision	0.7755	0.8510	0.7405	1.0000
Recall	0.8323	0.6852	0.8799	0.6856
F1	0.8029	0.7592	0.8042	0.8135

Table 3. Performance changes on training set for the Semantic Model (SEM) and the Priority Model (PM). FP indicates that learned false positives were removed from predictions.

3 Datasets

For experiments, we rigorously tested the proposed method on gene mention gold-standard sets and newly annotated sets. GENETAG (Smith et al., 2008) is the dataset released for BioCreative I and BioCreative II workshops. Since it is well-known for a gene mention gold-standard set, we used GENETAG as training data.

For test data, two previous gold-standard sets were selected and new test sets were also built for gene sentence classification. YAPEX (Franzen et al., 2002) and JNLPBA (Kim et al., 2004) are considered of moderate difficulty because they are both related to GENIA corpus, a well-known gold-standard set. However, Disease, Cell Line and Reptiles are considered as more difficult tasks because they represent new areas and contain recently published articles. The annotation guideline for new test sets basically followed those used in GENETAG (Tanabe et al., 2005), however domains, complexes, subunits and promoters were not included in new sets.

(1) "Disease" Set: This set of 60 PubMed documents was obtained from two sources. Fifty of the documents were obtained from the 793 PubMed documents used to construct the AZDC (Leaman et al., 2009). They are the fifty most recent among these records. In addition to these fifty documents, ten documents were selected from PubMed on the topic of maize to add variety to the set and because one of the curators who worked with the set had experience studying the maize genome. These ten were chosen as recent documents as of early March 2012 and which contained the text word maize and discussed genetics. The whole set of 60 documents were annotated by WJW to produce a gold standard.

(2) "CellLine" Set: This set comprised the most recent 50 documents satisfying the query "cell line[MeSH]" in PubMed on March 15, 2012. This query was used to obtain documents which discuss cell lines, but most of these documents also discuss genes and for this reason the set was expected to be challenging. The set was annotated by WJW and DC and after independently annotating the set they reconciled differences to produce a final gold standard.

(3) "Reptiles" Set: This set comprised the most recent 50 documents satisfying the query "reptiles AND genes [text]" in PubMed on March 15, 2012. This set was chosen because it would have little about human or model organisms and for this reason it was expected to be challenging. The set was annotated by WJW and DC and after independently annotating the set they reconciled differences to produce a final gold standard.

For both "CellLine" and "Reptiles" Sets, the most recent data was chosen in an effort to make the task more challenging. Presumably such documents will contain more recently created names and phrases that do not appear in the older training data. This will then pose a more difficult test for NER systems.

Table 4 shows all datasets used for training and testing. The new sets, "Disease", "CellLine" and "Reptiles" are also freely available at http://www.ncbi.nlm.nih.gov/CBBresearch/Wilbur/IRET/bionlp.zip

	Positives	Negatives	Total
GENETAG	10245	9755	20000
YAPEX	1298	378	1676
JNLPBA	17761	4641	22402
Disease	345	251	596
CellLine	211	217	428
Reptiles	179	328	507

Table 4. Datasets. "GENETAG" was used for training data and others were used for test data. "YAPEX" and "JNLPBA" were selected from previous gold-standard corpora. "Disease", "Cell Line" and "Reptiles" are newly created from recent publications and considered as difficult sets.

4 Results and Discussion

In this paper, our goal is to achieve higher-prediction performance on a wide range of gene sentences by combining multiple gene mention identifiers. The basic assumption here is that there is a strong predictor that performs well for previously known gold-standard datasets. For this strong predictor, we selected BANNER since it includes basic features that are known to give good performance.

	Accuracy	Precision	Recall	F1
GENETAG	0.9794	0.9817	0.9779	0.9799
YAPEX	0.9051	0.9304	0.9483	0.9392
JNLPBA	0.8693	0.9349	0.8976	0.9159
Disease	0.8591	0.9223	0.8261	0.8716
Cell Line	0.8925	0.9146	0.8626	0.8878
Reptiles	0.8994	0.8478	0.8715	0.8595

Table 5. Performance of BANNER on training and test datasets.

Table 5 presents the gene sentence classification performance of BANNER on training and test sets. We emphasize that performance here means that if BANNER annotates a gene/protein name in a sentence, that sentence is classified as positive, otherwise it is classified as negative. BANNER used GENETAG as training data, hence it shows excellent classification performance on the same set.

	Unigrams	Substrings	BANNER	Ensemble	Uni+Ensemble	Sub+Ensemble
YAPEX	0.9414	0.9491	0.9685	**0.9704**	0.9624	0.9678
JNLPBA	0.9512	0.9504	0.9584	**0.9651**	0.9625	0.9619
Disease	0.8255	0.8852	0.9238	0.9501	0.9573	**0.9610**
CellLine	0.8174	0.9004	0.9281	**0.9539**	0.9429	0.9496
Reptiles	0.6684	0.7360	0.8696	**0.9049**	0.9001	0.8937

Table 6. Average precision results on test sets for different feature combinations.

	Unigrams	Substrings	BANNER	Ensemble	Uni+Ensemble	Sub+Ensemble
YAPEX	0.8735	0.8819	0.9321	0.9196	0.9298	**0.9336**
JNLPBA	0.8902	0.8938	0.9111	0.9197	0.9262	**0.9264**
Disease	0.7449	0.7884	0.8479	0.8894	0.8957	**0.9043**
CellLine	0.7346	0.8057	0.8698	0.9017	**0.9052**	0.8957
Reptiles	0.6257	0.6816	0.8499	0.8199	**0.8547**	**0.8547**

Table 7. Breakeven results on test sets for different feature combinations.

- Just one fiber gene was revealed in this strain.
- This transcription factor family is characterized by a DNA-binding alpha-subunit harboring the Runt domain and a secondary subunit, beta, which binds to the Runt domain and enhances its interaction with DNA.

Figure 2. False positive examples including misleading words.

YAPEX and JNLPBA are gold-standard sets that partially overlap the GENIA corpus. Since BANNER utilizes features from previous research on GENETAG, YAPEX and JNLPBA, we expect good performance on these data sets. For that reason, we created the three additional gold-standard sets to use in this study, and we believe the performance on these sets is more representative of what could be expected when our method is applied to cases recently appearing in the literature.

Table 6 show average precision results for the different methods and all the test sets. GENETAG is left out because BANNER is trained on GENETAG. We observe improved performance of the ensemble methods over unigrams, substrings and BANNER. The improvement is small on YAPEX and JNLPBA, but larger for Disease, CellLine and Reptiles. We see that unigrams and substrings tend to add little to the plain ensemble.

The MAP (Mean Average Precision) values in Table 6 are in contrast to the breakeven results in Table 7, where we see that unigrams and substrings included with the ensemble generally give improved results. Some of the unigrams and substrings are specific enough to detect gene/protein names with high accuracy, and improve precision in top ranks in a way that cannot be duplicated by the annotations coming from Semantic or Priority Models or BANNER. In addition, substrings may capture more information than unigrams because of their greater generality.

Some of our errors are due to false positive NER identifications. By this we mean a token was classified as a gene/protein by BANNER or the Semantic or Priority Models. This often happens when the name indeed represents a gene/protein class, which is too general to be marked positive (Figure 2). A general way in which this problem could be approached is to process a large amount of literature discussing genes or proteins and look for names that are marked as positives by one of the NER identifiers, and which appear frequently in plural form as well as in the singular. Such names are likely general class names, and have a high probability to be false positives.

Another type of error will arise when unseen tokens are encountered. If such tokens have string similarity to gene/protein names already encountered in the SemCat data, they may be recognized by the Priority Model. But there will be completely new strings. Then one must rely on context and this may not be adequate. We think there is little that can be done to solve this short of better language understanding by computers.

There is a benefit in considering whole sentences as opposed to named entities. By considering whole sentences, name boundaries become a non-issue. For this reason, one can expect training data to be more accurate, i.e., human judges will tend to agree more in their judgments. This may allow for improved training and testing performance of machine learning methods. We believe it beneficial that human users are directed to sentences that contain the entities they seek without necessity of viewing the less accurate entity specific tagging which they may then have to correct.

5 Conclusions

We defined a new task for classifying gene/protein sentences as an aid to human curation and information retrieval. An ensemble approach was used to combine three different NER identifiers for improved gene/protein sentence recognition. Our experiments show that one can indeed find improved performance over a single NER identifier for this task. An additional advantage is that performance at this task is significantly more accurate than gene/protein NER. We believe this improved accuracy may benefit human users of this technology. We also make available to the research community three gold-standard gene mention sets, and two of these are taken from the most recent literature appearing in PubMed.

Acknowledgments

This work was supported by the Intramural Research Program of the National Institutes of Health, National Library of Medicine.

References

R. B. Altman, C. M. Bergman, J. Blake, C. Blaschke, A. Cohen, F. Gannon, L. Grivell, U. Hahn, W. Hersh, L. Hirschman, L. J. Jensen, M. Krallinger, B. Mons, S. I. O'donoghue, M. C. Peitsch, D. Rebholz-Schuhmann, H. Shatkay, and A. Valencia. 2008. Text mining for biology - the way forward: opinions from leading scientists. *Genome Biol*, 9 Suppl 2:S7.

C. N. Arighi, Z. Lu, M. Krallinger, K. B. Cohen, W. J. Wilbur, A. Valencia, L. Hirschman, and C. H. Wu. 2011. Overview of the BioCreative III workshop. *BMC Bioinformatics*, 12 Suppl 8:S1.

J. R. Curran, S. Clark, and J. Bos. 2007. Linguistically motivated large-scale NLP with C&C and boxer. In *Proceedings of the 45th Annual Meeting of the ACL on Interactive Poster and Demonstration Sessions*, pages 33-36.

K. Franzen, G. Eriksson, F. Olsson, L. Asker, P. Liden, and J. Coster. 2002. Protein names and how to find them. *Int J Med Inform*, 67:49-61.

Z. S. Harris. 1954. Distributional structure. *Word*, 10:146-162.

M. Huang, S. Ding, H. Wang, and X. Zhu. 2008. Mining physical protein-protein interactions from the literature. *Genome Biol*, 9 Suppl 2:S12.

J.-D. Kim, T. Ohta, Y. Tateisi, and J. Tsujii. 2003. GENIA corpus - semantically annotated corpus for bio-textmining. *Bioinformatics*, 19 Suppl 1:i180-i182.

J.-D. Kim, T. Ohta, Y. Tsuruoka, Y. Tateisi, and N. Collier. 2004. Introduction to the bio-entity recognition task at JNLPBA. In *Proceedings of the International Joint Workshop on Natural Language Processing in Biomedicine and its Applications*, pages 70-75.

S. Kim and W. J. Wilbur. 2011. Classifying protein-protein interaction articles using word and syntactic features. *BMC Bioinformatics*, 12 Suppl 8:S9.

R. Leaman and G. Gonzalez. 2008. BANNER: an executable survey of advances in biomedical named entity recognition. In *Proceedings of the Pacific Symposium on Biocomputing*, pages 652-663.

R. Leaman, C. Miller, and G. Gonzalez. 2009. Enabling recognition of diseases in biomedical text with machine learning: corpus and benchmark. In *2009 Symposium on Languages in Biology and Medicine.*

Z. Lu, H. Y. Kao, C. H. Wei, M. Huang, J. Liu, C. J. Kuo, C. N. Hsu, R. T. Tsai, H. J. Dai, N. Okazaki, H. C. Cho, M. Gerner, I. Solt, S. Agarwal, F. Liu, D. Vishnyakova, P. Ruch, M. Romacker, F. Rinaldi, S. Bhattacharya, P. Srinivasan, H. Liu, M. Torii, S. Matos, D. Campos, K. Verspoor, K. M. Livingston, and W. J. Wilbur. 2011. The gene normalization task in BioCreative III. *BMC Bioinformatics*, 12 Suppl 8:S2.

S. G. Nash and J. Nocedal. 1991. A numerical study of the limited memory BFGS method and the truncated-Newton method for large scale optimization. *SIAM Journal on Optimization*, 1:358-372.

P. Pantel and D. Lin. 2002. Discovering word senses from text. In *Proceedings of the Eighth ACM SIGKDD International Conference on Knowledge Discovery and Data Mining*, pages 613-619.

L. Smith, L. K. Tanabe, R. J. Ando, C. J. Kuo, I. F. Chung, C. N. Hsu, Y. S. Lin, R. Klinger, C. M. Friedrich, K. Ganchev, M. Torii, H. Liu, B. Haddow, C. A. Struble, R. J. Povinelli, A. Vlachos, W. A. Baumgartner, Jr., L. Hunter, B. Carpenter, R. T. Tsai, H. J. Dai, F. Liu, Y. Chen, C. Sun, S. Katrenko, P. Adriaans, C. Blaschke, R. Torres, M. Neves, P. Nakov, A. Divoli, M. Mana-Lopez, J. Mata, and W. J. Wilbur. 2008. Overview of BioCreative II gene mention recognition. *Genome Biol*, 9 Suppl 2:S2.

L. Tanabe, L. H. Thom, W. Matten, D. C. Comeau, and W. J. Wilbur. 2006. SemCat: semantically categorized entities for genomics. In *AMIA Annu Symp Proc*, pages 754-758.

L. Tanabe and W. J. Wilbur. 2006. A priority model for named entities. In *Proceedings of the Workshop on Linking Natural Language Processing and Biology: Towards Deeper Biological Literature Analysis*, pages 33-40.

L. Tanabe, N. Xie, L. H. Thom, W. Matten, and W. J. Wilbur. 2005. GENETAG: a tagged corpus for gene/protein named entity recognition. *BMC Bioinformatics*, 6 Suppl 1:S3.

P. D. Turney and P. Pantel. 2010. From frequency to meaning: vector space models of semantics. *Journal of Artificial Intelligence Research*, 37:141-188.

Z. Yang, H. Lin, and Y. Li. 2008. Exploiting the contextual cues for bio-entity name recognition in biomedical literature. *J Biomed Inform*, 41:580-587.

Effect of small sample size on text categorization with support vector machines

Paweł Matykiewicz
Biomedical Informatics
Cincinnati Children's Hospital
3333 Burnet Ave
Cincinnat, OH 45220, USA
pawel.matykiewicz@gmail.com

John Pestian
Biomedical Informatics
Cincinnati Children's Hospital
3333 Burnet Ave
Cincinnat, OH 45220, USA
john.pestian@cchmc.org

Abstract

Datasets that answer difficult clinical questions are expensive in part due to the need for medical expertise and patient informed consent. We investigate the effect of small sample size on the performance of a text categorization algorithm. We show how to determine whether the dataset is large enough to train support vector machines. Since it is not possible to cover all aspects of sample size calculation in one manuscript, we focus on how certain types of data relate to certain properties of support vector machines. We show that normal vectors of decision hyperplanes can be used for assessing reliability and internal cross-validation can be used for assessing stability of small sample data.

1 Introduction

Every patient visit generates data, some on paper, some stored in databases as structured form fields, some as free text. Regardless of how they are stored, all such data are to be used strictly for patient care and for billing, not for research. Patient health records are maintained securely according to the provisions of the Health Insurance Portability and Accountability Act (HIPAA). Investigators must obtain informed consent from patients whose data will be used for other purposes. This means defining which data will be used and how they will be used. In addition to writing protocols and obtaining consent from patients, medical experts must either manually codify important information or teach a machine how to do it. All of these labor-intensive tasks are expensive. No one wants to collect more data than is necessary.

Our research focuses on answering difficult neuropsychiatric questions such as, "Who is at higher risk of dying by suicide?" or "Who is a good candidate for epilepsy surgery evaluation?" Large amounts of data that might answer these questions exist in the form of text dictated by clinicians or written by patients and thus unavailable. Parallel to the collection of such data, we explored whether small datasets can be used to build reliable methods of making this information available. Here, we investigate how text classification training size relates to certain aspects of linear support vector machines. We hypothesize that *a sufficiently large training subset will generate stable and reliable performance estimates of a classifier*. On the other hand, *if the dataset is too small, then even small changes to the training size will change the performance of a classifier and manifest unstable and unreliable estimates*. We introduce quantitive definitions for stability and reliability and give empirical evidence on how they work.

2 Background

How much data is needed for reliable and stable analysis? This question has been answered for most univariate problems, and a few solutions exist for multivariate problems, but no widely accepted answer is available for sparse and high-dimensional data. Nonetheless, we will review the few sample size calculation methods that have been used for machine learning.

Hsieh et al. (1998) described a method for calculating the sample size needed for logistic and linear regression models. The multivariate problem was simplified to a series of univariate two-sample t-tests on the input variables. A variance inflation factor was used to correct for the multi-dimensionality which quantifies the severity of multicollinearity in the least squares regression: collinearity deflates and non-collinearity inflates sample size estimation. Computer simulations were done on low-dimensional and continuous data, so it is not known whether the method is applicable to text categorization.

Guyon et al. (1998) addressed the problem of determining what size test set guarantees statistically significant results in a character recognition task, as a function of the expected error rate. This method does not assume which learner will be used. Instead, it requires specific parameters that describe hand-writing data collection properties such as between-writers variance and within-writer variance. The downside of this method is that it must assume the worst-case scenario: a large variance in data and a low error rate for the classifier. For this reason larger datasets are recommended.

Dobbin et al. (2008) and Jianhua Hu (2005) focused only on sample size for a classifier that learns from gene expression data. No assumptions were made about the classifier, only about the data structure. All gene expressions were measured on a continuous scale that denotes some luminescence corresponding to the relative abundance of nucleic acid sequences in the target DNA strand. The data, regardless of size, can be qualified using just one parameter, fold change, which measures changes in the expression level of a gene under two different conditions. Furthermore, the fold change can be standardized for compatibility with other biological experiments: with a lower standardized fold change, more samples are needed, and with more genes, more samples are needed. There is a strong assumption about data makeup, but no assumption is made about the classifier. This solution allows for small sample sizes but does not generalize to text classification data.

Way et al. (2010) evaluated the performance of various classifiers and featured a selection technique in the presence of different training sample sizes. Experiments were conducted on synthetic data, with two classes drawn from multivariate Gaussian distributions with unequal means and either equal or unequal covariance matrices. The conclusion was that support vector machines with a radial kernel performed slightly better than the LDA when the training sample size was small. Only certain combinations of feature selection and classification methods work well with small sample sizes. We will use similar assumptions for sparse and high-dimensional data.

Most recently, Juckett (2012) developed a method for determining the number of documents needed for a gold standard corpus. The sample size calculation was based on the concept of capture probabilities. It is defined as the normalized sum of probabilities over all words of interest. For example, if the required capture probability is 0.95 for a set of medical words, when using larger corpora that contain these words, it must first be calculated how many documents are needed to capture the same probability in the target corpus. This method is specific to linguistic research on annotated corpora, where the probabilities of individual words in the sought corpora must match the probabilities of words in the target domain. This method focuses solely on the data structure and does not assume an algorithm or the task that it will serve. The downside is a higher sample size.

When reviewing various methods for sample size calculation, we found that as more assumptions can be made, fewer data are needed for meaningful analysis. Assumptions can be made about data structure and quality, the task the data serve, feature selection, and the classifier. Our approach exploits a scenario where the task, the feature selection, and the classifier are known.

3 Data

We used four data sets to test our hypothesis: versicolor and virginica samples from the Iris dataset (**VV**), newswires about corn and wheat from the ModApte split of the Reuters-21578 dataset (**WCT** and **WCE**), suicide notes reprinted in Shneidman and Farberow (1957) (**SN**), and ubiquitous questionnaire patient interviews (**UQ**). Properties of these data are summarized in Table 1.

The first dataset was created by Anderson (1935) and introduced to the world of statistics by Fisher (1936). Since then it has been used on countless occasions to benchmark machine learning algorithms. Each row of data has four variables to describe the shape of an iris calyx: sepal length, sepal width, petal length, and petal width. The dataset contains 50 measurements for each of three subspecies of the iris flower: setosa, versicolor, and virginica. All measurements of the setosa calyx are separable from the rest of the data and thus were not used in our experiments. Instead, we used data corresponding to versicolor and virginica (**VV**), which is more interesting because of a small class overlap. The noise is introduced mostly by sepal width and sepal length.

The second dataset was created by Lewis and Ringuette (1994) and is the one most commonly used to benchmark text classification algorithms. The collection is composed of 21,578 short news stories from the Reuters news agency. Some stories have manually assigned topics, like "earn," "acq," or "money-fx," and others do not. In order to make the dataset comparable across different uses, a "Modified Apte" ("ModApte") split was proposed by Apté et al. (1994). It has 9,603 training and 3,299 external testing documents, a total of 135 distinct topics, with at least one topic per document. The most frequent topic is "earn," which appears in 3,964 documents. Here, we used only the "wheat" and "corn" categories, which appear 212 and 181 times in the training set along with 71 and 56 cases in the test set. These topics are semantically related, so it is no surprise that 59 documents in the training set and 22 documents in test set have both labels. This gives a total of 335 unique training instances and 105 unique test instances. Interestingly, it is easier to distinguish "corn" news from "not corn just wheat" news than it is to distinguish "wheat" from "not wheat just corn." The latter seems to be a good dataset for benchmarking sample size calculation. We will refer to the "wheat" versus "not wheat" training set as **WCT** and the "wheat" versus "not wheat" external test set as **WCE**.

The third dataset was extracted from the appendix in Shneidman and Farberow (1957). It contains 66 suicide notes (**SN**) organized into two categories: 33 genuine and 33 simulated. The authors of the notes were matched in both groups by gender (male), race (white), religion (Protestant), nationality (native-born U.S. citizens), and age (25-59). Authors of the simulated suicide notes were screened for personality disorders or tendencies toward morbid thoughts that would exclude them from the study. Individuals enrolled in the study were asked to write a suicide note as if they were going to take their own life. Notes were anonymized, digitized, and prepared for text processing (Pestian et al., 2010).

The fourth dataset was collected in a clinical controlled trial at Cincinnati Children's Hospital Medical Center Emergency Department. Sixty patients were enrolled, 30 with suicidal behavior and 30 controls from the orthopedic service. The suicidal behavior group comprised 15 females and 15 males with an average age of $\approx$ 15.7 years (SD $\approx$ 1.15). The control group included 15 females and 15 males with an average age of $\approx$ 14.3 years (SD $\approx$ 1.21). The interview consisted of five open-ended ubiquitous questions (**UQ**): "Does it hurt emotionally?" "Do you have any fear?" "Are you angry?" "Do you have any secrets?" and "Do you have hope?" The interviews were recorded in an audio format, transcribed by a medical transcriptionist, and prepared for analysis by removing the sections of the interview where the questions were asked. To preserve the **UQ** structure, n-grams from each of the five questions were separated (Pestian et al., 2012).

	VV	SN	UQ	WCT	WCE
Samples (m)	100	66	60	335	105
Classes	2	2	2	2	2
Class balance	100%	100%	100%	58%	48%
Min row freq	100	2	2	3	0
Max row freq	100	66	60	335	105
Min cell value	1	0	0	0	0
Max cell value	7.9	102.045	64	117	892
Features (n)	4	60	7,282	7,132	7,132
Sparsity	0%	60%	92.3%	97%	98%

Table 1: Four very different benchmark data: versicolor and virginica (**VV**) from iris data, representing a dense, low-dimensional dataset; suicide notes (**SN**) from *Clues to Suicide* (Shneidman and Farberow, 1957), representing a mildly sparse, high-dimensional dataset; ubiquitous questionnaires, (**UQ**) representing a sparse, extremely high-dimensional dataset; and "wheat" versus "not wheat just corn" (**WCT** and **WCE**) from the "ModApte" split of Reuters-21578 data, representing an unbalanced, extremely sparse, high-dimensional dataset.

4 Methods

Feature extraction. Every text classification algorithm starts with feature engineering. Documents in the **UQ**, **WCT**, and **WCE** sets were represented by a bag-of-n-grams model (Manning and Schuetze, 1999; Manning et al., 2008). Every document was tokenized, and frequencies of unigrams, bigrams, and trigrams were calculated. All digit numbers that appeared in a document were converted to the same token ("NUMB"). Documents become row vectors and n-grams become column vectors in a large sparse matrix. Each n-gram has its own dimension, with the exception of **UQ** data, where n-grams are represented separately for each of the five questions. Neither stemming nor a stop word list were applied to the textual data. Suicide notes (**SN**) were not represented by n-grams. In previous studies, we found that the structure of the note and its emotional content are indicative of suicidality, not its semantic content. Hence, the **SN** dataset is represented by the frequency of 23 emotions assigned by mental health professionals, the frequency of 34 parts of speech, and by three readability scores: Flesch, Fog, and Kincaid.

Feature weighting. Term weighting was chosen *ad hoc*. **UQ**, **WCT**, and **WCE** had a logarithmic term frequency (log-tf) as local weighting and an inverse document frequency (idf) as global weighting but were derived only from the training data (Salton and Buckley, 1988; Nakov et al., 2001).

Feature selection. To speed up calculations, the least frequent features were removed from the **SN**, **UQ**, **WCT**, and **WCE** datasets (see minimum row frequency in Table 1). Further optimization of the feature space was done using an information gain filter (Guyon and Elisseeff, 2003; Yang and Pedersen, 1997). Depending on the experiment, some of the features with the lowest information gain were removed. For example, $IG = 0.4$ means that 40% of the features, those with a higher information gain, were kept, and the other 60%, those with a lower information gain, were removed. Lastly, all row vectors in **UQ**, **WCT**, and **WCE** were normalized to unit length (Joachims, 1998).

Learning algorithm. We used linear support vector machines (**SVM**) to learn from the data. Support vector machines are described in great detail in Schlkopf and Smola (2001). We will focus on just two aspects: properties of the normal vector of decision hyperplane (see Figure 1) and internal cross-validation (see Figure 2). **SVM** is in essence a simple linear classifier:

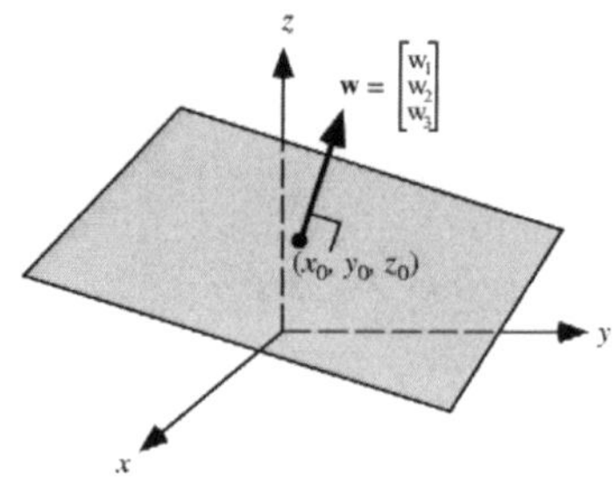

Figure 1: Normal vector **w** of a hyperplane.

$$f(\mathbf{x}) = \mathrm{sgn}(\langle \mathbf{w}, \mathbf{x} \rangle + b) \qquad (1)$$

where $\mathbf{x}$ is an input vector that needs to be classified, $\langle \cdot, \cdot \rangle$ is the inner product, $\mathbf{w}$ is a weight vector with the same dimensionality as $\mathbf{x}$, and b is a scalar. The function f outputs $+1$ if $\mathbf{x}$ belongs to the first class or -1 if $\mathbf{x}$ belongs to the second class. **SVM** differs from other linear classifiers on how $\mathbf{w}$ is computed. Contrary to other classifiers, it does not solve $\mathbf{w}$ directly. Instead, it uses convex optimization to find vectors from the training set that can be used for creating the largest margin between training examples from the first and second class. Hence, the solution to $\mathbf{w}$ is in the form of the linear combination of coefficients and training vectors:

$$\mathbf{w} = \sum_{i=1}^{m} \alpha_i y_i \mathbf{x}_i \qquad (2)$$

where m is the number of training vectors, $\alpha_i \geq 0$ are Lagrange multipliers, $y_i \in \{-1, 1\}$ are numerical codes for class labels, and $\mathbf{x}_i$ are training row vectors. Vector $\mathbf{w}$ is perpendicular to the decision boundary, and its proper name in the context of **SVM** is the normal vector of decision hyperplane[1] (see Figure 1). One of the properties of **SVM** is that outlying training vectors are not used in $\mathbf{w}$. These vectors have the corresponding coefficient $\alpha_i = 0$. In fact, these vectors can be removed from the training set and the convex optimization procedure will

[1] If *R* with **SVM** from the *e1071* package is used, the command to obtain the normal vector is `w = c(t(model$coefs)%*%model$SV)`.

result in exactly the same solution. We can use this property to probe how reliable training data are for the classification task. If we have enough data that we can randomly remove some, what is left will result in $\mathbf{w}^* \approx \mathbf{w}$. On the other hand, if we do not have enough data, then random removal of training data will result in a very different equation, because the decision boundary changes and $\mathbf{w}^* \neq \mathbf{w}$.

Reliability of performance. The relationship between $\mathbf{w}^*$ and $\mathbf{w}$ can be measured. We introduce the **SVM** reliability index (**SRI**):

$$\mathrm{SRI}(\mathbf{w}^*, \mathbf{w}) = |r(\mathbf{w}^*, \mathbf{w})| \tag{3}$$
$$= \frac{|\sum_{i=1}^{n}(w_i^* - \overline{\mathbf{w}}^*)(w_i - \overline{\mathbf{w}})|}{\sqrt{\sum_{i=1}^{n}(w_i^* - \overline{\mathbf{w}}^*)^2}\sqrt{\sum_{i=1}^{n}(w_i - \overline{\mathbf{w}})^2}}$$

which is the absolute value of the Pearson product-moment correlation coefficient between convex optimization solution $\mathbf{w}^*$ corresponding to a training subset and $\mathbf{w}$ corresponding to the full dataset[2]. Pearson's correlation coefficient discovers linear dependency between two normally distributed random variables and has its domain on a continuous segment between -1 and $+1$. In our case, we are looking for a strong linear dependency between constituents of the training weight vector w_i^* and constituents of the full dataset weight vector w_i. Some numerical implementations of **SVM** cause the output values for the class labels to switch. We corrected for this effect by applying absolute value to the Pearson's coefficient, resulting in $\mathrm{SRI} \in [0, 1]$. We did not have a formal proof on how **SRI** relates to **SVM** performance. Instead, we showed empirical evidence for the relationship based on a few small benchmark data. **Stability of performance.** **SVM** generalization performance is usually measured using cross-validation accuracy. In particular, we use balanced accuracy because it gives better evidence for a drop in performance when solving unbalanced problems. Following Guyon and Elisseeff (2003) and many others, we divided the data into three sets: test, training, and validation. Mean test balanced accuracy $\overline{a}^T$ is estimated using stratified Monte Carlo cross-validation (**MCCV**), where

[2]We experimented with Pearson's correlation, Spearman's correlation, one-way intraclass correlation, Cosine correlation, Cronbach's coefficient, and Krippendorff's coefficients and found that Pearson's correlation coefficient works well with both low-dimensional and high-dimensional spaces.

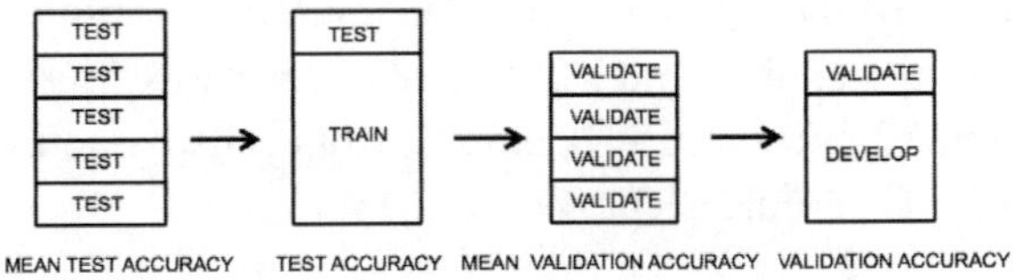

Figure 2: Estimation and resampling: mean test balanced accuracy and mean validation balanced accuracy should match. To prevent overfitting, tuning machine learning should be guided by mean validation accuracy and confirmed by mean test accuracy. This procedure requires the "develop" set to be large enough to give reliable and stable estimates.

the proportion of the training set to the test set is varied between 0.06 and 0.99. Mean validation balanced accuracy $\overline{a}^V$ (**MVA**) is estimated using K-fold cross-validation (also known as internal cross-validation), where $K = \frac{m}{2}$ and m is the number of training cases. In the case of the "wheat" versus "not wheat just corn" dataset, we have, in addition, the external validation set **WCE** and corresponding mean external balanced accuracy $\overline{a}^E$. Correct estimation of the learner's generalization performance should result in all three accuracies being equal: $\overline{a}^T \approx \overline{a}^V \approx \overline{a}^E$. Furthermore, we want all three accuracies to be the same regardless of the amount of data. If we have enough data that we can randomly remove some, what is left will result in $\overline{a}^{V^*} \approx \overline{a}^{V^{**}}$. On the other hand, if we do not have enough data, then random removal of training data will result in very different accuracy estimations: $\overline{a}^{V^*} \neq \overline{a}^{V^{**}}$.

Sample size calculation. We do not have a good way of predicting how much data will be needed to solve a problem with a small p-value, but this is a matter of convenience. Rather than looking to the future, we can simply ask if what we have now is enough. *If we can build a classifier that gives reliable and stable estimates of performance, we can stop collecting data.* Reliability is measured by **SRI**, while stability is measured by **MVA**, not as a single value but merely as a function of the training size:

$$SRI(t) = |r(\mathbf{w}^{tm}, \mathbf{w}^m)| \quad \text{and} \tag{4}$$
$$a^T(t) = a^{T^{tm}} \tag{5}$$

where t is a proportion of the training data, $t \in (0, 1)$, m is size of the full dataset, and tm is the actual number of training instances. To quantify the ability of the dataset to produce classification models with reliable and stable performance estimates, we need two more measures: sample dispersion of **SRI** and sample dispersion of **MVA**:

$$c_{SRI}(t \geq p) = \frac{s_{SRI(t \geq p)}}{SRI(t \geq p)} \quad \text{and} \tag{6}$$

$$c_{MVA}(t \geq p) = \frac{s_{a^T(t \geq p)}}{a^T(t \geq p)} \tag{7}$$

defined as the coefficient of variation of all **SRI** or **MVA** measurements for training data sizes greater than $p\dot{m}$. For example, we want to know if our 10-fold cross-validation (**CV**) for a dataset that has 400 training samples is reliable and stable. 10-fold CV is 0.9 of training data, so we need to measure **SRI** and **MVA** for different proportions of training data, $t = \{0.90, 0.91, \ldots, 0.99\}$, and then calculate dispersion for $c_{SRI}(t \geq 0.9)$ and $c_{MVA}(t \geq 0.9)$. Numerical calculations will give us sense of good and bad dispersion across different datasets.

5 Results

Do I have enough data? The first set of experiments was done with untuned algorithms. We set the **SVM** parameter to $C = 1$ and did not use any feature selection. Figure 3 shows four examples of how **SVM** performance depends on the training set size. The performance was measured using mean test balanced accuracy, **MVA**, and **SRI**. Numerical calculations showed that **VV** needs at least 30 randomly selected training examples to produce reliable and stable results with high accuracy. $c_{SRI}(t \geq 0.75)$ is 0.005 and $c_{MVA}(t \geq 0.75)$ is 0.016. **SN** was not encouraging regarding the estimated accuracy; **SRI** dropped, suggesting that the **SVM** decision hyperplanes are unreliable. Mental health professionals can distinguish between genuine and simulated notes about 63% of time. Machine learning does it correctly about 73% of time if text structure and emotional content are used. Even so, the sample size calculation yields high dispersion ($c_{SRI}(t \geq 0.75) = 0.134$ and $c_{MVA}(t \geq 0.75) = 0.082$). **UQ** is small and high-dimensional, and yet the results were reliable and stable ($c_{SRI}(t \geq 0.75) = 0.015$ and $c_{MVA}(t \geq 0.75) = 0.023$). Patients enrolled in the **UQ** study also received the Suicide Ideation Questionnaire (Raynolds, 1987) and the Columbia-Suicide Severity Rating Scale (Posner et al., 2011). We found that **UQ** was no different from the structured questionnaires. **UQ** detects suicidality mostly by emotional pain and hopelessness, which were mildly present in four control patients. Other instruments returned errors because the same few teenagers reported risky behavior and morbid thoughts. **WCT** produced reliable and stable accuracy estimates, but no large amounts of data could be removed ($c_{SRI}(t \geq 0.75) = 0.010$ and $c_{MVA}(t \geq 0.75) = 0.053$). It seems that **WCE** is somehow different from **WCT**, or it might be a case of overfitting, which causes the mean test accuracy to diverge from **MVA** as the training dataset gets smaller. **Algorithm tuning.** No results should be regarded as satisfactory until a thorough parameter space search has been completed. Each step of a text classification algorithm can be improved. To attempt a complete description of the dependency of a minimal viable sample size on text classification would be both impossible and futile, since new methods are discovered every day. However, to start somewhere, we focused only on the feature selection and **SVM** parameter C [3]. Feature selection removes noise from data. Parameter C informs the convex optimization process about the expected noise level. If both parameters are set correctly, we should see an improvement in the reliability and stability of the results. There are several methods for tuning **SVM**; the most commonly used but computationally expensive is internal cross-validation (Duan et al., 2003; Chapelle et al., 2002). Figure 5 shows the results of the parameter tuning procedure. **VV** and **SN** are not extremely high-dimensional, so we tuned just parameter C. **MVA** maxima were found at $C = 0.45$ with **VV**, $C = 0.05$ with **SN**, $C = 0.4$ and $IG = 0.1584$ with **UQ**, and $C = 2.5$ and $IG = 0.8020$ with **WCT**. **Do I have enough data after algorithm tuning?** Internal cross-validation (**MVA**) did not improve dispersion universally (see Table 2). **VV** improved on reliability but not stability. **SN** scored much better on both measures, but we do not yet know what the cutoff for having a low enough dispersion is. **UQ** did worse on all measures after tuning. **WCT** improved greatly on mean

[3] Please note that most **SVM** implementations do not allow for simultaneous feature selection and internal cross-validation.

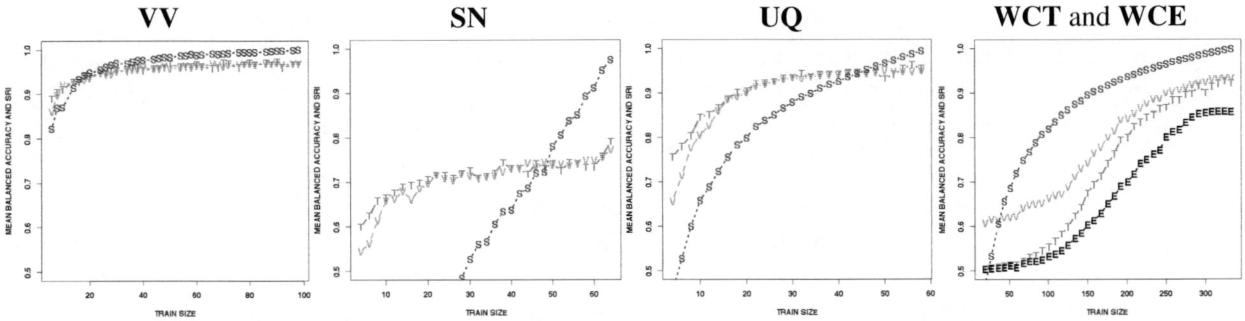

Figure 3: **SRI** index (**S**), **MVA** accuracy (**V**) and mean test accuracy (**T**) averaged over 120 repetitions and different training data sizes. Linear **SVM** with $C = 1$ and no feature selection. **VV** ($c_{SRI}(t \geq 0.75) = 0.005$ and $c_{MVA}(t \geq 0.75) = 0.016$), **UQ** ($c_{SRI}(t \geq 0.75) = 0.015$ and $c_{MVA}(t \geq 0.75) = 0.023$), and **WCT** ($c_{SRI}(t \geq 0.75) = 0.010$ and $c_{MVA}(t \geq 0.75) = 0.053$) gave stable and reliable estimates, but **SN** did not ($c_{SRI}(t \geq 0.75) = 0.134$ and $c_{MVA}(t \geq 0.75) = 0.082$).

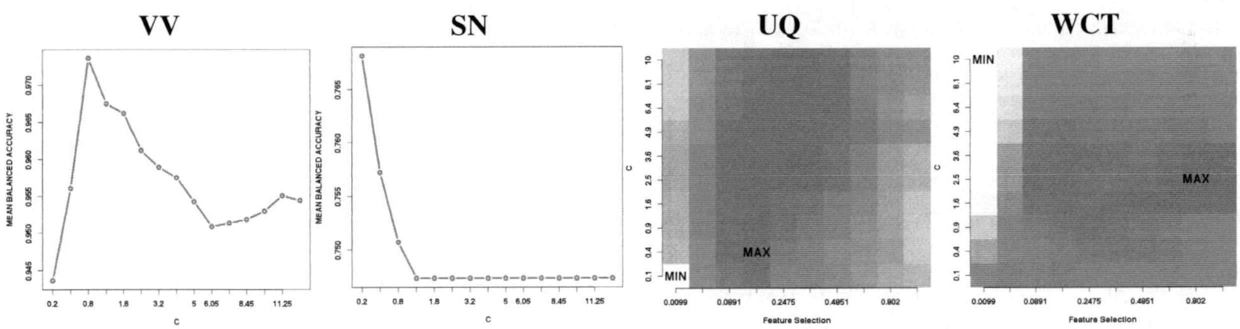

Figure 4: **MVA** (internal cross-validation) parameter tuning results. Maxima were found at $C = 0.45$ with **VV**, $C = 0.05$ with **SN**, $C = 0.4$ and $IG = 0.1584$ with **UQ**, and $C = 2.5$ and $IG = 0.8020$ with **WCT**.

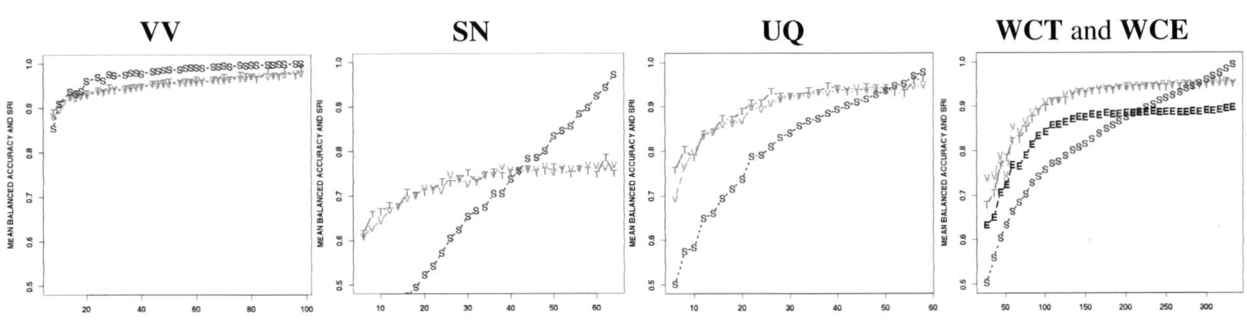

Figure 5: **SRI** index (**S**), **MVA** accuracy (**V**), and mean test accuracy (**T**) averaged over 60 repetitions and different training data sizes. Tuned classification algorithms: **VV** with $C = 0.45$ and no feature selection, **SN** with $C = 0.05$ and no feature selection, **UQ** with $C = 0.4$ and $IG = 0.1584$, and **WCT** with $C = 2.5$ and $IG = 0.8020$. Stability and reliability: **VV** had $c_{SRI}(t \geq 0.75) = 0.003$ and $c_{MVA}(t \geq 0.75) = 0.018$), **SN** had $c_{SRI}(t \geq 0.75) = 0.085$ and $c_{MVA}(t \geq 0.75) = 0.075$, **UQ** had $c_{SRI}(t \geq 0.75) = 0.025$ and $c_{MVA}(t \geq 0.75) = 0.024$, and **WCT** had $c_{SRI}(t \geq 0.75) = 0.025$ and $c_{MVA}(t \geq 0.75) = 0.011$.

test accuracy, mean external validation, and stability dispersion (see Figure 5). It would be interesting to see if improvement on both reliability dispersion and stability dispersion would bring mean test accuracy and mean external validation even closer together.

	$\overline{a^T(t \geq 0.75)}$	$c_{SRI}(t \geq 0.75)$	$c_{MVA}(t \geq 0.75)$
VV no tuning	0.965	0.005	**0.016**
SN no tuning	0.744	0.134	0.082
UQ no tuning	**0.946**	**0.015**	**0.023**
WCT no tuning	0.862	**0.010**	0.053
VV with tuning	**0.970**	**0.003**	0.018
SN with tuning	**0.755**	**0.085**	**0.075**
UQ with tuning	0.941	0.025	0.024
WCT with tuning	**0.946**	0.025	**0.011**

Table 2: Sample size calculation before and after tuning with internal cross-validation (**MVA**). Even though mean test accuracy ($\overline{a^T(t \geq 0.75)}$) improved for **VV**, **SN**, and **WCT**, reliability and stability did not improve universally. Internal cross-validation alone might not be adequate for tuning classification algorithms for all data.

6 Discussion

Sample size calculation data for a competition and for problem-solving. In general, there might be two conflicting objectives when calculating whether what we have collected is a large enough dataset. If the objective is to have a shared task with many participants and, thus, many unknowns, the best course of action is to assume the weakest classifier: unigrams with no feature weighting or selection trained using the simplest logistic regression. On the other hand, if the problem is to be solved with only one classifier and the least amount of data, then the strongest assumptions about the data and the algorithm are required.

The fallacy of untuned algorithms. After years of working with classification algorithms to solve difficult patient care problems, we have found that a large amount of data is not needed; usually samples measured in the hundreds will suffice, but this is only possible when a thorough parameter space search is conducted. It seems that reliability and stability dispersions are good measures of how well the algorithm is tuned to the data without overfitting. Moreover, we now have a new direction for thinking about optimizing classification algorithms: instead of focusing solely on accuracy, we can also measure the dispersion and see whether this is a better indicator of what would happen with unevaluated data. There is a great deal of data available, but very little that can be used for training.

What to measure? VC-bound, span-bound, accuracy, F_1, reliability, and stability dispersions are just a few examples of indicators of how well our models fit. What we have outlined here is how one of the many properties of **SVM**, the property of the normal vector, can be used to obtain insights into data. Normal vectors are constructed using Lagrangian multipliers and support vectors; accuracy is constructed using a sign function on decision values. It is feasible that other parts of **SVM** may be more suited to algorithm tuning and calculation of minimum viable training size.

7 Conclusion

Power and sample size calculations are very important in any domain that requires extensive expertise. We do not want to collect more data than necessary. There is, however, a scarcity of research in sample size calculation for machine learning. Nonetheless, the existing results are consistent: the more that can be assumed about the data, the problem and the algorithm, the fewer data are needed.

We proposed two independent measures for evaluating whether available datasets are sufficiently large: reliability and stability dispersions. Reliability dispersion measures indirectly whether the decision hyperplane is always similar and how much it varies, while stability dispersion measures how well we are generalizing and how much variability there is. If the sample size is large enough, we should always get the same decision hyperplane with the same generalization accuracy.

With little empirical evidence, we can conclude that classifier performance measured by just a single K in a cross-validation test is not sufficient. K must be be varied, and other measures must be present, such as the SVM reliability index, that support or contradict the generalization accuracy estimates. We suggest that other measures for sample size calculation and algorithm tuning may exist and there is still much to be learned about the mechanics of support vector machines.

References

Edgar Anderson. 1935. The irises of the gaspe peninsula. *Bulletin of the American Iris Society*, 59:2–5.

Chidanand Apté, Fred Damerau, and Sholom M. Weiss. 1994. Automated learning of decision rules for text categorization. *ACM Trans. Inf. Syst.*, 12(3):233–251, July.

Olivier Chapelle, Vladimir Vapnik, Olivier Bousquet, and Sayan Mukherjee. 2002. Choosing multiple parameters for support vector machines. *Machine Learning*, 46:131–159.

Kevin K. Dobbin, Yingdong Zhao, and Richard M. Simon. 2008. How large a training set is needed to develop a classifier for microarray data? *Clinical cancer research : an official journal of the American Association for Cancer Research*, 14(1):108–114, January.

Kaibo Duan, S. Sathiya Keerthi, and Aun Neow Poo. 2003. Evaluation of simple performance measures for tuning svm hyperparameters. *Neurocomputing*, 51:41–59.

Ronald A. Fisher. 1936. The use of multiple measurements in taxonomic problems. *Annals of Eugenics*, 7:179–188.

Isabelle Guyon and Andre Elisseeff. 2003. An introduction to variable and feature selection. *J. Mach. Learn. Res.*, 3:1157–1182, March.

Isabelle Guyon, John Makhoul, Richard Schwartz, and Vladimir Vapnik. 1998. What size test set gives good error rate estimates? *Pattern Analysis and Machine Intelligence, IEEE Transactions on*, 20(1):52–64, January.

Fushing Y. Hsieh, Daniel A. Bloch, and Michael D. Larsen. 1998. A simple method of sample size calculation for linear and logistic regression. *Statistics in Medicine*, 17(14):1623–1634, December.

Fred A. Wright Jianhua Hu, Fei Zou. 2005. Practical fdr-based sample size calculations in microarray experiments. *Bioinformatics*, 21(15):3264–3272, August.

Thorsten Joachims. 1998. Text categorization with support vector machines: Learning with many relevant features. In Claire Ndellec and Cline Rouveirol, editors, *Machine Learning: ECML-98*, volume 1398, pages 137–142. Springer-Verlag, Berlin/Heidelberg.

David Juckett. 2012. A method for determining the number of documents needed for a gold standard corpus. *Journal of Biomedical Informatics*, page In Press, January.

David D. Lewis and Marc Ringuette. 1994. A comparison of two learning algorithms for text categorization. In *Third Annual Symposium on Document Analysis and Information Retrieval*, pages 81–93.

Christopher D. Manning and Hinrich Schuetze. 1999. *Foundations of Statistical Natural Language Processing*. The MIT Press, 1 edition, June.

Christopher D. Manning, Prabhakar Raghavan, and Hinrich Schtze. 2008. *Introduction to Information Retrieval*. Cambridge University Press, 1 edition, July.

Preslav Nakov, Antonia Popova, and Plamen Mateev. 2001. Weight functions impact on lsa performance. In *EuroConference RANLP'2001 (Recent Advances in NLP)*, pages 187–193.

John Pestian, Henry Nasrallah, Pawel Matykiewicz, Aurora Bennett, and Antoon Leenaars. 2010. Suicide note classification using natural language processing: A content analysis. *Biomedical Informatics Insights*, pages 19–28, August.

John Pestian, Jacqueline Grupp-Phelan, Pawel Matkiewicz, Linda Richey, Gabriel Meyers, Christina M. Canter, and Michael Sorter. 2012. Suicidal thought markers: A controlled trail examining the language of suicidal adolescents. *To Be Determined*, In Preparation.

Kelly Posner, Gregory K. Brown, Barbara Stanley, David A. Brent, Kseniya V. Yershova, Maria A. Oquendo, Glenn W. Currier, Glenn A. Melvin, Laurence Greenhill, Sa Shen, and J. John Mann. 2011. The ColumbiaSuicide severity rating scale: Initial validity and internal consistency findings from three multisite studies with adolescents and adults. *The American Journal of Psychiatry*, 168(12):1266–1277, December.

William M. Raynolds, 1987. *Suicidal Ideation Questionnaire - Junior*. Odessa, FL: Psychological Assessment Resources.

Gerard Salton and Christopher Buckley. 1988. Term-weighting approaches in automatic text retrieval. *Information Processing & Management*, 24(5):513–523.

Bernhard Schlkopf and Alexander J. Smola. 2001. *Learning with Kernels: Support Vector Machines, Regularization, Optimization, and Beyond*. The MIT Press, 1st edition, December.

Edwin S. Shneidman and Norman Farberow. 1957. *Clues to Suicide*. McGraw Hill Paperbacks.

Ted W. Way, Berkman Sahiner, Lubomir M. Hadjiiski, and Heang-Ping Chan. 2010. Effect of finite sample size on feature selection and classification: a simulation study. *Medical Physics*, 37(2):907–920, February.

Yiming Yang and Jan O. Pedersen. 1997. A comparative study on feature selection in text categorization. In *Proceedings of the Fourteenth International Conference on Machine Learning*, ICML '97, pages 412–420, San Francisco, CA, USA. Morgan Kaufmann Publishers Inc.

PubAnnotation - a persistent and sharable corpus and annotation repository

Jin-Dong Kim and **Yue Wang**
Database Center for Life Science (DBCLS),
Research Organization of Information and Systems (ROIS),
2-11-16, Yayoi, Bunkyo-ku, Tokyo, 113-0032, Japan
{jdkim|wang}@dbcls.rois.ac.jp

Abstract

There has been an active development of corpora and annotations in the BioNLP community. As those resources accumulate, a new issue arises about the reusability. As a solution to improve the reusability of corpora and annotations, we present *PubAnnotation*, a persistent and sharable repository, where various corpora and annotations can be stored together in a stable and comparable way. As a position paper, it explains the motivation and the core concepts of the repository and presents a prototype repository as a proof-of-concept.

1 Introduction

Corpora with high-quality annotation is regarded indispensable for the development of *language processing technology (LT)*, e.g. *natural language processing (NLP)* or *textmining*. Biology is one of the fields which have strong needs for LT, due to the high productivity of new information, most of which is published in literature. There have been thus an active development of corpora and annotations for the *NLP for biology (BioNLP)*. Those resources are certainly an invaluable asset of the community.

As those resources accumulate, however, a new issue arises about the reusability: the corpora and annotations need to be sharable and comparable. For example, there are a number of corpora that claim to have annotations for protein or gene names, e.g, Genia (Kim et al., 2003), Aimed (Bunescu et al., 2004), and Yapex (Franzén et al., 2002). To reuse them, a user needs to be able to compare them so that they can devise a strategy on how to use them. It is however known that often the annotations in different corpora are incompatible to each other (Wang et al., 2010): while one is considered as a protein name in a corpus, it may not be the case in another.

A comparison of annotations in different corpora could be made directly or indirectly. If there is an overlap between two corpora, a direct comparison of them would be possible. For example, there are one[1], two[2] and three[3] PubMed abstracts overlapped between Genia - Yapex, Genia - Aimed, and Yapex - Aimed corpora, respectively. When there is no or insufficient overlap, an indirect comparison could be tried (Wang et al., 2010). In any case, there are a number of problems that make it costly and troublesome, though not impossible, e.g. different formats, different ways of character encoding, and so on.

While there have been a few discussions about the reusability of corpora and annotations (Cohen et al., 2005; Johnson et al., 2007; Wang et al., 2010; Campos et al., 2012), as a new approach, we present *PubAnnotation*, a persistent and sharable storage or repository, where various corpora and annotations can be stored together in a stable and comparable way. In this position paper, after the motivation and background are explained in section 1, the initial design and a prototype implementation of the storage are presented in section 2 and 3, respectively and future works are discussed in section 4.

2 Design

Figure 1 illustrates the current situation of corpus annotation in the BioNLP community, which we consider problematic. In the community, there

[1]PMID-10357818

[2]PMID-8493578, PMID-8910398

[3]PMID-9144171, PMID-10318834, PMID-10713102

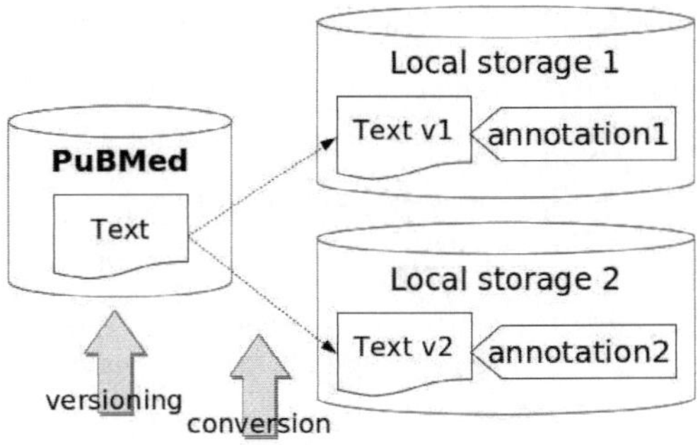

Figure 1: Usual setup of PubMed text annotation

are several central sources of texts, e.g. PubMed, PubMed Central (PMC), and so on. In this work, we consider only PubMed as the source of texts for brevity, but the same concept should be applicable to other sources. Texts from PubMed are mostly the title and abstract of literature indexed in PubMed. For an annotation project, text pieces from a source database (DB) are often copied in a local storage and annotations are attached to them.

Among others, the problem we focus on in this situation is the variations that are made to the texts. Suppose that there are two groups who happen to produce annotations to a same PubMed abstract. The abstract will be copied to the local storages of the two groups (illustrated as the local storage 1 and 2 in the figure). There are however at least two reasons that may cause the local copies to be different from the abstract in PubMed, and also to be different from each other even though they are copies of the same PubMed abstract:

Versioning This variation is made by PubMed. The text in PubMed is changed from time to time for correction, change of policy, and so on. For example, Greek letters, e.g., α, are spelled out, e.g., `alpha`, in old entries, but in recent entries they are encoded as they are in Unicode. For the reason, there is a chance that copies of the same entry made at different times (*snapshots*, hereafter) may be different from each other.

Conversion This variation is made by individual groups. The texts in a local storage are sometimes changed for local processing. For example, most of the currently available NLP tools (for English), e.g., POS taggers and parsers that are developed based on Penn Treebank, cannot treat Unicode characters appropriately. For such NLP tools to be used, all the Unicode characters need to be converted to ASCII character sequences in local copies. Sometimes, the result of some pre-processing, e.g. tokenization, also remains in local copies.

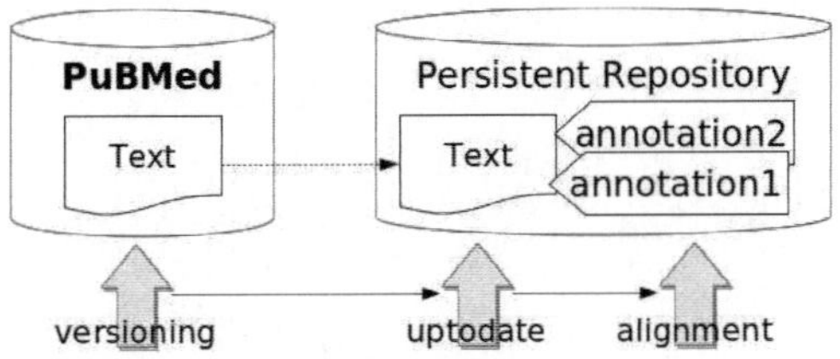

Figure 2: Persistent text/annotation repository

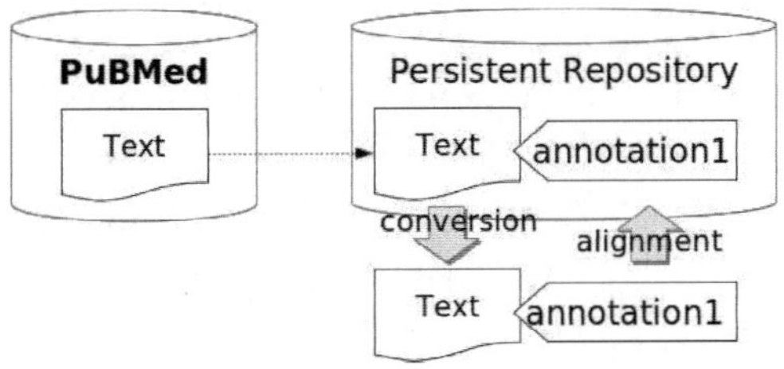

Figure 3: Text/annotation alignment for integration

The problem of text variation may not be such a problem that makes the reuse of corpora and annotations extremely difficult, but a problem that makes it troublesome, raising the cost of the entire community substantially.

To remedy the problem, we present, a persistent and sharable storage of corpora and annotations, which we call *PubAnnotation*. Figure 2 illustrates an improved situation we aim at with PubAnnotation. The key idea is to maintain all the texts in PubAnnotation in their canonical form, to which all the corresponding annotations are to be aligned. For texts from PubMed, the canonical form is defined to be exactly the same as in PubMed. With the definition, a text entry in PubAnnotation needs to be updated (*uptodate* in the figure) as the corresponding text in PubMed changes (*versioning*). Accordingly, the annotations belonging to the entry also need to be re-aligned (*alignment*).

There also would be a situation where a variation of a text entry is required for some reason, e.g. for

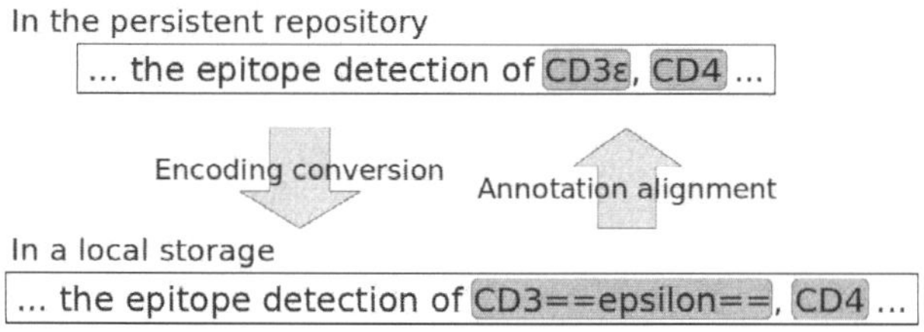

Figure 4: Text/annotation alignment example

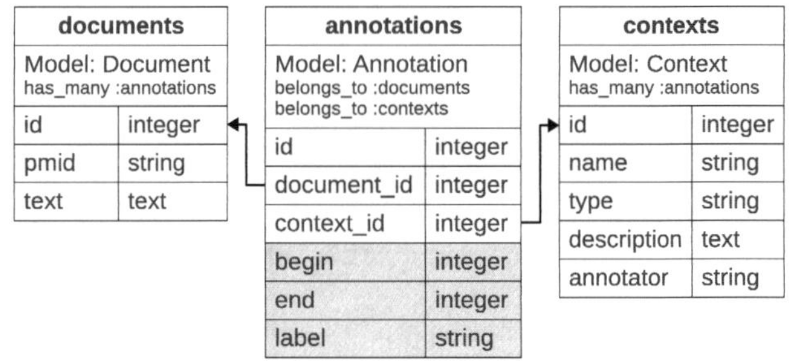

Figure 5: DB schema of persistent annotation repository

application of an NLP tool that cannot handle Unicode characters. Figure 3 illustrates a required process to cope with such a situation: first, the text is exported in a desired form (*conversion* in the figure); second, annotations are made to the text; and third, the annotations are aligned back to the text in its canonical form in the repository.

Figure 4 shows an example of text conversion and annotation alignment that are required when the Enju parser (Miyao and Tsujii, 2008) needs to be used for the annotation of protein names. The example text includes a Greek letter, ε, which Enju cannot properly handle. As Enju expects Greek letters to be spelled out with double equal signs on both sides, the example text is converted as so when it is exported into a local storage. Based on the pre-processing by Enju, the two text spans, `CD==epsilon==` and `CD4`, are annotated as protein names. When they are imported back to PubAnnotation, the annotations are re-aligned to the canonical text in the repository. In this way, the texts and annotations can be maintained in their canonical form and in alignment respectively in PubAnnotation. In the same way, existing annotations, e.g. Genia, Aimed, Yapex, may be imported in the repository, as far as their base texts are sufficiently similar to the canonical entries so that they can be aligned reliably. In this way, various existing annotations may be integrated in the repository,

To enable all the processes described so far, any two versions of the same text need to be aligned, so that the places of change can be detected. Text alignment is therefore a key technology of PubAnnotation. In our implementation of the prototype repository, the Hunt-McIlroy's longest common subsequence (LCS) algorithm (Hunt and McIlroy, 1976), as implemented in the `diff-lcs` ruby gem package, is used for the alignment.

3 Prototype implementation

As a proof-of-concept, a prototype repository has been implemented. One aspect considered seriously is the *scalability*, as repository is intended to be "persistent". Therefore it is implemented on a relational database (Ruby on Rails with PostgreSQL 9.1.3), instead of relying on a plain file system.

Figure 5 shows the database schema of the repository.[4] Three tables are created for *documents*, *annotations*, and (annotation) *contexts*, respectively. The annotations are stored in a stand-off style, each of which belongs to a *document* and also to an *annotation context* (context, hereafter). A context represents a set of annotations sharing the same set of meta-data, e.g., the type of annotation and the annotator. For brevity, we only considered PubMed as the source DB, and named entity recognition (NER)-type annotations, which may be simply represented by the attributes, `begin`, `end`, and `label`.

The prototype repository provides a RESTful interface. Table 1 shows some example which can be accessed with the standard HTTP GET method. A new entry can be created in the repository using a HTTP POST method with data in JSON format. Figure 6 shows an example of JSON data for the creation of annotations in the repository. Note that, the base text of the annotations needs to be passed together with the annotations, so that the text can be compared to the canonical one in the repository. If a difference is detected, the repository will try to align the annotations to the text in the repository.

[4]Although not shown in the figure, all the records are stored with the date of creation.

`http://server_url/pmid/8493578` to retrieve the document record of a specific PMID
`http://server_url/pmid/8493578.ascii` same as above, but in US-ASCII encoding (Unicode characters are converted to HTML entities).
`http://server_url/pmid/8493578/annotations` to retrieve all the annotations to the specific document.
`http://server_url/pmid/8493578/contexts` to retrieve all the annotation contexts created to the specific document.
`http://server_url/pmid/8493578/annotations?context=genia-protein` to retrieve all the annotations that belong to genia-protein context.
`http://server_url/pmid/8493578/annotations.json?context=genia-protein` the same as above, but in JSON format.

Table 1: Examples of RESTful interface of the prototype repository

```
{
 "document":
    {"pmid":"8493578",
     "text":"Regulation ..."},
 "context":
    {"name":"genia-protein"},
 "annotations":
    [
      {"begin":51,"end":56,
       "label":"Protein",
      {"begin":75,"end":97,
       "label":"Protein",
    ]
}
```

Figure 6: The JSON-encoded data for the creation of two protein annotations to the document of PMID:8493578.

4 Discussions and conclusions

The current state of the design and the prototype implementation are largely incomplete, and there is a much room for improvement. For example, the database schema has to be further developed to store texts from various source DBs, e.g., PMC, and to represent various types of annotations, e.g., relations and events. The issue of governance is yet to be discussed. We, however, hope the core concepts presented in this position paper to facilitate discussions and collaborations of the community and the remaining issues to be addressed in near future.

Acknowledgments

This work was supported by the "Integrated Database Project" funded by the Ministry of Education, Culture, Sports, Science and Technology of Japan.

References

Razvan Bunescu, Ruifang Ge, Rohit J. Kate, Edward M. Marcotte, Raymond J. Mooney, Arun K. Ramani, and Yuk Wah Wong. 2004. Comparative experiments on learning information extractors for proteins and their interactions. *Artificial Intelligence in Medicine*, 33(2):139–155.

David Campos, Srgio Matos, Ian Lewin, Jos Lus Oliveira, and Dietrich Rebholz-Schuhmann. 2012. Harmonization of gene/protein annotations: towards a gold standard medline. *Bioinformatics*, 28(9):1253–1261.

K. Bretonnel Cohen, Philip V Ogren, Lynne Fox, and Lawrence Hunter. 2005. Empirical data on corpus design and usage in biomedical natural language processing. In *AMIA annual symposium proceedings*, pages 156–160.

Kristofer Franzén, Gunnar Eriksson, Fredrik Olsson, Lars Asker, Per Lidén, and Joakim Cöster. 2002. Protein names and how to find them. *International Journal of Medical Informatics*, 67(13):49 – 61.

James W. Hunt and M. Douglas McIlroy. 1976. An Algorithm for Differential File Comparison. Technical Report 41, Bell Laboratories Computing Science, July.

Helen Johnson, William Baumgartner, Martin Krallinger, K Bretonnel Cohen, and Lawrence Hunter. 2007. Corpus refactoring: a feasibility study. *Journal of Biomedical Discovery and Collaboration*, 2(1):4.

Jin-Dong Kim, Tomoko Ohta, Yuka Tateisi, and Jun'ichi Tsujii. 2003. GENIA corpus - a semantically annotated corpus for bio-textmining. *Bioinformatics*, 19(suppl. 1):i180–i182.

Yusuke Miyao and Jun'ichi Tsujii. 2008. Feature forest models for probabilistic hpsg parsing. *Computational Linguistics*, 34(1):35–80, March.

Yue Wang, Jin-Dong Kim, Rune Sætre, Sampo Pyysalo, Tomoko Ohta, and Jun'ichi Tsujii. 2010. Improving the inter-corpora compatibility for protein annotations. *Journal of Bioinformatics and Computational Biology*, 8(5):901–916.

Using Natural Language Processing to Identify Pharmacokinetic Drug-Drug Interactions Described in Drug Package Inserts

Richard Boyce, PhD
University of Pittsburgh
5607 Baum Avenue
Pittsburgh, PA 15206, USA
rdb20@pitt.edu

Gregory Gardner, MS
University of Pittsburgh
5607 Baum Avenue
Pittsburgh, PA 15206, USA
gag30@pitt.edu

Henk Harkema, PhD
University of Pittsburgh
6425 Penn Ave.
Pittsburgh, PA 15206
hendrik.harkema@nuance.com

Abstract

The package insert (aka drug product label) is the only publicly-available source of information on drug-drug interactions (DDIs) for some drugs, especially newer ones. Thus, an automated method for identifying DDIs in drug package inserts would be a potentially important complement to methods for identifying DDIs from other sources such as the scientific literature. To develop such an algorithm, we created a corpus of Federal Drug Administration approved drug package insert statements that have been manually annotated for pharmacokinetic DDIs by a pharmacist and a drug information expert. We then evaluated three different machine learning algorithms for their ability to 1) identify pharmacokinetic DDIs in the package insert corpus and 2) classify pharmacokinetic DDI statements by their modality (i.e., whether they report a DDI or no interaction between drug pairs). Experiments found that a support vector machine algorithm performed best on both tasks with an F-measure of 0.859 for pharmacokinetic DDI identification and 0.949 for modality assignment. We also found that the use of syntactic information is very helpful for addressing the problem of sentences containing both interacting and non-interacting pairs of drugs.

1 Introduction

Package inserts (PIs, aka drug product label) are the primary source of information for newly approved drugs and a potentially authoritative source of drug information from a medical-legal standpoint (Marroum & Gobburu 2002). Among the information provided by PIs are *drug-drug interactions* (DDIs): known and predicted drug combinations that could lead to a clinically meaningful alteration in the effect of one of the drugs. The United States Federal Drug Administration (FDA) mandates that PIs for FDA-approved drugs include both observed and predicted clinically significant DDIs, as well as the results of pharmacokinetic studies that establish the absence of effect (FDA. 2010). Moreover, the PI is the only publically-available source of information on DDIs for some drugs, especially newer ones (Dal-Ré et al. 2010). Hence, an automated method for identifying DDIs from drug PIs would be an important complement to methods for identifying DDIs from other sources such as the scientific literature. In this paper we describe the creation of a new corpus of FDA-approved drug package insert statements that have been manually annotated for pharmacokinetic DDIs. We then discuss how three different machine learning algorithms were evaluated for their ability to 1) identify pharmacokinetic DDIs in drug package inserts and 2) classify pharmacokinetic DDI statements by their modality (i.e., whether they report a DDI or that a drug pair does not interact).

2 Materials and Methods

2.1 The DDI Corpus and Schema

A corpus of annotated statements derived from FDA-approved drug PIs was created for use as training and test data while developing automated DDI extraction algorithms. The statements were derived from PIs using a strategy that ensured there would be a representative sample of statements that 1) unambiguously identified interacting drug pairs, 2) unambiguously identified non-interacting drug pairs, and 3) included no mention of interacting drug pairs. Previous experience by our research group suggested that the manner in which DDI statements are described in PIs has changed over time in response to changing FDA regulations. Most notably, an FDA guidance document issued in 1999 was (to our knowledge) the first to explicitly suggest the inclusion of brief descriptions of pharmacokinetic DDI studies within specific sections of drug PIs (FDA. 1999). To account for this, investigators selected 64 PIs using a strategy that ensured the corpus would have a balanced sample of statements from drugs marketed before and after 2000. For the purpose of this study we designated all PIs for drugs marketed prior to 2000 as "older" and those for drugs marketed in or after 2000 as "newer." PIs were downloaded from the DailyMed website,[1] and the entire "Drug Interactions" and "Clinical Pharmacology" sections were selected as text sources from "newer" PIs. For "older" PIs, which often lacked these two sections, investigators chose a section containing an apparent interaction statement and one randomly-selected section.

DDIs are typically classified as occurring by either pharmacodynamic or pharmacokinetic mechanisms. A pharmacodynamic DDI involves the additive or synergistic amplification of a drug's effect. In a pharmacokinetic (PK) DDI, one drug, called a *precipitant*, affects (inhibits or induces) the absorption, distribution, metabolism, or excretion of another drug, called the *object*. To simplify our task, we decided to focus specifically on PK DDIs. Prior to annotating the PI statements, a schema was created for the entities that the investigators considered important components of a PK DDI. The schema modeled drugs as having two characteristics, *type* and *role*. The *type* of drug could be `active ingredient` (e.g., simvastatin), `drug product` (e.g., Zocor), or `metabolite` (e.g., beta-OH-simvastatin). Drugs annotated as `metabolite` also referred to the `active ingredient` parent compound. The *role* of a drug could be either an `object` or a `precipitant`. Two other properties were provided to model each PK DDI: 1) whether the statement from which the DDI was identified suggested an observed effect or a lack of an observed effect between two coadministered drugs (i.e., `positive` vs `negative` *modality* statements), and 2) whether the statement included quantitative or qualitative data in describing an interaction or non-interaction between a drug pair (i.e., `quantitative` vs `qualitative` statements). Finally, the segment of text in which the interaction claim was made was annotated as an `interaction phrase`. With the corpus and schema in place, drugs and PK DDIs present in the PI statements were then annotated by two independent reviewers using Knowtator, an annotation tool integrated with the Protégé ontology editor (Ogren 2006).

One annotator was a pharmacist and DDI expert, and the other a librarian specializing in drug information retrieval. To help the annotators, co-investigator RB ran the NCBO Annotator (Jonquet, Shah & Musen 2009) over the corpus using the RxNorm drug terminology (Nelson et al. 2011) to pre-annotate as many active ingredients and drug products as possible. The annotators reviewed these "pre-annotations" while identifying entities that missed during the pre-annotation process. Co-investigator HH used Knowtator to calculate inter-annotator agreement statistics from the annotators' initial annotation sets. RB then worked with the two annotators to achieve consensus on the final corpus of annotated DDI statements.

2.2 Setting up the DDI statement extraction experiment

Once the set of DDI annotations was compiled, we devised two machine learning tasks. The first task was to determine whether two drugs mentioned in a statement taken from a PI are noted as either interacting or not interacting with each other by pharmacokinetic mechanisms (i.e., *does the statement report a PK DDI with the drug pair of either a positive or negative modality?*). The second task was to determine the modality of a given PK DDI. The first task *did not* include determining the roles of the drugs if an interaction is found, i.e., which member of the pair of drug mentions is the precipitant and which one is the object. To enable the exploration of the performance of multiple machine learning methods, we divided two-thirds of the annotated PI statements into a *development* set and one-third into a blind *test* set. PI statements annotated as reporting DDIs were stratified within the

[1] http://dailymed.nlm.nih.gov/

two sets using a random selection method that ensured a representative balance of sentence distance between drug mentions, DDI modality, DDI type, and drug PI age designation (see above). Statements not containing an interaction were stratified by sentence distance between drug mentions, and PI age designation. Stratification was done on the level of statements. Thus, statements taken from the same package insert may have been distributed over the development and test set.

We observed that 99% of corpus statements annotated as a PK DDI mentioned an interacting drug pair within a three sentence region. Thus, we created a *baseline* dataset by iterating through PI statements in the development set and identifying all drug pair mentions that occurred within a three-sentence span. Throughout the remainder of this paper we refer to the statements identified by this process as *instances*.

Instances containing drug pairs that were manually annotated as participating in an interaction (either with positive or negative modality) were labeled as *positive* instances for the extraction task; all other pairs were labeled as *negative* instances. Prior to generating features for machine learning, each instance was pre-processed. Numbers (e.g. "1", "34", "5.2", etc.) were replaced by the string "`num`" to make them more meaningful to a learning algorithm across instances. This allowed the algorithm to associate numerical references with each other using a general pattern, instead of learning phrases with specific numbers (e.g. the phrase "`num` mg" may be significant, whereas "10 mg" may be less significant). Similarly, to abstract away from specific names, the names of drug products, active ingredients, and metabolites in each statement were replaced by the string "drugname". This forces the learning algorithm to generalize over the participants of interactions, preventing it from identifying interactions based on the identity of the participants.

In the baseline dataset, each instance's pre-processed sentence text was translated to bigrams using TagHelper, a text analysis program written on top of the Weka machine learning software (Hall et al. 2009; Rosé et al. 2008). Bigrams are a comprehensive set of consecutive word pairs that appear in a sentence. Words in bigrams were stemmed by TagHelper to facilitate learning more general concepts conveyed by phrases. For example, the commonly occurring phrases "increases auc" and "increased auc" are stemmed to "increase auc" and then merged to the bigram. The baseline set of instances was loaded into Weka and three models were built using three different machine learning algorithms. The three algorithms were a rule learner ("JRip"), a decision tree ("J48"), and an SVM algorithm ("SMO"). Algorithm parameters were left at Weka defaults and 10-fold cross-validation was used to develop each model.

Exploration of Weka predictions from the baseline dataset showed that a major source of confusion for the machine learning algorithms was an inability to distinguish between pairs of drugs that do and do not interact within the same sentence. A frequent source of this kind of occurrence in the package insert text was *coordinate structures* such as "Drug A interacts with Drugs B and C", where "B and C" is a coordinate structure. For such sentences, the baseline dataset contains the interacting pairs (A,B) and (A,C), along with the non-interacting pair (B,C). However, because all three pairs are represented by the same set of bigrams, it is obvious that information from bigrams alone is insufficient to distinguish which pairs interact and which simply co-occur within the sentence.

Another problem was that of multiple mentions of the same drug within an instance's sentence span, as, for example, in the sentence "Co-administration of A and B leads to increased AUC levels for B." Because the annotators had identified only one drug mention per annotated interaction, the algorithms incorrectly considered other mentions of the same drug as part of a non-interacting pair. Two solutions were implemented to help alleviate these problems. First, the dataset was condensed to a set of instances with unique drug pairs and sentence spans. If any of the baseline instances contributing to the condensed instance contained interactions, the condensed instance was said to contain an interaction. In this way, multiple drug mentions within a sentence span containing an interaction would translate to a single instance representing an interaction between the two drugs.

Second, two natural language dependency parsers were used to extract extra features from the sentence text for each instance: the Stanford NLP Parser (Klein & Manning 2003) and ClearParser (Choi 2011). Following approaches to relation extraction proposed in other domains e.g., (Bunescu & Mooney 2005), the dependency structure produced by each parser was searched for the shortest path between the pair of drug mentions of the instance. The words on this path were stemmed using the Stanford NLP Tools stemmer (Stanford NLP 2011), and added to the dataset as the instance's "syntactic path".

Once a statement is classified as describing a PK DDI between two drugs, it is important to know if there is an observed effect or a lack of effect between two coadministered drugs (i.e., *positive* vs *negative* modality statements). To present the learning algorithms with the most relevant training data, modality prediction was treated as a separate task from interaction prediction. Development and test sets were created in the same manner as for interaction prediction, however instances that did not represent interactions were excluded. Only bigram features were used for modality prediction. Model training and testing proceeded in the same manner as for interaction prediction.

3 Results

A total of 208 multi-sentence sections were extracted from 64 PIs. Prior to consensus, inter-annotator agreement between the two annotators on PK DDI, active ingredient, drug product, metabolite mentions and was found to be 60%, 96.3%, 99.5%, and 60.8% respectively. The majority of disagreements about DDIs were due to a tendency of one annotator to incorrectly annotate some pharmacodynamic DDIs as PK DDIs. Also, one annotator incorrectly assumed that all metabolites had been pre-annotated and so did not actively attempt to annotate metabolite entities. These and other minor issues were corrected and full consensus was reached by both annotators. The final drug package insert PK DDI corpus contains 592 PK DDIs, 3,351 active ingredient mentions, 234 drug product mentions, and 201 metabolite mentions.[2]

Tables 1 and 2 provide more details on the modality and drug types present in the 592 consensus PK DDI statements. Table 1 shows that 388 statements indicated that a PK DDI would occur between a drug pair, while 204 statements indicated that an interaction would not occur. The table also shows that 204 statements reported quantitative measures while 388 did not. Table 2 shows that the majority (86%) of PK DDI statements reported interactions by stating the two active ingredients involved in the DDI, with a much smaller proportion using a drug product in the description. Also, 35 DDI statements reported an effect on a drug metabolite.

	Interaction Type		
Modality	Qualitative	Quantitative	Total
Negative	202	2	204
Positive	186	202	388
Total	388	204	592

Table 1. PK DDI statement modality shown by interaction type.

A total of 11,048 PI instances were generated for the baseline dataset. This was reduced to 5,015 instances after condensing the instances down to unique drug pairs and sentence spans. In the final dataset, about a third of instances were drug pairs within the same sentence (1,583). The rest were split between drug pairs in adjacent sentences (1,717), and drug pairs with two sentences of separation (1,715). The dataset included 542 interactions of which 493 included the drug pair within a single sentence. 355 interactions were positive modality and 187 negative; 360 were qualitative, 182 quantitative. 1,636 instances were categorized as "new" based on drug release data while 3,379 were classified as "old".

Results for interaction and modality prediction are shown in Table 3. For both the interaction and modality prediction tasks, the SVM algorithm (SMO) outperformed the rule learner (Jrip) and decision tree (J48). On the test set which was not used in training, the SVM classifier identified PK DDIs with an F-measure of 0.859 vs 0.762 for the rule learner and 0.802 for the decision tree algorithm. All algorithms performed quite well on the modality classification task but the SVM algorithm performed best with an F-measure of 0.949 vs 0.929 (rule learner) and 0.917 (decision tree).

4 Discussion

The automatic identification of DDIs in unstructured text is a topic that is gaining much interest. This work makes an important contribution to the field by being the first to demonstrate that machine learning can be applied quite effectively to the task of extracting PK DDIs from FDA-approved PIs.

[2] http://purl.org/NET/nlprepository/PI-PK-DDI-Corpus

	Object Type			
Precipitant Type	Active ingredient	Drug product	Metabolite	Total
Active ingredient	506	14	34	554
Drug product	37	-	1	38
Total	543	14	35	592

Table 2. A summary of consensus annotated PK DDIs by precipitant and object type.

As our work focuses on extracting PK DDIs, it is most similar to that of Karnik *et al.* (Karnik et al. 2011) who explored the performance of an "all paths" graph kernel (Airola et al. 2008) on a corpora of PK DDIs derived from 219 MEDLINE abstracts. The best performing algorithm in their experiments had an F-measure of 0.658 which is considerably less than the F-measure of 0.859 that our SVM achieved. However, the two results are not directly comparable because of unknown differences between the corpora. For example, it may be that PIs use more standard language patterns to report PK DDIs than what is found in MEDLINE abstracts. In future work we will explore how well the SVM algorithm performs over MEDLINE abstracts and contrast any differences between the two DDI sources that might affect NLP.

The only other project we are aware of that focused explicitly on extracting PK DDIs from unstructured text is that of Tari *et al.* (Tari et al. 2010), who evaluated a rule-based algorithm for extracting PK DDIs from papers and abstracts in the scientific literature. In this study the authors distinguished between *explicit* DDIs (statements indicating a direct observation of a PK effect from a give drug combination) and *implicit* DDIs (DDIs that can be inferred based on claims about drug metabolic properties extracted from scientific texts). The algorithm was ran over ~17 million MEDLINE abstracts and the output DDIs were compared with a reference standard set of 494 DDIs identified manually from 265 DrugBank drug pages. The algorithm's recall of DrugBank interactions was only 12%. However, a manual inspection of the results found that 78% of the DDIs extracted by the algorithm were valid based on the source texts, even though they were not present in their reference standard. These results are important because they suggest that the set of DDIs present in DrugBank are *incomplete* and highlight the need for corpora derived from other text sources such as the one we developed from drug PIs for this study.

A larger body of research exists for the task of extracting DDIs of any type (i.e., PK *or* pharmacodynamic DDIs). Ten research papers were presented at the recent "Challenge Task on Drug-Drug Interaction Extraction" held at the 2011 SemEval Conference (Segura-Bedmar, Martinez & Sanchez-Cisneros 2011). All systems in this challenge were tested against the "DrugDDI corpus"; a set of 579 documents from the DrugBank database with 3,160 manually-annotated DDIs (Segura-Bedmar, Martinez & Pablo-Sanchez 2010). The best performing system in this challenge utilized an ensemble learning approach (Thomas et al. 2011) and produced an F-measure of 0.657. The second best performing method utilized composite kernels, a method that combines feature-based and kernel-based methods, and was found to perform with an F-measure of 0.64 (Chowdhurry et al. 2011). Airola *et al*'s "all paths" graph kernel (mentioned above) performed much more poorly on the DrugDDI corpora than on the Karnik's PK-DDI corpus (F-measure 0.16 vs 0.658). The authors note that there were significant differences between in the two corpora with regards to the length and complexity of the sentences reporting DDIs .

To the best of our knowledge, only one other NLP study that has focused specifically on drug interactions reported in drug product labeling (Rubrichi & Quaglini 2012). The investigators compared the ability of an SVM classifier and a conditional random fields (CRF) classifier for assigning 13 semantic labels to Italian language text present in the interaction section of "Summary of Product Characteristics" documents (the Italian equivalent of PIs). The investigators explored the influence of a range of features on classifier performance, including orthographical, neighboring word, syntactic, parts of speech, and dictionary features. When all features were employed, the SVM had slightly better performance than the CRF classifier (micro-averaged F-measure: 91.41 vs 91.13, macro-averaged F-measure: 84.99 vs 80.83).

	Jrip			*J48*			*SMO*		
Model (dataset)	Prec	Recall	F	Prec	Recall	F	Prec	Recall	F
Baseline (development)	0.588	0.656	0.62	0.584	0.573	0.578	0.639	0.677	0.658
Stanford Parser (development)	0.762	0.68	0.719	0.809	0.804	0.807	0.851	0.815	0.833
ClearParser (development)	0.787	0.793	0.79	0.822	0.791	0.806	0.828	0.887	0.856
Stanford Parser (test)	0.778	0.665	0.717	0.828	0.832	0.83	0.843	0.838	0.84
ClearParser (test)	0.764	0.76	0.762	0.85	0.76	0.802	0.836	0.883	0.859
Modality (test)	0.963	0.897	0.929	0.887	0.948	0.917	0.941	0.957	0.949

Table 3. Results for interaction prediction on the baseline, development, and blind test set. Also shown are results for modality prediction for the blind test set (results over the development set are similar but not shown).

One key difference between the Rubrichi study and ours is that the task of tagging unstructured text with semantic elements that describe a DDI is not the same as classifying whether or not a statement containing a drug pair is reporting a DDI between the drugs. The difference is especially apparent when considering coordinate structures such as "Drug A interacts with Drugs B and C", Semantic tagging would be useful for identifying the drug entities but is not useful (on its own) for identifying which of the three drug pairs interact with each other.

It is interesting to note that most recent work on DDI extraction had not made the distinction between PK and pharmacodynamic DDIs that is standard in the fields of pharmacology and pharmacy. This distinction might be relevant to DDI extraction because the two types of interactions are discovered in distinct ways that might lead to significant differences in how they are described in scientific documents. For example, there is a fairly standard set of *in vitro* experiments and clinical trials that have been a routine part of drug development for more than a decade (FDA. 1999). The same is not true for pharmacodynamic DDIs, which are more challenging to study because they involve additive and synergistic effects that are not necessarily related to a drug's dose or clearance. Since it is reasonable that the methods used to investigate a DDI strongly influences its description, we think future work should examine if PK and pharmacodynamic DDI descriptions are different enough to warrant distinct DDI extraction efforts.

An error analysis of the final dataset suggested some reasons for cases where the machine learning algorithms misclassified instances. Instances that were not interactions, but were classified as such, contained a large number of sentences with descriptions of studies or biochemical processes and measurements. These types of statements may share a number of features with actual interactions (e.g. numerical data, changing levels of drug, etc.) without containing an interaction. There also remain cases where several drug names occur and the classifiers were unable to differentiate between the interacting pair and non-interacting pairs. Unfortunately, no such clear pattern was apparent for instances that descrived interactions, but were classified as containing no interaction statement. A number of large sentences were observed in these instances, suggesting sentence complexity may play a role, increasing the difficulty of natural language parsing.

Analysis of the attribute weights assigned by the SVM algorithm (SMO) after training for interaction prediction shows some commonality regardless of whether the data was processed by the Stanford Parser or the ClearParser. For example, 19 out of the 20 most significant features identified by the algorithm from the dataset when processed by the Stanford Parser were words on the syntactic path; one less than when the dataset was processed by the ClearParser. Common significant features include words such as "coadminister", "auc", "pharmacokinetic", and "absorption". The algorithm placed greater importance on the words "increase" and "decrease" when the dataset was processed by the Stanford Parser, while the words "reduce" and "enhance" received greater attribute weights when the data was processed by the ClearParser. A similar analysis of the SVM algorithm developed for PK DDI modality prediction shows that bigrams with the words "no" or "not" are clearly the features of most importance to the model.

We also note that the algorithm's performance on the test set of PI statements is very similar to the algorithm's performance over the development set (see Table 3). We think that this finding is largely due to the careful stratification approach we used when creating the development and test sets. It might also be possible that the features in the unstructured PI text do not vary greatly between PIs regardless of their age. However, Table 2 shows that our PK DDI corpora had considerable variation in terms of quantitative vs qualitative and positive vs negative DDI statements. Thus, we anticipate that the SVM algorithm's performance will be maintained when ran against a much larger PI corpus and future work will test how well the algorithm generalizes to other sets of PIs.

5 Conclusion

We created a new, publically available, corpus of FDA-approved drug PI statements that have been manually annotated for PK DDIs by a pharmacist and a drug information expert. Also, we evaluated three different machine learning algorithms for their ability to 1) identify PK DDIs in drug PIs and 2) classify PK DDI statements by their modality (i.e., whether they report a DDI or no interaction between drug pairs). Experiments found that an SVM algorithm performed best on both tasks with an F-measure of 0.859 for PK DDI identification and 0.949 for modality assignment. We found that the use of syntactic information is very helpful for addressing the problem of sentences containing both interacting and non-interacting pairs of drugs. The strong performance of our algorithm for PK DDIs suggests that approaching pharmacokinetic and pharmacodynamic interactions as different NLP tasks is a potentially promising approach for advancing automated DDI extraction. Given the marked difference in performance between our extraction methods and previous work, we are planning further experiments to establish whether this difference reflects the comparative simplicity of the extraction task represented by our corpus, some specific strength of the applied extraction methods, or some other factor.

Acknowledgement

This project was funded by grant K12-HS019461 from the Agency for Healthcare Research and Quality (AHRQ). The content is solely the responsibility of the authors and does not represent the official views of AHRQ. We also thank John Horn, PharmD (University of Washington) and Mr. Rob Guzman (University of Pittsburgh) for their work annotating the corpus and identifying related research.

References

Airola, Antti, Sampo Pyysalo, Jari Björne, Tapio Pahikkala, Filip Ginter & Tapio Salakoski. 2008. All-paths graph kernel for protein-protein interaction extraction with evaluation of cross-corpus learning. *BMC Bioinformatics* 9(Suppl 11). S2. doi:10.1186/1471-2105-9-S11-S2 (3 May, 2012).

Bunescu, Razvan C. & Raymond J. Mooney. 2005. A shortest path dependency kernel for relation extraction. *Proceedings of the conference on Human Language Technology and Empirical Methods in Natural Language Processing*, 724–731. (HLT '05). Stroudsburg, PA, USA: Association for Computational Linguistics. doi:10.3115/1220575.1220666. http://dx.doi.org/10.3115/1220575.1220666 (2 May, 2012).

Choi, Jinho. 2011. ClearParser GoogleCode page. *clearparser*. http://code.google.com/p/clearparser/ (10 December, 2011).

Chowdhurry, Md. Faisal Mahbub, Asma Ben Abacha, Alberto Lavelli & Pierre Zweigenbaum. 2011. Two Different Machine Learning Techniques for Drug-Drug Interaction Extraction. *1st Challenge task on Drug-Drug Interaction Extraction (DDIExtraction 2011)*, 19–26. Huelva, Spain.

Dal-Ré, R., A. Pedromingo, M. García-Losa, J. Lahuerta & R. Ortega. 2010. Are results from pharmaceutical-company-sponsored studies available to the public? *European Journal of Clinical Pharmacology* 66(11). 1081–1089. doi:10.1007/s00228-010-0898-y (5 August, 2011).

FDA. 1999. *FDA Guideline: In Vivo Drug Metabolism/Drug Interaction Studies – Study Design, Data Analysis, and Implications for Dosing and Labeling*. Rockville, MD: Food and Drug Administration. http://www.fda.gov/downloads/Drugs/GuidanceComplianceRegulatoryInformation/Guidances/ucm072119.pdf.

FDA. 2010. CFR - Code of Federal Regulations Title 21.

http://www.accessdata.fda.gov/scripts/cdrh/cfdocs/cfcfr/CFRSearch.cfm?fr=201.57 (7 June, 2011).

Hall, Mark, Eibe Frank, Geoffrey Holmes, Bernhard Pfahringer, Peter Reutemann & Ian H Witten. 2009. The WEKA data mining software: an update. *SIGKDD Explorations* 11(1). 10–18.

Jonquet, Clement, Nigam H Shah & Mark A Musen. 2009. The open biomedical annotator. *Summit on Translational Bioinformatics* 2009. 56–60. (10 December, 2011).

Karnik, Shreyas, Abhinita Subhadarshini, Zhiping Wang, Luis M Rocha & Lang Li. 2011. Extraction Of Drug-Drug Interactions Using All Paths Graph Kernel. *1st Challenge task on Drug-Drug Interaction Extraction (DDIExtraction 2011)*. Huelva, Spain.

Klein, Dan & Christopher D Manning. 2003. Fast Exact Inference with a Factored Model for Natural Language Parsing. (Ed.) S Thrun S Becker & Keditors Obermayer. *Science* 15. 3–10.

Marroum, P.J. & J. Gobburu. 2002. The product label: how pharmacokinetics and pharmacodynamics reach the prescriber. *Clinical Pharmacokinetics* 41(3). 161–169. (7 June, 2011).

Nelson, Stuart J, Kelly Zeng, John Kilbourne, Tammy Powell & Robin Moore. 2011. Normalized names for clinical drugs: RxNorm at 6 years. *Journal of the American Medical Informatics Association: JAMIA* 18(4). 441–448. doi:10.1136/amiajnl-2011-000116 (10 December, 2011).

Ogren, Philip V. 2006. Knowtator: a Protégé plug-in for annotated corpus construction. *Proceedings of the 2006 Conference of the North American Chapter of the Association for Computational Linguistics on Human Language Technology*, 273–275. Morristown, NJ, USA: Association for Computational Linguistics. doi:http://dx.doi.org/10.3115/1225785.1225791.

Rosé, Carolyn, Yi-Chia Wang, Yue Cui, Jaime Arguello, Karsten Stegmann, Armin Weinberger & Frank Fischer. 2008. Analyzing collaborative learning processes automatically: Exploiting the advances of computational linguistics in computer-supported collaborative learning. *International Journal of Computer-Supported Collaborative Learning* 3(3). 237–271. doi:10.1007/s11412-007-9034-0 (10 December, 2011).

Rubrichi, S & S Quaglini. 2012. Summary of Product Characteristics content extraction for a safe drugs usage. *Journal of Biomedical Informatics* 45(2). 231–239. doi:10.1016/j.jbi.2011.10.012 (3 May, 2012).

Segura-Bedmar, Isabel, Paloma Martinez & Cesar Pablo-Sanchez. 2010. Extracting drug-drug interactions from biomedical texts. *Workshop on Advances in Bio Text Mining*, vol. 11 Suppl 5, 9. Madrid, Spaim: BMC Bioinformatics. http://www.biomedcentral.com/1471-2105/11/S5/P9.

Segura-Bedmar, Isabel, Paloma Martinez & Daniel Sanchez-Cisneros (eds.). 2011. *Proceedings of the First Challenge Task: Drug-Drug Interaction Extraction 2011*. Huelva, Spain. http://sunsite.informatik.rwth-aachen.de/Publications/CEUR-WS/Vol-761/ (9 December, 2011).

Stanford NLP. 2011. The Stanford NLP (Natural Language Processing) Group. http://nlp.stanford.edu/software/ (10 December, 2011).

Tari, Luis, Saadat Anwar, Shanshan Liang, James Cai & Chitta Baral. 2010. Discovering drug-drug interactions: a text-mining and reasoning approach based on properties of drug metabolism. *Bioinformatics (Oxford, England)* 26(18). i547–553. doi:10.1093/bioinformatics/btq382 (9 December, 2011).

Thomas, Philippe, Mariana Neves, Illes Solt, Domonkos Tikk & Ulf Leser. 2011. Relation Extraction for Drug-Drug Interactions using Ensemble Learning. *1st Challenge task on Drug-Drug Interaction Extraction (DDIExtraction 2011)*, 11–18. Huelva, Spain.

Automatic Approaches for Gene-Drug Interaction Extraction from Biomedical Text: Corpus and Comparative Evaluation

Nate Sutton, Laura Wojtulewicz, Neel Mehta, Graciela Gonzalez
Department of Biomedical Informatics
Arizona State University, Tempe, Arizona, USA
{nate.sutton, whitz, nbmehta2, graciela.gonzalez}@asu.edu

Abstract

Publications that report genotype-drug interaction findings, as well as manually curated databases such as DrugBank and PharmGKB are essential to advancing pharmacogenomics, a relatively new area merging pharmacology and genomic research. Natural language processing (NLP) methods can be very useful for automatically extracting knowledge such as gene-drug interactions, offering researchers immediate access to published findings, and allowing curators a shortcut for their work.

We present a corpus of gene-drug interactions for evaluating and training systems to extract those interactions. The corpus includes 551 sentences that have a mention of a drug and a gene from about 600 journals found to be relevant to pharmacogenomics through an analysis of gene-drug relationships in the PharmGKB knowledgebase.

We evaluated basic approaches to automatic extraction, including gene and drug co-occurrence, co-occurrence plus interaction terms, and a linguistic pattern-based method. The linguistic pattern method had the highest precision (96.61%) but lowest recall (7.30%), for an f-score of 13.57%. Basic co-occurrence yields 68.99% precision, with the addition of an interaction term precision increases slightly (69.60%), though not as much as could be expected. Co-occurrence is a reasonable baseline method, with pattern-based being a promising approach if enough patterns can be generated to address recall. The corpus is available at http://diego.asu.edu/index.php/projects

1 Introduction

Pharmacogenomics is a relatively new area of biomedical research that merges pharmacology and molecular genomics, among other disciplines, and focuses on studying the effects of genetic variability on drug toxicity and efficacy, on the discovery of novel genomic targets for drug development, and on the identification and functional characterization of polymorphisms relevant to drug action. Thus, publications that report genotype-drug findings and manually curated databases that collect such findings, like PharmGKB and DrugBank (Hewett et al., 2002; Wishart, 2006) are of paramount importance to the field. However, manual curation is expensive and time consuming, and cannot keep up with the ever increasing number of publications. Natural language processing (NLP) methods can be very useful for automatically extracting such gene-drug interactions, offering researchers immediate access to published findings, and allowing curators a shortcut for their work.

Consider for example a sentence containing an interaction NLP can help extract: “Only the epsilon4 allele of APOE was found to make a significant (P = 0.002) but small contribution to warfarin dose requirement.” (PMID: 16847429). We can easily see that in the sentence, an APOE allele interacts with the drug warfarin in its dose requirement. Furthermore, at a higher level of abstraction, the sentence can help researchers infer that APOE affects the metabolic processes targeted by the drug warfarin.

NLP researchers attacking an interaction extraction project such as this one, will usually start by identifying the entities involved in the extractions and the terms that indicate such interactions. Assuming named entity recognition (NER) systems exist for the entities in question (or a dictionary is available for direct match), the main concern becomes extracting true interactions. A gold standard corpus would then need to be identified or created in order to evaluate and develop interaction extraction approaches, starting with the simplest ones. We aim to support advancement in the area of gene-drug interaction extraction through the construction of a corpus for that task that offers advantages not available in another similar corpus. Also for that support we report on a study of the capabilities of different methods for that form of extraction.

To achieve our aim, we describe a new corpus of gene-drug interactions, and compare the performance of two basic approaches plus the re-implementation of a more advanced pattern-based approach measured against this corpus. We do not seek in this publication to advance the extraction methods themselves, but allow a side-to-side comparison of approaches on a single corpus.

The sentences in the corpus (a total of 551) were randomly selected from sentences that include both a gene and a drug mention from the abstracts published on a selection of journals that have articles relevant to pharmacogenomics. In general, annotations include interactions evident from the sentence, if any, also noting when mentioned genes or drugs are *not* involved in interactions. All sentences were annotated by the main author, with a second and third annotator verifying 26% of the corpus. The corpus is publicly available online along with other supplementary materials including the annotation guide[1].

The extraction methods evaluated include co-occurrence of a gene and a drug, co-occurrence of a gene and a drug plus a recognized interaction term, and one that uses specific linguistic patterns for classification based on (Coulet, Shah, Garten, Musen, & Altman, 2010). The linguistic pattern method had the highest precision (96.61%) but lowest recall (7.30%), for an f-score of 13.57%. Basic co-occurrence yields 68.99% precision, with the addition of an interaction term increasing precision slightly (69.60%), though not as much as could be expected. Analysis of our results show that performance could be immediately improved by improving the fundamental entity-recognition of drugs and genes.

2 Related Work

A good portion of the work presented here follows prior approaches to high quality protein-protein interaction (PPI) corpora development and extraction. Given that our corpus contains genes and proteins as entities, procedures used to create PPI corpora were a useful resource. A variety of annotation decisions made were informed by the work of Pyysalo et. al. on their BioInfer corpus (Pyysalo et al., 2007). A detailed annotation guide used in their work was referenced in annotation rules in this work. Other corpora, such as the ones used in Biocreative challenges, have also made valuable contributions to PPI extraction progress (Hakenberg et al., 2010; Krallinger, Leitner, Rodriguez-Penagos, & Valencia, 2008).

Unlike for PPI interaction extraction, there are very limited currently available corpora that can be used for automatic gene-drug interaction extraction system development and evaluation. One corpus that contains those interactions is a 300 sentence corpus by Ahlers et al. (Ahlers, Fiszman, Demner-Fushman, Lang, & Rindflesch, 2007). The Ahlers et. al. corpus include the biological interaction categories of inhibit, and stimulate in addition to interaction annotations for genes and drugs. Our corpus does not contain those additional categories directly, but the interaction words that are annotated in our corpus can indicate such categories as well as others. All in all, our focus was on creating a corpus that could be used for evaluation of basic as well as complex approaches, and allow machine-learning based systems to be trained on it.

Current systems for extracting gene-drug interactions are based on entity co-occurrence and some include matching of relationship terms. Those systems commonly use statistical formulas for ranking the relevance of results. Polysearch, Pharmspresso, and others are examples of such systems (Cheng et al., 2008; Garten & Altman, 2009). Some systems integrate linguistic patterns into their methods, such as those by Coulet et. al. and Tari et. al. (Luis Tari, Jörg Hakenberg, Graciela Gonzalez, & Baral, 2009). The system by Coulet et al. explores the value of dependency graph information for relationship extraction. Another result of Coulet et. al.'s work was the *Phare* ontology that includes concepts relevant to those relationships, which we utilize in this work. The value of such collections of interaction-indicating terms has been highlighted before in the biomedical relationship extraction context (Bui, Nualláin, Boucher, & Sloot, 2010; Chowdhary, Zhang, & Liu, 2009).

[1] http://diego.asu.edu/index.php/projects

3 Materials and Methods

3.1 Corpus design.

The purpose for the creation of the new corpus was to create a resource that NLP developers can use to train and test gene-drug interaction extraction systems. The corpus was based on articles from journals that are known to contain pharmacogenomic relationships. Genes and drugs were automatically tagged and then 551 sentences that contain both a gene and drug were randomly selected for annotation. The corpus and sentence selection process is described in the following subsections.

Journal Selection. A list of journals relevant to pharmacogenomics was generated by extracting the journal names from articles that have been curated in PharmGKB as containing evidence of gene-drug relationships. This list was generated from their downloadable "relationships" file, which contains the abstract IDs of articles with manually curated gene-drug relationships. 591 journal names were obtained this way. The goal of using only those journals is to make the corpus representative of typical sentences containing a gene and drug from literature known to report pharmacogenomic findings.

Sentence processing. All abstracts in PubMed from the relevant journal names were downloaded. A sentence splitter program from OpenNLP was used to extract sentences from the abstracts ("The OpenNLP Homepage," n.d.). A total of 22,601,402 sentences were processed.

Identification of entites. Previous work in pharmacogenomics relationship extraction has shown effective results by classifying relationships after identifying sentences with entities of interest through dictionary matching techniques (Garten & Altman, 2009; Rebholz-Schuhmann et al., 2007). Our work takes a similar approach, but utilizes a machine-learning based method, BANNER, for gene recognition, as it was shown to have better performance than a dictionary-based method (Leaman & Gonzalez, 2008). Drugs were recognized through the use of dictionary matching. The dictionaries used for drugs were based on drug names available at DrugBank. Exact full token matching of drug terms was used to identify them in sentences. Although incorrectly tagged (false entity) genes and drugs were corrected by annotators, they did not add entities missed by NER recognition. A second round of annotation will correct this when we shift focus to NER.

Terms indicative of an interaction for adding to basic co-occurrence relationship extraction were extracted from the *Phare* ontology. The terms acquired were from rdfs labeled text in the "object properties" in the ontology. Object properties are elements of the ontology that describe relationships between classes such as gene and drugs, yielding 168 unique terms after stemming.

Sentence selection. The initial annotation effort that is the focus of this paper was aimed at completing around 500 sentences as a proof of concept, with a total of 1,500 to be completed in the second phase of this project. Random selection of sentences that include a gene and a drug, in contrast to balanced positive and negative selection, was used to make the corpus reflect typical sentences potentially containing an interaction that can be easily extracted from the source articles after simple (drug and gene) concept tagging, which is the most basic approach to interaction extraction. The randomized ratio of positive and negative interactions in the corpus is useful for training classification systems that operate on similarly pre-processed sentences to account for that naturally occurring ratio.

3.2 Annotation.

An annotation tool named STAV was used to create annotations ("stav," n.d.). Customization of the tool was performed to match the types of annotations needed for the corpus. The identified entities were formatted for use with the tool. Annotations created with the tool were stored in the BioNLP shared task file format. That format is compatible with a variety of existing systems for relationship extraction.

Annotation guidelines. Based on a review of literature on related annotation guidelines for relationships such as PPIs, an initial annotation guideline was created based on a small sample of sentences. The guide was iteratively refined through annotation of additional sentences, until considered sufficiently stable for release to additional annotators.

The guideline was refined to achieve a balance of complexity and clarity to assist annotators.

Only a few (5-10) example sentences per annotator have been discussed in person. The explicit written instructions in the guide were relied on more than in-person example sentence discussions to train annotators to handle the complicated content of the corpus and avoid over-influencing the annotators, as noted that is possible with the overuse of those examples (Hovy & Lavid, 2008).

The first annotator, a student with a Bachelor of Science (BS) in Biology, was the main annotator and author of the guidelines. The second and third annotators are PhD students in Biomedical Informatics, the second with a BS in Biology and 10 years nursing experience, and the other with a Bachelor of Technology in Bioinformatics. Weekly annotation meetings were done on individual bases. A short checklist of things to look for in annotations was distributed in addition to the guidelines.

Annotations. The following describes major annotation categories and subcategories in the corpus:

- **Interaction** Genes and drugs are annotated simply as "having an interaction" broadly understood as having an "action, effect, or influence" on each other. All gene-drug interactions annotated must have at least one interaction term that helps explain the interaction. Additional properties that were annotated and a brief explanation of their purpose include:
    - **Direct/Indirect:** Describes the complexity in the interaction statements. An "indirect" interaction is one where the presence of an intermediary entity is needed for semantic understanding of the interaction.
    - **Explicit/Inferred:** Records if an inference had to be made on whether the interaction was present because an interaction was not explicitly stated.
- **Non-interaction**
    - **Shared Entity:** An entity connected to both a gene and a drug that don't interact with each other. In contrast to an intermediary entity.
- **Interaction Term** Terms that are descriptive of the interaction (as defined earlier). These terms are helpful for capturing more specifically the type of interaction present.
- **Intermediary Entity** These are non-gene, non-drug entities that are closely connected to the interaction. They are entities that are needed for understanding of the full semantic meaning of gene-drug interactions. These entities are not annotated themselves but they are used to determine the indirectness property.

Examples of these categories can be seen in the sentence: "Using standard steady-state kinetic analysis, it was demonstrated that *paclitaxel* was a possible uncompetitive inhibitor to NAT activity in cytosols based on the decrease in apparent values of K(m) and V(max)." (PMID: 11955677). This sentence includes an interaction between the drug *paclitaxel* and gene *NAT*. An interaction term that helps establish that the interaction is present is "inhibitor". "Cytosols" is where the NAT inhibition activity can occur and represents an intermediary entity that is needed in the semantic meaning of the interaction.

The broad definition of interaction was used to make progress toward annotations including, and in turn being representative of, the most general form of gene-drug interaction that is described in the source abstracts. We chose to first concentrate on getting good inter-annotator agreement using the general definition before considering additionally annotating specific biological interaction types. Annotated interactions are required to have at least one annotated interaction term (although terms do not have to be from the predefined list) to ensure that specific and identifiable language is present that justifies the annotation.

The subcategories included were added to record the linguistic complexity in which the interactions and non-interactions are described. Recording that complexity can help system developers handle its presence when trying to automatically recognize interaction statements. Additionally, the annotation properties of speculation, negation, and nesting were allowed but not separately annotated in interaction annotations.

Each annotator reported annotation time estimates. Total time spent on annotations including meetings but not other work (e.g. guideline development) was approximately 80 hours for the primary annotator and 20 hours combined for other annotators. Hard sentences to annotate required research into source articles and entities described.

Evaluation of the Corpus. Around 26% of the corpus was annotated by a second and third annotator. A program was created for IAA scoring, accounting for nested entities and equivalent entities including abbreviations. Manual review was used to verify the program's scores. Example sentences from the corpus discussed with annotators were not used for IAA scoring.

3.3 Relationship Extraction methods.

Three basic methods for extracting interactions were implemented for evaluation. The basic method, co-occurrence, is inherent to the corpus as all sentences are selected based on both entities being present in them. Thus, in co-occurrence, any mention of a gene and a drug together in a sentence represents an interaction between those entities.

Co-occurrence plus interaction terms, the second method tried, identifies that interactions are present only when sentences contain an interaction word from a predefined list. The list of interaction terms obtained from the *Phare* ontology was filtered by removing common stop words. Also, a filter was applied to only use terms greater than two letters in size. Those filters were used to avoid unneeded matches from common words.

The linguistic pattern based extraction method developed for this evaluation was based on the work by Coulet et. al. Specific linguistic patterns described in that work were used to classify the presence of interactions between genes and drugs. A program named Graph Spider was used to match the specified patterns within sentences (Shepherd & Clegg, 2008). The Stanford Parser was used to generate dependency graphs for use with the pattern recognition in Graph Spider.

The dependency rules designed by Coulet. et. al. were entered into Graph Spider using the metapattern language (MPL) designed by the Graph Spider authors. MPL is a pattern formalism that can be used to match dependency subgraph patterns in dependency parsed text. After dependency graphs were generated for processing in Graph Spider, text representing genes and drugs in the graphs were converted to general tags for those entity types. Those conversions were made to allow the patterns in MPL to be generalizable.

Java programs were created to reformat and score the subgraph pattern match results made by Graph Spider. Scoring used text character positions (spans) of entities included in annotations. True positives were recorded when pairs of entity spans in Graph Spider subgraph results matched annotated pairs of entity spans labeled as having interactions. False positives and false negatives were similarly assessed using entity spans. A manual evaluation of pattern matched output compared to annotations was performed to ensure accuracy.

A condition applied in the pattern based system was that the patterns can match up to four modifier words for each individual gene and drug in interaction pattern matches. Those words are additional words that modify the meaning of the gene or drug in the interaction. The limit was included for practical reasons, as hand coding of patterns in MPL is complex. The rules described by Coulet et. al. did not specify any limit on modifier words but the difference in results by including a realistic limit is predicted to be negligible.

4 Results

A total of 551 sentences are annotated, with 781 interactions present in them. There are 351 instances of non-interactive entities in the same set. The average length of sentences is 28.1 words. Table 1 describes further properties of the corpus.

Annotation Analysis. The inter-annotator agreement scores are reported as accuracy and Cohen's kappa. Kappa was chosen due to its widespread use and therefore comparability with other work in corpus creation. Accuracy is found by the number of instances agreed on divided by the total instances annotated. A total of 144 sentences were used for the scoring. Annotators 1 and 2, 1 and 3, and 2 and 3 were compared using 92, 52, and 61 sentences respectively. IAA results with the main categories of interaction vs. non-interaction are shown in Table 2.

	1 & 2	1 & 3	2 & 3
Accuracy	81.1%	74.2%	73.0%
Kappa	45.7%	30.5%	11.4%

Table 2. Inter-annotator agreement results.

Sentences	Tokens (with punctuation)	Words (tokens with no punctuation)
551	18,585	15,464

Table 1. Statistics describing corpus properties.

IAA scores were found for all annotated subcategories. Those subcategories are DirectExplicit, IndirectExplicit, IndirectInferred for interactions and SharedEntity for non-interactions. Their ranges of scores with all annotator pair groups using accuracy scores are 72-79%, 40-69%, 62-82%, 50-60% and kappa scores are 31-58%, 1-27%, -4-31%, 0-4% respectively. Those scores are created by selecting main category inter-annotator matches (e.g. interaction) and calculating the IAA between the annotated subcategories.

In some sentences, annotators missed doing annotations for gene-drug instances that the other annotator added. IAA scores did not include annotations made by only one annotator. Confirmation with annotators was made that annotations not made were not intended to represent non-interactions. The percentage of missed inter-annotator instances was approximately 20%. Future work will be to improve the inter-annotator annotation process so that those instances are not missed for IAA scoring. While some annotations were missed in IAA scoring, annotations by the primary annotator that are included in the corpus contain all instances (none missed) from the source text to our knowledge.

ID	Contents	Agreement	Sentence text
A	One direct explicit interaction	Y	This suggests that galantamine (GAL), a cholinesterase inhibitor, could be effective when seeking to prolong abstinence in recently detoxified alcoholics. (PMID: 16328375)
B	One indirect explicit and four shared entity non-interactions	Y	They are widely distributed and mediate all of the known biologic effects of angiotensin II (AngII) through a variety of signal transduction systems, including activation of phospholipases C and A2, inhibition of adenylate cyclase, opening of calcium channels, and activation of tyrosine kinases. (PMID: 9892138)
C	One indirect explicit interaction	N	The results of studies of perfused rat hearts with completely inhibited creatine kinase show significantly decreased work capacity and respectively, energy fluxes, in these hearts in spite of significant activation of adenylate kinase system (Dzeja et al. this volume). (PMID: 9746326)

Table 3. Example sentences from the corpus.

Interaction Extractor Type	Precision (TP/TP+FP)	Recall (TP/TP+FN)	F1-Score (2*((P*R)/(P+R)))
Co-occurrence	68.99% (781/1132)	100.00% (781/781)	81.65%
Co-occurrence plus int. terms	69.60% (664/954)	85.02% (664/781)	76.54%
Pattern-based	96.61% (57/59)	7.30% (57/781)	13.57%

Table 4. Extraction system performances. Note that sentences were selected based on co-occurrence of a gene and a drug, thus recall is 100% for that method, as it essentially defines the corpus.

The scoring methods used were instance level scoring instead of sentence level scoring. In the instance level scoring each gene-drug instance counted in performance scores.

A caveat about the pattern-based system scoring should be noted. That caveat was that the Graph Spider software used was unable to process approximately 10% (around 50) of the sentences in the corpus due to errors. The pattern-based system is likely to have scored slightly higher if it could have processed those sentences.

5 Discussion

5.1 Analyses of interaction extraction methods performance.

The f-score of co-occurrence with and without interaction terms showed better performance than the pattern-based interaction extractions, which was expected. Pattern based methods, particularly those where the patterns were manually created, are typically very high in precision and very low in recall, as they are highly dependant on the specific patterns included for recognition. Although recall was low, users who want very high confidence interaction predictions or interactions of a very specific type can benefit from the pattern-based system's demonstrated high precision. Co-occurrence can suit users who want to focus on recall.

Coulet et al. reported their system scored a precision of 70% for exact match and 87.7% for exact or incomplete match but true classification. Our results are similar to their 87.7% results in both percentage and scoring method. The method that allows incompleteness accepts matches that accurately identify core pharmacogenomic relationships but don't need to correctly match modifier words. Our scoring is similar in not needing to match modifier words. The similarity in results indicates that we correctly implemented the system that Coulet et al. designed. That indication does have the limitation that the 10% of sentences unable to be processed may have affected the results.

An example of a more complex interaction that was matched by co-occurrence with an interaction term but not the pattern-based method was "Moreover, S-nitrosylation of thioredoxin was also significantly augmented after atorvastatin treatment." (PMID: 15289372). In that sentence, an interaction occurred where thioredoxin's (gene) S-nitrosylation was augmented by atorvastatin (drug). Analysis of the dependency graphs used by the pattern-based system revealed some reasons why it was unable to identify the interaction.

The pattern-based system uses a rule that applies to that sentence: a potential pattern sequence match can be "interrupted" by a dependency that does not fit accepted patterns. In the non-classified sentence, the entities "was" and "augmented" were terms that caused the pattern matching to be interrupted. Both "was" and "augmented" are not nouns or prepositions. They both also are needed in the dependency subgraph that connects the gene and drug together. Those parts of speech are not allowed to be chained together in the pattern-based system's patterns. That deviation from the allowed patterns caused the system to miss that interaction.

Adding patterns with more diversity in allowed parts of speech in series of interaction terms that connect genes and drugs in interactions can improve recall performance. A review of parts of speech (POS) in missed matches showed that some misses were due to no verb POS tags being present in interaction descriptions. That can occur when verbs are in their nominalized form or other situations. Mining the corpus for both part of speech and dependency graph patterns can identify patterns that are able to correct those misses. Also, the POS tagger included with the parser mis-tagged a variety of words. Using a higher performance tagger or one trained on biomedical text may help with pattern matches.

Ahlers et. al. also reported relationship extraction performance from a new system with their gene-drug corpus. That system achieved a precision of 73% and recall of 50% extracting an annotation category including gene-drug relationships. The system is built upon an earlier system and an important part of its capabilities comes from specialized linguistic rules it uses. The corpus included in this work can be useful for further development of systems that integrate such rules with other methods to improve extraction performances.

Some characteristics were notable about the results of the methods using co-occurrence with and without interaction terms. The performances found of those methods may be specific to an increased amount of gene-drug interactions found in the journals used compared to other journals. Also, the use of interaction terms from the Phare ontology was expected to increase precision because they were found from predicted pharmacogenomic relationships. The co-occurrence with interaction terms method resulted in only approximately equaling the precision of basic co-occurrence. One possible reason for that is the terms were originally found partly with disease relationships. They therefore can be less relevant to gene-drug interactions.

5.2 Analyses of annotations

Table 2 includes that the general interaction annotations had the kappa values 46%, 30%, 11% which are considered only moderate to low scores by common rating methods. Some IAA scores, such as kappa, include a correction for chance agreement probability. An intentional design choice was made in the corpus to allow an unbalanced but natural ratio of interactions to non-interactions. That imbalance increased kappa's correction. Although our reasonably high IAA scores with accuracy helped increase the kappa score, they were not enough to offset the correction and bring kappa above the moderate score.

An article by Strijbos et. al. states that kappa can have a strict chance agreement correction in the case of few categories (Strijbos, Martens, Prins, & Jochems, 2006). Given that general interaction scores were only based on the categories of present or absent, kappa may have been overly strict with the correction. If that correction in our data is not strict, but justified, than that indicates how further improving our annotation process can be valuable. Further investigation will go into understanding what statistics may be useful for scoring given the corpus properties. Exploration will also continue with talking to annotator s about what may be causing disagreement. That exploration will help reveal ways to improve IAA.

Subcategories showed mixed results in their IAA performances. The subcategories with the highest IAA scores may indicate that those subcategories are more clearly defined than others in the annotation guide.

Reviewing some annotated sentences can help clarify how the IAA results occurred. All annotators agreed the drug galantamine has a direct explicit interaction with cholinesterase in sentence A in Table 3. Such an interaction description is simply described and an annotator has reported that type of interaction being the easiest to identify.

Agreement was found with all annotators for annotations in sentence B in Table 3. It was readily understandable to annotators that calcium and other signal transduction systems do not have an interaction simply for all being a part of those types of systems.

An example of a sentence with annotator disagreement was sentence C in table 3. Although endogenously produced in this case, the nested entity creatine was considered a drug due to being relevant to creatine in its exogenous drug form.

The occurrence of multiple properties, such as inhibition and effects on hearts can make it difficult to follow the logic of the interaction between creatine and adenylate kinase (enzyme). The interaction annotation can be hard for annotators to find due to that complexity and the subtleness of the "in spite of" phrase describing the negated effect between the drug and gene. The interaction is negated but that still is considered an interaction by the annotation rules used.

5.3 Future Work

As mentioned before, the corpus will grow from around 500 sentences that it has right now to around 1,500. The larger the corpus expands to be, the more representative it will become of gene-drug interactions. Other future work includes work with more advanced interaction extraction systems.

Along with this publication, a version of the corpus with high confidence in annotations will be released. Given that this is an initial work, a relatively modest amount of annotation revisions may occur with a few periodic later version releases of the corpus to improve its quality.

Unfortunately no tagger is perfect so as annotations proceed, drugs or genes that were missed by the tagger can be investigated to further understand why that occurred. An example of a commonly missed drug was acetylcholine. Acetylcholine was picked up as a drug if it was spelled out, but not if it was abbreviated as ACh and it is commonly abbreviated.

6 Conclusion

The extraction results indicated that the systems tested can be utilized and built upon according to user preferences in precision, recall, or specific interaction terms. The corpus presented here offers valuable utility to system developers working toward achieving favorable balances of precision and recall in gene-drug interaction extractions. The growth of that corpus will also increasingly benefit the developers working on those extractions. That type of extraction is important to advancing work in pharmacogenomics by retrieving knowledge for individuals working in the field.

Acknowledgements

The authors wish to thank Ehsan Emadzadeh for his help with the annotation tool and Robert Leaman for his help with annotation methods.

References

Ahlers, C., Fiszman, M., Demner-Fushman, D., Lang, F.-M., & Rindflesch, T. (2007). Extracting semantic predications from Medline citations for pharmacogenomics. *Pacific Symposium on Biocomputing. Pacific Symposium on Biocomputing*, 209–220.

Bui, Q.-C., Nualláin, B. O., Boucher, C. A., & Sloot, P. M. A. (2010). Extracting causal relations on HIV drug resistance from literature. *BMC Bioinformatics*, *11*, 101. doi:10.1186/1471-2105-11-101

Cheng, D., Knox, C., Young, N., Stothard, P., Damaraju, S., & Wishart, D. S. (2008). PolySearch: a web-based text mining system for extracting relationships between human diseases, genes, mutations, drugs and metabolites. *Nucleic Acids Research*, *36*(Web Server issue), W399–405. doi:10.1093/nar/gkn296

Chowdhary, R., Zhang, J., & Liu, J. S. (2009). Bayesian inference of protein-protein interactions from biological literature. *Bioinformatics (Oxford, England)*, *25*(12), 1536–1542. doi:10.1093/bioinformatics/btp245

Coulet, A., Shah, N. H., Garten, Y., Musen, M., & Altman, R. B. (2010). Using text to build semantic networks for pharmacogenomics. *Journal of Biomedical Informatics*, *43*(6), 1009–1019. doi:10.1016/j.jbi.2010.08.005

Garten, Y., & Altman, R. B. (2009). Pharmspresso: a text mining tool for extraction of pharmacogenomic concepts and relationships from full text. *BMC Bioinformatics*, *10 Suppl 2*, S6. doi:10.1186/1471-2105-10-S2-S6

Hakenberg, J., Leaman, R., Vo, N. H., Jonnalagadda, S., Sullivan, R., Miller, C., Tari, L., et al. (2010). Efficient extraction of protein-protein interactions from full-text articles. *IEEE/ACM Transactions on Computational Biology and Bioinformatics / IEEE, ACM*, *7*(3), 481–494. doi:10.1109/TCBB.2010.51

Hewett, M., Oliver, D. E., Rubin, D. L., Easton, K. L., Stuart, J. M., Altman, R. B., & Klein, T. E. (2002). PharmGKB: The Pharmacogenetics Knowledge Base. *Nucleic Acids Research*, *30*(1), 163–165. doi:10.1093/nar/30.1.163

Krallinger, M., Leitner, F., Rodriguez-Penagos, C., & Valencia, A. (2008). Overview of the protein-protein interaction annotation extraction task of BioCreative II. *Genome Biology*, *9 Suppl 2*, S4. doi:10.1186/gb-2008-9-s2-s4

Leaman, R., & Gonzalez, G. (2008). BANNER: an executable survey of advances in biomedical named entity recognition. *Pacific Symposium on Biocomputing. Pacific Symposium on Biocomputing*, 652–663.

Luis Tari, Jörg Hakenberg, Graciela Gonzalez, & Baral, C. (2009). Querying parse tree database of medline text to synthesize user-specific biomolecular networks. CiteSeerX. Retrieved from http://citeseerx.ist.psu.edu/viewdoc/summary?doi=10.1.1.140.8574

Pyysalo, S., Ginter, F., Heimonen, J., Björne, J., Boberg, J., Järvinen, J., & Salakoski, T. (2007). BioInfer: a corpus for information extraction in the biomedical domain. *BMC Bioinformatics*, *8*, 50. doi:10.1186/1471-2105-8-50

Rebholz-Schuhmann, D., Kirsch, H., Arregui, M., Gaudan, S., Riethoven, M., & Stoehr, P. (2007). EBIMed—text Crunching to Gather Facts for Proteins from Medline. *Bioinformatics*, *23*(2), e237–e244. doi:10.1093/bioinformatics/btl302

Sconce, E. A., Daly, A. K., Khan, T. I., Wynne, H. A., & Kamali, F. (2006). APOE genotype makes a small contribution to warfarin dose requirements. *Pharmacogenetics and Genomics*, *16*(8), 609–611. doi:10.1097/01.fpc.0000220567.98089.b5

Shepherd, A. J., & Clegg, A. B. (2008). Syntactic pattern matching with GraphSpider and MPL. *Proceedings of the Third International Symposium on Semantic Mining in Biomedicine SMBM 2008 Turku Finland*, 129–132.

stav. (n.d.).*GitHub*. Retrieved March 26, 2012, from https://github.com/TsujiiLaboratory/stav

Strijbos, J.-W., Martens, R. L., Prins, F. J., & Jochems, W. M. G. (2006). Content analysis: What are they talking about? *Computers & Education*, *46*(1), 29–48. doi:10.1016/j.compedu.2005.04.002

T1.pdf. (n.d.). Retrieved from http://www.lrec-conf.org/proceedings/lrec2008/workshops/T1.pdf

The OpenNLP Homepage. (n.d.). Retrieved March 26, 2012, from http://opennlp.sourceforge.net/projects.html

Wishart, D. S. (2006). DrugBank: a comprehensive resource for in silico drug discovery and exploration. *Nucleic Acids Research*, *34*(90001), D668–D672. doi:10.1093/nar/gkj067

A Preliminary Work on Symptom Name Recognition from Free-Text Clinical Records of Traditional Chinese Medicine using Conditional Random Fields and Reasonable Features

Yaqiang Wang, Yiguang Liu, Zhonghua Yu*, Li Chen
Department of Computer Science
Sichuan University
Chengdu, Sichuan 610064, China
yaq.wang@yahoo.com, lygpapers@yahoo.com.cn, yuzhonghua@scu.edu.cn, cl@scu.edu.cn

Yongguang Jiang
Department of Preclinical Medicine
Chengdu University of TCM
Chengdu, Sichuan 610075, China
cdtcm@163.com

Abstract

A preliminary work on symptom name recognition from free-text clinical records (FCRs) of traditional Chinese medicine (TCM) is depicted in this paper. This problem is viewed as labeling each character in FCRs of TCM with a pre-defined tag ("B-SYC", "I-SYC" or "O-SYC") to indicate the character's role (a beginning, inside or outside part of a symptom name). The task is handled by Conditional Random Fields (CRFs) based on two types of features. The symptom name recognition F-Measure can reach up to 62.829% with recognition rate 93.403% and recognition error rate 52.665% under our experiment settings. The feasibility and effectiveness of the methods and reasonable features are verified, and several interesting and helpful results are shown. A detailed analysis for recognizing symptom names from FCRs of TCM is presented through analyzing labeling results of CRFs.

1 Introduction

Traditional Chinese medicine (TCM), a complementary medical theory to western medicine, provides a distinct way to view our human life (Pal, 2002; Barnes, et al., 2004; Molassiotis, et al., 2005). Moreover, it has shown that TCM knowledge, which is accumulated in clinical practice, has become one of the most important sources of modern biomedical research (Zhou, et al., 2010).

In recent years, Data Mining and Machine Learning have been more than ever before applied to TCM clinical research, such as establishing TCM diagnosis expert systems for supporting decision making (Wang, et al., 2004; Huang and Chen, 2007; Zhang, et al., 2008). However, most of the works are based on manually well-structured datasets.

Because of the high cost of manually structuring and maintaining free-text clinical records (FCRs) of TCM, large volume of such datasets has not been exploited effectively (Zhou, et al., 2010), although they are significant for discovering new knowledge or capturing medical regularities. Therefore, developing appropriate information extraction methods for handling FCRs of TCM is an urgent need to reduce the manual labor for researchers.

Automatically extracting meaningful information and knowledge from FCRs of TCM is challenging in Data Mining and Machine Learning fields (Zhou, et al., 2010). As the basis, symptom name recognition or extraction from FCRs of TCM is in an early stage. To the best of our knowledge, there has little work to solve this problem (Wang, et al., 2010; Wang, et al., 2012). Symptom name recognition from FCRs of TCM was firstly attempted in (Wang, et al., 2010) through normalizing the symptom names in clinical records based on literal similarity and remedy-based similarity methods but not directly recognizing original clinical symptom names from FCRs of TCM. In 2012, Wang, et al. proposed a framework of automatic diagnosis of TCM for practice. Symptom name recognition is one part of the framework and simp-

*Corresponding author

ly attempted through a literal similarity method without detailed analysis (summarized procedures for the previous wok are shown in figure 1).

Wang, Y., et al., 2010:

-> Input FCR's clauses "昨日肠鸣", "失气多", "心中不适"

-> Output the standard symptom names that are most similar to these clauses (the similarity measured by literal similarity metrics, remedy-based similarity metrics or hybrid similarity metrics)

Wang, Y., et al., 2012:

(1) Dictionary-based method:

-> Input a FCR "昨日肠鸣, 失气多, 心中不适"

-> Based on symptom name dictionary matching symptom name

-> Output symptom names: "肠鸣", "失气多", "心中不适"

(2) Bigram-based method & literal similarity method:

-> Input a FCR "昨日肠鸣, 失气多, 心中不适"

-> Segment the FCR into bigrams "昨日, 日肠, 肠鸣, 失气, ..., 不适"

-> Generating all possible bigram combination lists through combining and merging neighbor bigrams

-> Output the combination list that has the highest *CombValue* defined in the paper

Figure 1. Simple Conclusions of the Previous Work.

Named Entity Recognition (NER) has been widely studied. There have been lots of methods for Chinese NER (Zhang, et al., 2003; Wu, et al., 2003; Gao, et al., 2005; Fu and Luke, 2005; Zhou, 2006; Duan and Zhang, 2011). However, these methods cannot be directly applied on symptom name recognition from FCRs of TCM due to big differences of characteristics of the corpus (Wang, et al., 2012). There are also several related work on English NER, but Chinese NER has more challenges because of the distinct characteristics of Chinese (Wu, et al., 2003).

In this paper, the task of symptom name recognition from FCRs of TCM is studied. The symptom names are recognized through finding their description boundaries from FCRs of TCM, and the method is described in section 2. Several reasonable and helpful features are introduced for CRFs to label the characters in FCRs of TCM with pre-defined boundary tags to indicate their roles (a beginning, inside or outside part of a symptom name) (presented in section 3). At last, several interesting and valuable experimental results are shown in section 4 and a conclusion is given in section 5.

2 Symptom Name Recognition from FCRs of TCM

The task of symptom name recognition from FCRs of TCM can be treated as detecting the boundaries of the symptom name descriptions in the sentences of FCRs of TCM. Therefore, this task can be viewed as labeling each tagging unit (e.g. word) in the sentences with a pre-defined tag indicating whether the unit is a beginning, inside, or outside part of a symptom name.

Generally, the tagging unit is word (Ramshaw and Marcus, 1995). However, there is no natural segmentation for words in Chinese sentences. Therefore, Chinese word segmentation problem has to face up firstly (Gao, et al., 2005). Because of the characteristics of FCRs of TCM (Wang, et al., 2012), automatically segmenting FCRs of TCM into words is not trivial and common Chinese word segmentation methods are not suitable. In order to tackle this problem, Chinese character is settled as the basic tagging unit. An example sentence of the labeling task is shown in figure 2.

昨 日 肠 鸣 , 失 气 多 , 心 中 不 适

O-SYC O-SYC [B-SYC I-SYC] [B-SYC I-SYC I-SYC] [B-SYC I-SYC I-SYC I-SYC]

Figure 2. An Example Sentence of the Symptom Name Recognition Task.

In figure 2, each character is labeled with a pre-defined tag ("B-SYC", "I-SYC" or "O-SYC"). The meaning of each tag is defined in table 1.

Tag	Meaning
B-SYC	Beginning of a TCM symptom name
I-SYC	Inside a TCM symptom name
O-SYC	Outside the TCM symptom names

Table 1. Meanings of the Pre-defined Tags.

Consequently, a recognized symptom name should start with a character labeled with "B-SYC" and end before the character whose corresponding label changes from "I-SYC" to "B-SYC" or "O-SYC" for the first time. The labeling task can be formulated as follows:

Given a FCR $\mathbf{x} = x_1, x_2, ..., x_n$, where x_i is a Chinese character, the goal is to build a annotator p to accurately label $\mathbf{x}$ with the credible corresponding tag sequence $\mathbf{y} = p(\mathbf{x})$, where $\mathbf{y} = y_1, y_2, ..., y_n$ and $y_n \in \{\text{B-SYC,I-SYC,O-SYC}\}$. This task can be effectively done by CRFs (Sha and Pereira, 2003) based on a training dataset which is consisted of pairs of sequences $(\mathbf{x,y})$.

3 Conditional Random Fields for Symptom Name Recognition

3.1 Conditional Random Fields

A Conditional Random Field can be defined as an undirected graphical model (see figure 3) which consists of a sequence of vertices representing random variables $\mathbf{Y} = (Y_1, Y_2, ..., Y_n)$ and edges representing conditional dependencies, conditioned on $\mathbf{X} = (X_1, X_2, ..., X_n)$. The random variable Y_i only has edges with its predecessor Y_{i-1} and successor Y_{i+1}, thus, random variables $Y_1, Y_2, ..., Y_n$ obey the Markov property and form a linear Markov chain.

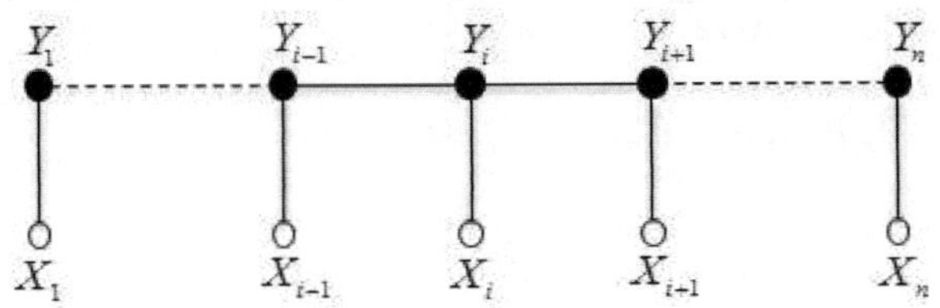

Figure 3. An Undirected Graphical Structure for a Conditional Random Field.

Then the conditional probability of a label sequence given an input sequence can be defined as:

$$p_{\boldsymbol{\lambda}}(\mathbf{y}, \mathbf{x}) = \frac{\exp \boldsymbol{\lambda} \cdot \sum_{i=1}^{n} \boldsymbol{f}(\mathbf{y}, \mathbf{x}, i)}{Z_{\boldsymbol{\lambda}}(\mathbf{x})}$$

Where $\boldsymbol{f}$ is a *global feature vector* (Sha and Pereira, 2003) and each element of $\boldsymbol{f}$ is an arbitrary feature selection function f_k ($k \in [1, K]$, where K is the number of feature functions). $\boldsymbol{\lambda}$ is a weight vector comprised by the learned weight λ_k for each feature function. More detailed description is that,

$$p(\mathbf{y} \mid \mathbf{x}) = \frac{\exp\left(\sum_{i=1}^{n} \sum_{k=1}^{K} \lambda_k f_k (\mathrm{y}_{i-1}, \mathrm{y}_i, \mathbf{x}, i) \right)}{Z(\mathbf{x})}$$

$Z(\mathbf{x})$ in the equation is a normalization factor which is the sum over all possible label sequences S:

$$Z(\mathbf{x}) = \sum_{S} \exp\left(\sum_{i=1}^{n} \sum_{k=1}^{K} \lambda_k f_k (\mathrm{y}_{i-1}, \mathrm{y}_i, \mathbf{x}, i) \right)$$

The most likely label sequence for an input sequence $\mathbf{x}$ is:

$$\mathbf{y} = \arg\max_{\mathrm{y}} p(\mathbf{y} \mid \mathbf{x})$$

It can be found with the Viterbi algorithm. We use the CRF++ tool in the experiments, which provides an efficient implementation for CRFs by using the limited-memory quasi-Newton algorithm for training the models (Sha and Pereira, 2003; Lafferty, et al., 2001) and the default settings of CRF++ are used.

3.2 Features for Labeling

It is difficult to analyze the syntactic structure of the content in FCRs of TCM which has narrative form, concise style and nonstandard description characteristics. Therefore, no higher level syntactic features, such as POS tags or NP chunks, can be used at the moment. Through analyzing FCRs of TCM, two types of representative and reasonable features (i.e. literal features and positional features) are exploited. The features are introduced and their reasonableness is explained by examples as follows.

Literal Features: the simplest and the most obvious features for determining the boundaries of symptom name descriptions are literal features. For example, according to the observation that after a word which is used to specify time (e.g. "昨日" (yesterday)) there would usually follow a symptom name description, such as "肠鸣" (borborygmus).

The best approach to get such features is to divide the content of FCRs of TCM into words. However, as described before, Chinese word segmentation is not trivial work. Fortunately, segmenting the content into n-grams is considerable and reasonable, because the indicating words would be mixed in the n-gram segments and could be helpful to determine the boundaries of symptom name descriptions.

Furthermore, the FCRs of TCM have a concise style, i.e. the length of the clauses in FCRs of TCM is short and words are usually used in their brief form. Therefore, the n-grams as the literal features need not be too long. In general, the average length of a Chinese word approximates 2 (Nie, et al., 2000). Consequently, the value of n should set to range from 1 to 3. Moreover, according to the intuition that "the distance between current character and its related n-grams in FCRs of TCM would not be too far", the context window size, which is the fragment scope picking up literal features (i.e. n-grams (see examples in table 2)) in FCRs of TCM, would not be too large. Otherwise it would bring about noisy information, thereby reducing the labeling precision. The context window size in our experiment is specified smaller then 4.

Feature Type	Context Window Size (CWS)	Literal feature examples under different CWS
Unigram Features (Uni)	1	C_{i-1}, C_i, C_{i+1}
	2	C_{i-2}, C_{i-1}, C_i, C_{i+1}, C_{i+2}
	3	C_{i-3}, C_{i-2}, C_{i-1}, C_i, C_{i+1}, C_{i+2}, C_{i+3}
	4	C_{i-4}, C_{i-3}, C_{i-2}, C_{i-1}, C_i, C_{i+1}, C_{i+2}, C_{i+3}, C_{i+4}
Bigram Features (Big)	1	$C_{i-1}C_i$, $C_i\ C_{i+1}$
	2	$C_{i-2}C_{i-1}$, $C_{i-1}C_i$, C_iC_{i+1}, $C_{i+1}C_{i+2}$
	3	$C_{i-3}C_{i-2}$, $C_{i-2}C_{i-1}$, $C_{i-1}C_i$, C_iC_{i+1}, $C_{i+1}C_{i+2}$, $C_{i+2}C_{i+3}$
	4	$C_{i-4}C_{i-3}$, $C_{i-3}C_{i-2}$, $C_{i-2}C_{i-1}$, $C_{i-1}C_i$, C_iC_{i+1}, $C_{i+1}C_{i+2}$, $C_{i+2}C_{i+3}$, $C_{i+3}C_{i+4}$
Trigram Features (Tri)	1	$C_{i-1}C_iC_{i+1}$
	2	$C_{i-2}C_{i-1}C_i$, $C_{i-1}C_iC_{i+1}$, $C_iC_{i+1}C_{i+2}$
	3	$C_{i-3}C_{i-2}C_{i-1}$, $C_{i-2}C_{i-1}C_i$, $C_{i-1}C_iC_{i+1}$, $C_iC_{i+1}C_{i+2}$, $C_{i+1}C_{i+2}C_{i+3}$
	4	$C_{i-4}C_{i-3}C_{i-2}C_{i-1}$, $C_{i-3}C_{i-2}C_{i-1}C_i$, $C_{i-2}C_{i-1}C_iC_{i+1}$, $C_{i-1}C_iC_{i+1}C_{i+2}$, $C_iC_{i+1}C_{i+2}C_{i+3}$, $C_{i+1}C_{i+2}C_{i+3}C_{i+4}$

Table 2. Literal Feature Examples Used in the Experiments. C_i is the character at current position i in one clause.

Positional Features: positions of characters in FCRs of TCM are also helpful. They are assistant features to determine the boundaries of symptom name descriptions.

The start of a sentence would be usually a common character (i.e. its corresponding label is "O-SYC") rather than the beginning of a symptom name description. On the contrary, the starting positions of the following clauses have higher probabilities to be labeled with "B-SYC". Taking the FCR "昨日肠鸣, 失气多, 心中不适" (Yesterday, the patient had borborygmus and more farting, and his/her heart was uncomfortable) as an example, it starts with a common word "昨日" (yesterday) followed by a symptom name "肠鸣" (borborygmus). And at the same time, following clauses all start with symptom name descriptions.

The example of positional features is shown in figure 4.

Original record:

昨 日 肠 鸣 , 失 气 多 , 心 中 不 适

Transformed positional features:

[1-1] [1-2] [1-3] [1-4] [2-1] [2-2] [2-3] [3-1] [3-2] [3-3] [3-4]

Figure 4. Example of Positional Features.

In figure 4, one "[*SubSID-POS*]" represents a positional feature, and *SubSID* is the index of current clause in a FCR and *POS* indicates the position of a character in current clause.

4 Experiments

In this section, the proposed method for symptom name recognition from TCM FCRs is evaluated, and the usefulness of the introduced features is verified based on a TCM clinical dataset. The results are depicted bellow.

4.1 Experimental Datasets

In this paper, a clinical record dataset (CRD) is used. It contains 11613 FCRs of TCM and was collected by TCM doctors during their routine diagnostic work. The Chinese characters in FCRs of CRD are annotated with tags "B-SYC", "I-SYC", and "O-SYC". The number of each type of tags is 69193, 104243 and 142860, respectively. There are 4235 unique symptom names in CRD, and the amount of annotated symptom names is 69193.

	Training Data	Test Data
Number of Unique Symptom Names	1813	3463
Amount of Symptom Names	17339	51854
Number of Each Type of Tags ("B-SYC", "I-SYC", "O-SYC")	17339, 25738, 35995	51854, 78505, 106865

Table 3. Detailed Information of the Training and Test Datasets.

CRD is divided into two sub-datasets (i.e. a training dataset (3483 FCRs, 25% of CRD) and a test dataset (8130 FCRs, 75% of CRD)). For convenience, all numbers (e.g. integers, decimals and fractions, etc.) in CRD are uniformly replaced by a English character "*N*" in advance. Detailed information of training and test datasets is listed in table 3.

4.2 Evaluation Metrics

A new method for symptom name recognition from FCRs of TCM is proposed and two types of features are introduced. To evaluate the feasibility and effectiveness of the method and features, two groups of evaluation metrics are designed: (1) for assessing the ability of symptom name recognition, symptom name recognition rate, recognition error rate and recognition F-Measure are defined; (2) for giving a detailed analysis, the labeling precision, recall, and F-Measure are exercised. The detailed explanations of these metrics are described below.

Symptom name recognition rate (RR_{det}), recognition error rate (RER_{det}) and recognition F-Measure (RFM_{det}): these metrics are designed for assessing capability of the proposed method for symptom name recognition from TCM FCRs. If and only if the boundary of a symptom name is labeled accurately (i.e. starting with "B-SYC" and ending with the first change from "I-SYC" to "B--SYC" or "O-SYC"), the recognized symptom name is correct. Higher RR_{det} and lower RER_{det} are achieved; better symptom name recognition performance RFM_{det} would be obtained. RR_{det}, RER_{det} and RFM_{det} are formulated as follows.

$$RR_{det} = \frac{|NSDC|}{|NCS|}$$

$$RER_{det} = \frac{|SD| - |NSDC|}{|SD|}$$

$$RFM_{det} = \frac{2 \cdot DR_{det} \cdot (1 - DER_{det})}{DR_{det} - DER_{det} + 1}$$

Where $|NSDC|$ is the number of symptom name recognized correctly from the test dataset, $|NCS|$ is the number of clinical symptom names in the test dataset, and $|SD|$ is the number of symptom name recognized.

Labeling precision (Pre_{lab}), recall (Rec_{lab}) and F-Measure (FM_{lab}): the metrics (Pre_{lab}, Rec_{lab} and FM_{lab}) are used to evaluate the performance of labeling Chinese character sequences of FCRs of TCM for giving a detailed analysis. They are defined below.

$$Pre_{lab} = \frac{|NCLC|}{|NCL|}$$

$$Rec_{lab} = \frac{|NCL|}{|NC|}$$

$$FM_{lab} = \frac{2 \cdot Pre_{lab} \cdot Rec_{lab}}{Pre_{lab} + Rec_{lab}}$$

Where $|NCLC|$ is the number of characters labeled correctly with their corresponding tags, $|NCL|$ is the number of characters labeled with tags, and $|NC|$ is the number of characters should be labeled.

4.3 Evaluation of Symptom Name Recognition Ability

Comprehensive evaluations of symptom name recognition ability using CRFs with reasonable features are shown in figure 5, 6 and 7. These figures show that CRFs with reasonable features for symptom name recognition from FCRs of TCM is feasible. The best RFM_{det} 62.829% (RR_{det} 93.403% and RER_{det} 52.665%) is achieved under settings CWS = 3 and features Uni+Big+Tri used.

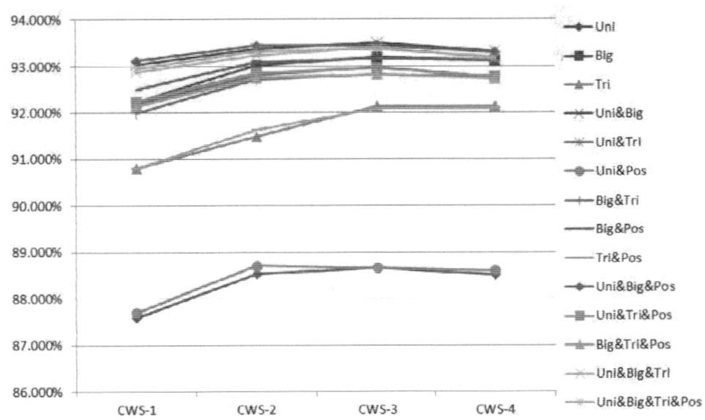

Figure 5. Symptom Name Recognition Rate.

It obviously shows in figures 5, 6 and 7 that literal features and positional features are helpful to symptom name recognition from FCRs of TCM. More types of features are used; better recognition performance would be obtained in most cases. When CWS = 1 and referred features changed from unigram literal features to the combination of unigram and bigram literal features, the highest growth about 3.925% of RFM_{det} is achieved (the

RR_{det} increases from 87.586% to 93.034% and the RER_{det} decreases from 56.173% to 53.118%).

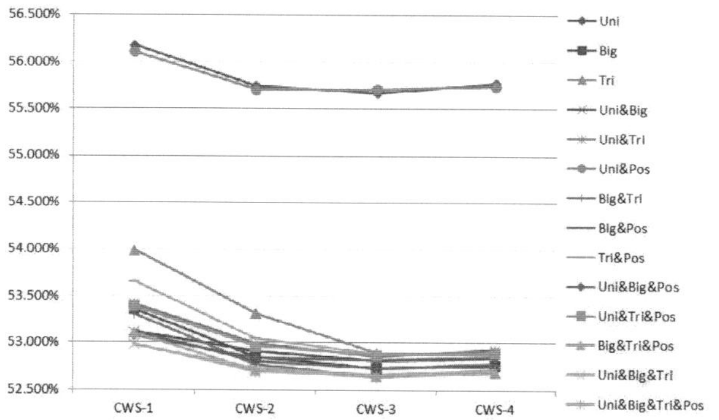

Figure 6. Symptom Name Recognition Error Rate.

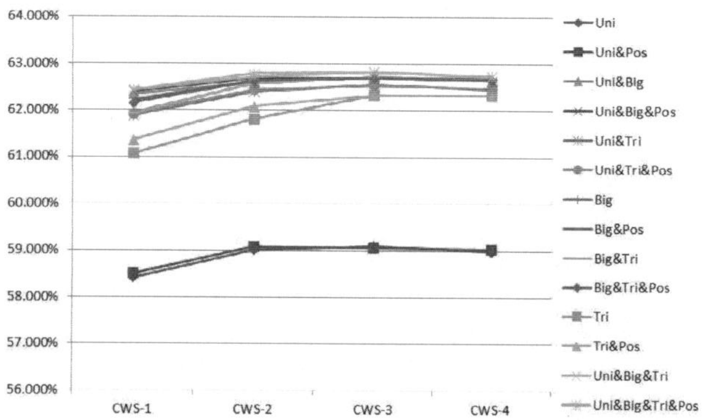

Figure 7. Symptom Name Recognition F-Measure.

As described previously, the context information is helpful to symptom name recognition. However, the context window size should not be too large. In figures 5, 6 and 7, it clearly shows that when CWS increase RR_{det} and RFM_{det} are improved and RFM_{det} is reduced. When CWS grows too large (larger than 3 here), RR_{det} and RFM_{det} begin, nevertheless, to descend and RER_{det} is raised in most every cases.

Moreover, positional features are complementary features to literal features for symptom name recognition from FCRs of TCM. It vividly shows in figures 5, 6 and 7 that RR_{det} and RFM_{det} would be improved and RER_{det} would be reduced more or less when literal features combined with positional features. The highest elevation can reach 0.297% if the combination features of trigram literal features and positional features are used and CWS = 1.

4.4 Evaluation of Labeling Performance and Detailed Analysis for Symptom Name Recognition

In this part, firstly, an evaluation for labeling performance is given, and then a detailed analysis for symptom name recognition from FCRs of TCM using CRFs with reasonable features would be described.

The results of Pre_{lab} and FM_{lab} under different situations are shown in figure 8 and 9, respectively. The Rec_{lab} here are all 100%. It can be seen from these figures that the FM_{lab} can reach nearly up to 97.596% with corresponding Pre_{lab} 95.305%. The results can also demonstrate the feasibility of the proposed method for symptom name recognition from FCRs of TCM and the worth of the representative and reasonable features introduced in this paper. The properties of literal features and positional features, which are just described in section 4.3, are also reflected in figures 8 and 9.

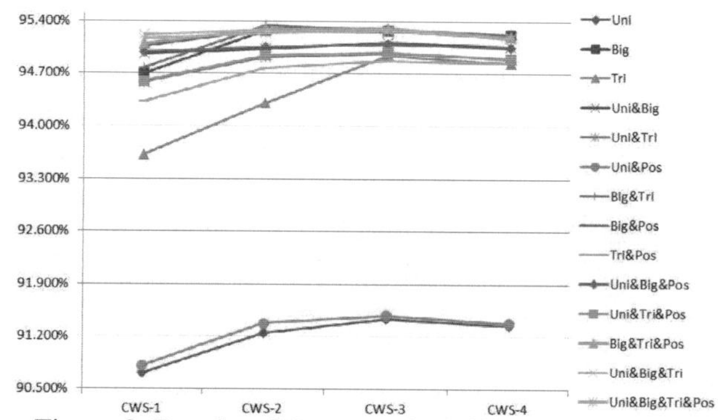

Figure 8. Results of Pre_{lab} under Different Situations.

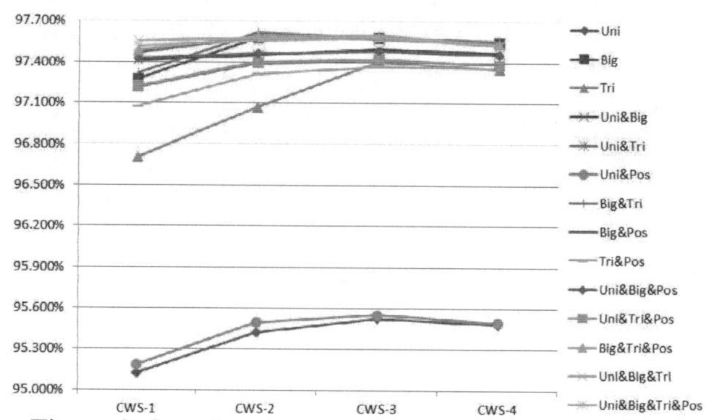

Figure 9. Results of FM_{lab} under Different Situations.

Although RR_{det} can achieve a very high performance, however, RER_{det} is also too high. In figures 8 and 9, high labeling results was gotten. It implies that the probable position of the symptom name can be found in TCM FCRs, but the exact boundaries of the symptom name descriptions cannot be detected accurately yet.

More careful results are listed in table 4. In this table, the average labeling Pre_{lab} of labels "B-

SYC" and "O-SYC" are always higher than the global average precision, but the average Pre_{lab} of "I-SYC" is lower than the global average precision. It implies that the performance of labeling the end position of a symptom name description is worse than the other position's. In other words, the judgment on whether "I-SYC" or "O-SYC" is more difficult. Therefore, as the future work, how to accurately determine the end of a symptom name description should be paid more attention to.

		CWS = 1	CWS = 2	CWS = 3	CWS = 4
Global P		94.186%	94.526%	94.616%	94.540%
B	P	95.184%	95.472%	95.519%	95.429%
	R	94.135%	94.243%	94.238%	94.113%
	F	94.656%	94.853%	94.873%	94.765%
I	P	93.085%	93.586%	93.772%	93.713%
	R	93.791%	94.181%	94.267%	94.201%
	F	93.434%	93.879%	94.016%	93.953%
O	P	94.533%	94.781%	94.819%	94.738%
	R	94.501%	94.916%	95.056%	94.996%
	F	94.514%	94.845%	94.934%	94.864%

Table 4. Detailed Results of Average Pre_{lab}, Rec_{lab} and FM_{lab} for Each Type of Labels. "B", "I" and "O" are short forms of "B-SYC", "I-SYC" and "O-SYC", respectively.

5 Conclusion

In this paper, a preliminary work on symptom name recognition from FCRs of TCM is described, and a feasible method based on CRFs with reasonable features is investigated. Through the experiments, the specialties, usage and effectiveness of the introduced features are verified.

In future, particular syntactic structure and grammatical rules for FCRs of TCM need to be defined and studied based on the characteristics of FCRs of TCM. On the one hand, they can help the TCM doctors and researchers to understand the clinical records deeper (Spasic, et al., 2005; Zhou, et al., 2010), and on the other hand, technically, they are good for filtering and reducing feature size and providing basics and adequate evidence for symptom name normalization process and automatic diagnosis procedure.

Acknowledgments

The authors would like to thank M.S. Xuehong Zhang and M.S. Shengrong Zhou for their helpful suggestions to this work and their valuable work on manually structuring the clinical records for us. The authors are grateful to Ms. Fang Yu and B.S. Yuheng Karen Chen for their helpful paper revising. The authors are also pleased to acknowledge the National Natural Science Foundation of China (Grant No. 61173182 and 61179071), the Provincial Science and Technology Foundation of Sichuan Province (Grant No. 2008SZ0049), the Specialized Research Fund for the Doctoral Program (Grant No. 20090181110052), and the New Century Excellent Talents Fund (Grant No. NCET-08-0370) for their supporting to this work.

References

P.M. Barnes, E. Powell-Griner, K. McFann, R.L. Nahin. 2004. Complementary and alternative medicine use among adults: United States, 2002. *Seminars in Integrative Medicine*, 2(2):54-71.

H. Duan, Y. Zheng. 2011. A study on features of the CRFs-based Chinese Named Entity Recognition. *International Journal of Advanced Intelligence*, 3(2):287-294.

G. Fu, K.K. Luke. 2005. Chinese named entity recognition using lexicalized HMMs. *SIGKDD Explorations*, 7(1):19-25.

J. Gao, M. Li, A. Wu, C.-N. Huang. 2005.Chinese word segmentation and named entity recognition: a pragmatic approach. *Computational Linguistics*, 31(4):531-574.

M. Huang, M. Chen. 2007. Integrated design of the intelligent web-based Chinese medical system (CMDS)-systematic development for digestive health. *Expert System with Applications*, 32:658-673.

J. Lafferty, A. McCallum, F. Pereira. 2001. Conditional random fields: probabilistic models for segmenting and labeling sequence data. *In Proceedings of the 18th International Conference on Machine Learning*.

D. Li, K. Kipper-Schuler, G. Savova. 2008. Conditioal Random Fields and Support Vector Machine for disorder named entity recognition in clinical texts. *In BioNLP 2008: Current Trends in Biomedical Natural Language Processing*, pp:94-95.

A. McCallum, W. Li. 2003. Early results for named entity recognition with conditional random fields, feature induction and web-enhanced lexicons. *In Proceedings of the 7th Conference on Natural Language Learning (CoNLL) at HLT-NAACL*.

M. Molassiotis, P. Fernadez-Ortega, D. Pud, G. Ozden, J.A. Scott, V. Panteli, A. Margulies, M. Browall, M.

Magri, S. Selvekerova, E. Madsen, L. Milovics, I. Bruyns, G. Gudmundsdottir, S. Hummerston, A. M.-A. Ahmad, N. Platin, N. Kearney, E. Pariraki. 2005. Use of complementary and alternative medicine in cancer patients: a European survey. *Annals of Oncology*, 16(4):655-663.

J.-J. Nie, J. Gao, J. Zhang, M. Zhou. 2000. On the use of words and n-grams for Chinese information retrieval. *In Proceedings of the fifth international workshop on Information Retrieval with Asian Languages.*

S.K. Pal. 2002. Complementary and alternative medicine: an overview. *Current Science*, 82(5):518-524.

L.A. Ramshaw, M.P. Marcus. 1995. Text chunking using transformation-based learning. *In Proceedings of the Third Workshop on Very Large Corpora. ACL.*

F. Sha, F. Pereira. 2003. Shallow parsing with conditional random fields. Proceedings of the 2003 Conference of the North American Chapter of the Association of Computer Linguistics on Human Language Technology.

I. Spasic, S. Ananiadou, J. McNaught, A. Kumar. 2005. Text mining and ontologies in biomedicine: making sense of raw text. *Briefings in Bioinformatics*, 6(3):239-251.

X. Wang, H. Qu, P. Liu. 2004. A self-learning expert system for diagnosis in traditional Chinese medicine. *Expert System with Applications*, 26:557-566.

Y. Wang, Z. Yu, Y. Jiang, K. Xu, X. Chen. 2010. Automatic symptom name normalization in clinical records of traditional Chinese medicine. *BMC Bioinformatics*, 11:40.

Y. Wang, Z. Yu, Y. Jiang, Y. Liu, L. Chen, Y. Liu. 2012. A framework and its empirical study of automatic diagnosis of traditional Chinese medicine utilizing raw free-text clinical records. *Journal of Biomedical Informatics*, 45:210-223.

Y. Wu, J. Zhao, B. Xu. 2003. Chinese named entity recognition combining a statistical model with human knowledge. *In Proceedings of the ACL 2003 Workshop on Multilingual and Mixed-Language Named Entity Recognition* (*MultiNER'03*), pp:65-72.

K. Yoshida, J. Tsujii. 2007. Reranking for biomedical named-entity recognition. *In BioNLP 2007: Biological, translational, and clinical language processing*, pp:209-216.

H.-P. Zhang, Q. Liu, H.-K. Yu, X.-Q. Cheng, S. Bai. 2003. Chinese named entity recognition using role model. Computational Linguistics and Chinese Language Processing, 8(2):29-60.

N.L. Zhang, S. Yuan, Y. Wang. 2008. Latent tree models and diagnosis in traditional Chinese medicine. *Artificial Intelligence in Medicine*, 42:229-245.

J. Zhou, L. He, X. Dai, J. Chen. 2006. Chinese named entity recognition with a multi-phase model. *In Proceedings of the fifth Workshop on Chinese Language Processing*, pp:213-216.

X. Zhou, Y. Peng, B. Liu. 2010. Text mining for traditional Chinese medical knowledge discovery: a survey. *Joural of Biomedical Informatics*, 43:650-660.

G.D. Zhou, J. Su. 2002. Named entity recognition using an HMM-based Chunk Tagger. *In Proceedings of the 40th Annual Meeting of the Association for Computational Linguistics.*

Scaling up WSD with Automatically Generated Examples

Weiwei Cheng, Judita Preiss and **Mark Stevenson**
Department of Computer Science,
Sheffield University,
Regent Court, 211 Portobello,
Sheffield, S1 4DP
United Kingdom
{W.Cheng, J.Preiss, M.Stevenson}@dcs.shef.ac.uk

Abstract

The most accurate approaches to Word Sense Disambiguation (WSD) for biomedical documents are based on supervised learning. However, these require manually labeled training examples which are expensive to create and consequently supervised WSD systems are normally limited to disambiguating a small set of ambiguous terms. An alternative approach is to create labeled training examples automatically and use them as a substitute for manually labeled ones. This paper describes a large scale WSD system based on automatically labeled examples generated using information from the UMLS Metathesaurus. The labeled examples are generated without any use of labeled training data whatsoever and is therefore completely unsupervised (unlike some previous approaches). The system is evaluated on two widely used data sets and found to outperform a state-of-the-art unsupervised approach which also uses information from the UMLS Metathesaurus.

1 Introduction

The information contained in the biomedical literature that is available in electronic formats is useful for health professionals and researchers (Westbrook et al., 2005). The amount is so vast that it is difficult for researchers to identify information of interest without the assistance of automated tools (Krallinger and Valencia, 2005). However, processing these documents automatically is made difficult by the fact that they contain terms that are ambiguous. For example, "culture" can mean "laboratory procedure" (e.g. "In peripheral blood mononuclear cell *culture*") or "anthropological culture" (e.g. "main accomplishments of introducing a quality management *culture*"). These lexical ambiguities are problematic for language understanding systems.

Word sense disambiguation (WSD) is the process of automatically identifying the meanings of ambiguous terms. Some WSD systems for the biomedical domain are only able to disambiguate a small number of ambiguous terms (see Section 2). However, for WSD systems to be useful in applications they should be able to disambiguate all ambiguous terms. One way to create such a WSD system is to automatically create the labeled data that is used to train supervised WSD systems. Several approaches (Liu et al., 2002; Stevenson and Guo, 2010; Jimeno-Yepes and Aronson, 2010) have used information from the UMLS Metathesaurus[1] to create labeled training data that have successfully been used to create WSD systems.

A key decision for any system that automatically generates labeled examples is the number of examples of each sense to create, known as the *bias* of the data set. It has been shown that the bias of a set of labeled examples affects the performance of the WSD system it is used to train (Mooney, 1996; Agirre and Martínez, 2004b). Some of the previous approaches to generating labeled data relied on manually annotated examples to determine the bias of the data sets and were therefore not completely unsupervised.

This paper describes the development of a large scale WSD system that is able to disambiguate all

[1] http://www.nlm.nih.gov/research/umls/

terms that are ambiguous in the UMLS Metathesaurus. The system relies on labeled examples that are created using information from UMLS. Various bias options are explored, including ones that do not make use of information from manually labeled examples, and thus we can create a completely unsupervised system. Evaluation is carried out on two standard datasets (the NLM-WSD and MSH-WSD corpora). We find that WSD systems can be created without using any information from manually labeled examples and that their performance is better than a state-of-the-art unsupervised approach.

The remainder of this paper is organized as follows. Previous approaches to WSD in biomedical documents are described in the next Section. Section 3 presents the methods used to identify bias in the labeled examples and WSD system. Experiments in which these approaches are compared are described in Section 4 and their results in Section 5.

2 Background

Many WSD systems for the biomedical domain are based on supervised learning (McInnes et al., 2007; Xu et al., 2007; Stevenson et al., 2008; Yepes and Aronson, 2011). These systems require labeled training data, examples of an ambiguous term labeled with the correct meaning. Some sets of labeled data have been developed for the biomedical domain (Weeber et al., 2001; Savova et al., 2008; Jimeno-Yepes et al., 2011). However, these data sets only contain examples for a few hundred terms and can only be used to develop WSD systems to identify the meanings of those terms. The process of creating labeled examples is extremely time-consuming and difficult (Artstein and Poesio, 2008), making it impractical to create labeled examples of all possible ambiguous terms found in biomedical documents.

Two alternative approaches have been explored to develop systems which are able to disambiguate all ambiguous terms in biomedical documents. The first makes use of unsupervised WSD algorithms (see Section 2.1) and the second creates labeled data automatically and uses it to train a supervised WSD system (see Section 2.2).

2.1 Unsupervised WSD

Unsupervised WSD algorithms make use of information from some knowledge source, rather than relying on training data.

Humphrey et al. (2006) describe an unsupervised system which uses semantic types in UMLS to distinguish between the possible meanings of ambiguous words. However, it cannot disambiguate between senses with the same semantic type, i.e., it is not possible for the system to recognise all sense distinctions.

The *Personalised Page Rank (PPR)* system (Agirre et al., 2010; Jimeno-Yepes and Aronson, 2010) relies on a a graph-based algorithm similar to the Page Rank algorithm originally developed for use in search engines (Brin, 1998). It performs WSD by converting the UMLS Metathesaurus into a graph in which the possible meanings of ambiguous words are nodes and relations between them are edges. Disambiguation is carried out by providing the algorithm with a list of senses that appear in the text that is being disambiguated. This information is then combined with the graph and a ranked list of the possible senses for each ambiguous word generated.

Unsupervised systems have the advantage of being able to disambiguate all ambiguous terms. However, the performance of unsupervised systems that have been developed for biomedical documents is lower than that of supervised ones.

2.2 Automatic Generation of Labeled Data

Automatic generation of labeled data for WSD combines the accuracy of supervised approaches with the ability of unsupervised approaches to disambiguate all ambiguous terms. It was first suggested by Leacock et al. (1998). Their approach is based on the observation that some terms in a lexicon occur only once and, consequently, there is no doubt about their meaning. These are referred to as being *monosemous*. Examples for each possible meaning of an ambiguous term are generated by identifying the closest monosemous term (the *monosemous relative*) in the lexicon and using examples of that term. Variants of the approach have been applied to the biomedical domain using the UMLS Metathesaurus as the sense inventory.

Liu et al. (2002) were the first to apply the monosemous relatives approach to biomedical WSD and use it to disambiguate a set of 35 abbreviations. They reported high precision but low recall, indicating that labeled examples could not be created for many of the abbreviations. Jimeno-Yepes and Aronson (2010) applied a similar approach and found that it performed better than a number of alternative approaches on a standard evaluation resource (the NLM-WSD corpus) but did not perform as well as supervised WSD. Stevenson and Guo (2010) compared two techniques for automatically creating labeled data, including the monosemous relatives approach. They found that the examples which were generated were as good as manually labeled examples when used to train a supervised WSD system. However, Stevenson and Guo (2010) relied on labeled data to determine the number of examples of each sense to create, and therefore the bias of the data set. Consequently their approach is not completely unsupervised since it could not be applied to ambiguous terms that do not have labeled training data available.

3 Approach

3.1 WSD System

The WSD system is based on a supervised approach that has been adapted for the biomedical domain (Stevenson et al., 2008). The system was tested on the NLM-WSD corpus (see Section 4.1) and found to outperform alternative approaches.

The system can exploit a wide range of features, including several types of linguistic information from the context of an ambiguous term, MeSH codes and Concept Unique Identifiers (CUIs) from the UMLS Metathesaurus. However, computing these features for every example is a time consuming process and to make the system suitable for large scale WSD it was restricted to using a smaller set of features. Previous experiments (Stevenson et al., 2008) showed that this only leads to a small drop in disambiguation accuracy while significantly reducing the computational cost of generating features.

3.1.1 Features

Two types of context words are used as features: the lemmas of all content words in the same sentence as the ambiguous word and the lemmas of all content words in a ±4-word window around the ambiguous term. A list of corpus-specific stopwords was created containing terms that appear frequently in Medline abstracts but which are not useful for disambiguation (e.g. "abstract", "conclusion"). Any lemmas found in this list were not used as features.

3.1.2 Learning algorithm

Disambiguation is carried out using the *Vector Space Model*, a memory-based learning algorithm in which each occurrence of an ambiguous word is represented as a vector created using the features extracted to represent it (Agirre and Martínez, 2004a). The Vector Space Model was found to outperform other learning algorithms when evaluated using the NLM-WSD corpus (Stevenson et al., 2008).

During the algorithm's training phase a single centroid vector, $\vec{C_{s_j}}$, is generated for each possible sense, s_j. This is shown in equation 1 where T is the set of training examples for a particular term and $sense(\vec{t})$ is the sense associated with the vector $\vec{t}$.

$$\vec{C_{s_j}} = \frac{\sum_{\vec{t_i}\,\epsilon\, T:sense(\vec{t_i})=s_j} \vec{t_i}}{|\vec{t_i}\,\epsilon\, T : sense(\vec{t_i}) = s_j|} \quad (1)$$

Disambiguation is carried out by comparing the vector representing the ambiguous word, $\vec{a}$, against the centroid of each sense using the cosine metric, shown in equation 2, and choosing the one with the highest score.

$$score(s_j, \vec{a}) = cos(\vec{C_{s_j}}, \vec{a}) = \frac{\vec{C_{s_j}}.\vec{a}}{|\vec{C_{s_j}}||\vec{a}|} \quad (2)$$

Note that the learning algorithm does not explicitly model the prior probability of each possible sense, unlike alternative approaches (e.g. Naive Bayes), since it was found that including this information did not improve performance.

3.2 Automatically generating training examples

The approaches used for generating training examples used here are based on the work of Stevenson and Guo (2010), who describe two approaches:

1. Monosemous relatives

2. Co-occurring concepts

Both approaches are provided with a set of ambiguous CUIs from the UMLS Metathesaurus, which represent the possible meanings of an ambiguous term, and a target number of training examples to be generated for each CUI. Each CUI is associated with at least one term and each term is labeled with a lexical unique identifier (LUI) which represents a range of lexical variants for a particular term. The UMLS Metathesaurus contains a number of data files which are exploited within these techniques, including:

`AMBIGLUI`: a list of cases where a LUI is linked to multiple CUIs.

`MRCON`: every string or concept name in the Metathesaurus appears in this file.

`MRCOC`: co-occuring concepts.

For the monosemous relatives approach, the strings of monosemous LUIs of the target CUI and its relatives are used to search Medline to retrieve training examples. The monosemous LUIs related to a CUI are defined as any LUIs associated with the CUI in the `MRCON` table and not listed in `AMBIGLUI` table.

The co-occurring concept approach works differently. Instead of using strings of monosemous LUIs of the target CUI and its relatives, the strings associated with LUIs of a number of co-occurring CUIs of the target CUI and its relatives found in MRCOC table are used. The process starts by finding the LUIs of the top n co-occurring CUIs of the target CUI. These LUIs are then used to form search queries. The query is quite restrictive in the beginning and requires all terms appear in the Medline citations files. Subsequently queries are made less restrictive by reducing the number of required terms in the query.

These techniques were used to generate labeled examples for all terms that are ambiguous in the 2010 AB version of the UMLS Metathesaurus.[2] The set of all ambiguous terms was created by analysing the `AMBIGLUI` table, to identify CUIs that are associated with multiple LUIs. The Medline Baseline Repository (MBR)[3] was also analysed and it was found that some terms were ambiguous in this resource, in the sense that more than one CUI had been assigned to an instance of a term, but could not be identified from the `AMBIGLUI` table. The final list of ambiguous CUIs was created by combining those identified from the `AMBIGLUI` table and those find in the MBR. This list contained a total of 103,929 CUIs.

Both techniques require large number of searches over the Medline database and to carry this out efficiently the MBR was indexed using the Lucene Information Retrieval system[4] and all searches executed locally.

Examples were generated using both approaches. The monosemous relatives approach generated examples for 98,462 CUIs and the co-occurring concepts for 98,540. (Examples generated using the monosemous relatives approach were preferred for the experiments reported later.) However, neither technique was able to generate examples for 5,497 CUIs, around 5% of the total. This happened when none of the terms associated with a CUI returned any documents when queried against the MBR and that CUI does not have any monosemous relatives. An example is C1281723 "Entire nucleus pulposus of intervertebral disc of third lumbar vertebra". The lengthy terms associated with this CUI do not return any documents when used as search terms and, in addition, it is only related to one other CUI (C0223534 "Structure of nucleus pulposus of intervertebral disc of third lumbar vertebra") which is itself only connected to C1281723. Fortunately there are relatively few CUIs for which no examples could be generated and none of them appear in the MBR, suggesting they refer to UMLS concepts that do not tend to be mentioned in documents.

3.3 Generating Bias

Three different techniques for deciding the number of training examples to be generated for each CUI (i.e. the bias) were explored.

Uniform Bias (UB) uses an equal number of training examples to generate centroid vectors for each of the possible senses of the ambiguous term.

Gold standard bias (GSB) is similar to the uniform bias but instead of being the same for all possible CUIs the number of training examples for each CUI is determined by the number of times it appears

[2]Stevenson and Guo (2010) applied them to a small set of examples from the NLM-WSD data set (see Section 4.1).

[3]`http://mbr.nlm.nih.gov`

[4]`http://lucene.apache.org/`

in a manually labeled gold standard corpus. Assume t is an ambiguous term and C_t is the set of possible meanings (CUIs). The number of training examples used to generate the centroid for that CUI, E_c, is computed according to equation 3 where G_c is the number of instances in the gold standard corpus annotated with CUI c and n is a constant which is set to 100 for these experiments.[5]

$$E_c = \frac{G_c}{\sum_{c_i \, \epsilon \, C_t} G_{c_{i,t}}} .n \tag{3}$$

The final technique, **Metamap Baseline Repository Bias (MBB)**, is based on the distribution of CUIs in the MBR. The number of training examples are generated in a similar way to the gold standard bias with MBR being used instead of a manually labeled corpus and is shown in equation 4 where M_c is the number of times the CUI c appears in the MBR.

$$E_c = \frac{M_c}{\sum_{c_i \, \epsilon \, C_t} M_{c_i}} .n \tag{4}$$

For example, consider the three possible CUIs associated with term "adjustment" in the NLM-WSD corpus: C0376209, C0456081 and C0683269[6]. The corpus contains 18 examples of C0376209, 62 examples of C0456081 and 13 of C0683269. Using equation 3, the number of training examples when GSB is applied for C0376209 is 20, 67 for C0456081 and 14 for C0683269. In the Metamap Baseline Repository files, C0376209 has a frequency count of 98046, C0456081 a count of 292809 and C0683269 a count of 83530. Therefore the number of training examples used for the three senses when applying MBB is: 21 for C0376209, 62 for C0456081 and 18 for C0683269.

4 Evaluation

4.1 Data sets

We evaluate our system on two datasets: the NLM-WSD and MSH-WSD corpora.

The NLM-WSD corpus[7] (Weeber et al., 2001) has been widely used for experiments on WSD in the biomedical domain, for example (Joshi et al., 2005; Leroy and Rindflesch, 2005; McInnes et al., 2007; Savova et al., 2008). It contains 50 ambiguous terms found in Medline with 100 examples of each. These examples were manually disambiguated by 11 annotators. The guidelines provided to the annotators allowed them to label a senses as "None" if none of the concepts in the UMLS Metathesaurus seemed appropriate. These instances could not be mapped onto UMLS Metathesaurus and were ignored for our experiments.

The larger MSH-WSD corpus (Jimeno-Yepes et al., 2011) contains 203 strings that are associated with more than one possible MeSH code in the UMLS Metathesaurus. 106 of these are ambiguous abbreviations, 88 ambiguous terms and 9 a combination of both. The corpus contains up to 100 examples for each possible sense and a total of 37,888 examples of ambiguous strings taken from Medline. Unlike the NLM-WSD corpus, all of the instances can be mapped to the UMLS Metathesaurus and none was removed from the dataset for our experiments.

The two data sets differ in the way the number of instances of each sense was determined. For the NLM-WSD corpus manual annotation is used to decide the number of instances that are annotated with each sense of an ambiguous term. However, the NLM-MSH corpus was constructed automatically and each ambiguous term has roughly the same number of examples of each possible sense.

4.2 Experiments

The WSD system described in Section 3 was tested using each of the three techniques for determining the bias, i.e. number of examples generated for each CUI. Performance is compared against various alternative approaches.

Two supervised approaches are included. The first, most frequent sense (MFS) (McCarthy et al., 2004), is widely used baseline for supervised WSD systems. It consists of assigning each ambiguous term the meaning that is more frequently observed in the training data. The second supervised approach

[5]Small values for E_c are rounded up to ensure that any rare CUIs have at least one training example.

[6]These CUIs are obtained using the mappings from NLM-WSD senses to CUIs available on the NLM website: `http://wsd.nlm.nih.gov/collaboration.shtml`

[7]`http://wsd.nlm.nih.gov`

is to train the WSD system using manually labeled examples from the NLM-WSD and MSH-WSD corpora. 10-fold cross validation is applied to evaluate this approach.

Performance of the Personalised Page Rank approach described in Section 2.1 is also provided to allow comparison with an unsupervised algorithm. Both Personalised Page Rank and the techniques we employ to generate labeled data, base disambiguation decisions on information from the UMLS Metathesaurus.

The performance of all approaches is measured in terms of the percentage of instances which are correctly disambiguated for each term with the average across all terms reported. Confidence intervals (95%) computed using bootstrap resampling (Noreen, 1989) are also shown.

5 Results

Results of the experiments are shown in Table 1 where the first three rows show performance of the approach described in Section 3 using the three methods for computing the bias (UB, MMB and GSB). MFS and Sup refer to the Most Frequent Sense supervised baseline and using manually labeled examples, respectively, and PPR to the Personalised PageRank approach.

When the performance of the approaches using automatically labeled examples (UB, MMB and GSB) is compared it is not surprising that the best results are obtained using the gold standard bias since this is obtained from manually labeled data. Results using this technique for computing bias always outperform the other two, which are completely unsupervised and do not make use of any information from manually labeled data. However, the improvement in performance varies according to the corpus, for the NLM-WSD corpus there is an improvement of over 10% in comparison to UB while the corresponding improvement for the MSH-WSD corpus is less than 0.5%.

A surprising result is that performance obtained using the uniform bias (UB) is consistently better than using the bias obtained by analysis of the MBR (MMB). It would be reasonable to expect that information about the distribution of CUIs in this corpus would be helpful for WSD but it turns out that making no assumptions whatsoever about their relative frequency, i.e., assigning a uniform baseline, produces better results.

The relative performance of the supervised (MFS, Sup and GSB) and unsupervised approaches (UB, MMB and PPR) varies according to the corpus. Unsurprisingly using manually labeled data (Sup) outperforms all other approaches on both corpora. The supervised approaches also outperform the unsupervised ones on the NLM-WSD corpus. However, for the MSH-WSD corpus all of the unsupervised approaches outperform the MFS baseline.

A key reason for the differences in these results is the different distributions of senses in the two corpora, as shown by the very different performance of the MFS approach on the two corpora. This is discussed in more detail later (Section 5.2).

Comparison of the relative performance of the unsupervised approaches (UB, MMB and PPR) shows that training a supervised system with the automatically labeled examples using a uniform bias (UB) always outperforms PPR. This demonstrates that this approach outperforms a state-of-the-art unsupervised algorithm that relies on the same information used to generate the examples (the UMLS Metathesaurus).

5.1 Performance by Ambiguity Type

The MSH-WSD corpus contains both ambiguous terms and abbreviations (see Section 4.1). Performance of the approaches on both types of ambiguity are shown in Table 2.

	MSH-WSD Ambiguity Type	
Approach	Abbreviation	Term
UB	91.40 [91.00, 91.75]	72.68 [72.06, 73.32]
MMB	84.43 [83.97, 84.89]	69.45 [68.86, 70.10]
GSB	90.82 [90.45, 91.22]	73.96 [73.40, 74.62]
MFS	52.43 [51.73, 53.05]	51.76 [51.11, 52.36]
Sup.	97.41 [97.19, 97.62]	91.54 [91.18, 91.94]
PPR	86.40 [86.00, 86.85]	68.40 [67.80, 69.14]

Table 2: WSD evaluation results for abbreviations and terms in the MSH-WSD data set.

The relative performance of the different approaches on the terms and abbreviations is similar to the entire MSH-WSD data set (see Table 1). In par-

Approach	Type	Corpus NLM-WSD	MSH-WSD
UB	Unsup.	74.00 [72.80, 75.29]	83.19 [82.87, 83.54]
MMB	Unsup.	71.18 [69.94, 72.38]	78.09 [77.70, 78.46]
GSB	Sup.	84.28 [83.12, 85.36]	83.39 [83.08, 83.67]
MFS	Sup.	84.70 [83.67, 85.81]	52.01 [51.50, 52.45]
Sup	Sup.	90.69 [89.87, 91.52]	94.83 [94.63, 95.02]
PPR	Unsup.	68.10 [66.80, 69.23]	78.60 [78.23, 78.90]

Table 1: WSD evaluation results on NLM-WSD and MSH-WSD data sets.

ticular using automatically generated examples with a uniform bias (UB) outperforms using the bias derived from the Medline Baseline Repository (MBR) while using the gold standard baseline (GSB) improves results slightly for terms and actually reduces them for abbreviations.

Results for all approaches are higher when disambiguating abbreviations than terms which is consistent with previous studies that have suggested that in biomedical text abbreviations are easier to disambiguate than terms.

5.2 Analysis

An explanation of the reason for some of the results can be gained by looking at the distributions of senses in the various data sets used for the experiments. Kullback-Leibler divergence (or KL divergence) (Kullback and Leibler, 1951) is a commonly used measure for determining the difference between two probability distributions. For each term t, we define S as the set of possible senses of t, the sense probability distributions of t as D and D'. Then the KL divergence between the sense probability distributions D and D' can be calculated according to equation 5.

$$KL(D||D') = \sum_{s \in S} D(s). \log \frac{D(s)}{D'(s)} \quad (5)$$

The three techniques for determining the bias described in Section 3.3 each generate a probability distribution over senses. Table 2 shows the average KL divergence when the gold standard distribution obtained from the manually labeled data (GSB) is compared with the uniform bias (UB) and bias obtained by analysing the Medline Baseline Repository (MMB).

Avg. KL Divergence	Corpus NLM-WSD	MSH-WSD
$KL(GSB\|\|MMB)$	0.5649	0.4822
$KL(GSB\|\|UB)$	0.4600	0.0406

Table 3: Average KL divergence of sense probability distributions in the NLM-WSD and MSH-WSD data sets.

The average KL divergence scores in the table are roughly similar with the exception of the much lower score obtained for the gold-standard and uniform bias for the MSH-WSD corpus (0.0406). This is due to the fact that the MSH-WSD corpus was designed to have roughly the same number of examples for each sense, making the sense distribution close to uniform (Jimeno-Yepes et al., 2011). This is evident from the MFS scores for the MSH-WSD corpus which are always close to 50%. This also provides as explanation of why performance using automatically generated examples on the MSH-WSD corpus only improves by a small amount when the gold standard bias is used (see Table 1). The gold standard bias simply does not provide much additional information to the WSD system. The situation is different in the NLM-WSD corpus, where the MFS score is much higher. In this case the additional information available in the gold standard sense distribution is useful for the WSD system and leads to a large improvement in performance.

In addition, this analysis demonstrates why performance does not improve when the bias generated from the MBR is used. The distributions which are obtained are different from the gold standard and are therefore mislead the WSD system rather than providing useful information. The difference between these distributions would be expected for the MSH-WSD corpus, since it contains roughly the same number of examples for each possible sense and does not attempt to represent the relative frequency of the different senses. However, it is surprising to observe a similar difference for the NLM-WSD corpus, which does not have this constraint. The difference suggests the information about CUIs in the MBR, which is generated automatically, has some limitations.

Table 4 shows a similar analysis for the MSH-WSD corpus when abbreviations and terms are considered separately and supports this analysis. The figures in this table show that the gold standard and uniform distributions are very similar for both abbreviations and terms, which explains the similar results for UB and GSB in Table 2. However, the gold standard distribution is different from the one obtained from the MBR. The drop in performance of MMB compared with GBS in Table 2 is a consequence of this.

	Ambiguity Type	
Avg. KL Divergence	Abbreviation	Term
$KL(GSB\|\|MMB)$	0.4554	0.4603
$KL(GSB\|\|UB)$	0.0544	0.0241

Table 4: Average KL divergence for abbreviations and terms in the MSH-WSD data set.

6 Conclusion

This paper describes the development of a large scale WSD system based on automatically labeled examples. We find that these examples can be generated for the majority of CUIs in the UMLS Metathesaurus. Evaluation on the NLM-WSD and MSH-WSD data sets demonstrates that the WSD system outperforms the PPR approach without making any use of labeled data.

Three techniques for determining the number of examples to use for training are explored. It is found that a supervised approach (which makes use of manually labeled data) provides the best results. Surprisingly it was also found that using information from the MBR did not improve performance. Analysis showed that the sense distributions extracted from the MBR were different from those observed in the evaluation data, providing an explanation for this result.

Evaluation showed that accurate information about the bias of training examples is useful for WSD systems and future work will explore other unsupervised ways of obtaining this information. Alternative techniques for generating labeled examples will also be explored. In addition, further evaluation of the WSD system will be carried out, such as applying it to an all words task and within applications.

Acknowledgements

This research has been supported by the Engineering and Physical Sciences Research Council and a Google Research Award.

References

E. Agirre and D. Martínez. 2004a. The Basque Country University system: English and Basque tasks. In Rada Mihalcea and Phil Edmonds, editors, *Proceedings of Senseval-3*, pages 44–48, Barcelona, Spain.

E. Agirre and D. Martínez. 2004b. Unsupervised WSD Based on Automatically Retrieved Examples: The Importance of Bias. In *Proceedings of EMNLP-04*, Barcelona, Spain.

E. Agirre, A. Sora, and M. Stevenson. 2010. Graph-based word sense disambiguation of biomedical documents. *Bioinformatics*, 26(22):2889–2896.

R. Artstein and M. Poesio. 2008. Inter-Coder Agreement for Computational Linguistics. *Computational Linguistics*, 34(4):555–596.

S. Brin. 1998. Extracting Patterns and relations from the Word-Wide Web. In *Proceedings of WebDB'98*.

S. Humphrey, W. Rogers, H. Kilicoglu, D. Demner-Fushman, and T. Rindflesch. 2006. Word Sense Disambiguation by Selecting the Best Semantic Type Based on Journal Descriptor Indexing: Preliminary experiment. *Journal of the American Society for Information Science and Technology*, 57(5):96–113.

A. Jimeno-Yepes and A. Aronson. 2010. Knowledge-based biomedical word sense disambiguation: comparison of approaches. *BMC Bioinformatics*, 11(1):569.

A. Jimeno-Yepes, B. McInnes, and A. Aronson. 2011. Exploiting MeSH indexing in MEDLINE to generate a data set for word sense disambiguation. *BMC Bioinformatics*, 12(1):223.

M. Joshi, T. Pedersen, and R. Maclin. 2005. A Comparative Study of Support Vector Machines Applied to the Word Sense Disambiguation Problem for the Medical Domain. In *Proceedings of IICAI-05*, pages 3449–3468, Pune, India.

M. Krallinger and A. Valencia. 2005. Text mining and information retrieval services for molecular biology. *Genome Biology*, 6(7):224.

S. Kullback and R. A. Leibler. 1951. On Information and Sufficiency. *The Annals of Mathematical Statistics*, 22(1):79–86.

C. Leacock, M. Chodorow, and G. Miller. 1998. Using Corpus Statistics and WordNet Relations for Sense Identification. *Computational Linguistics*, 24(1):147–165.

G. Leroy and T. Rindflesch. 2005. Effects of Information and Machine Learning algorithms on Word Sense Disambiguation with Small Datasets. *International Journal of Medical Informatics*, 74(7-8):573–585.

H. Liu, S. Johnson, and C. Friedman. 2002. Automatic Resolution of Ambiguous Terms Based on Machine Learning and Conceptual Relations in the UMLS. *Journal of the American Medical Informatics Association*, 9(6):621–636.

D. McCarthy, R. Koeling, J. Weeds, and J. Carroll. 2004. Finding Predominant Word Senses in Untagged Text. In *Proceedings of ACL-2004*, pages 280–287, Barcelona, Spain.

B. McInnes, T. Pedersen, and J. Carlis. 2007. Using UMLS Concept Unique Identifiers (CUIs) for Word Sense Disambiguation in the Biomedical Domain. In *Proceedings of the AMIA Symposium*, pages 533–537, Chicago, IL.

R. Mooney. 1996. Comparative Experiments on Disambiguating Word Senses: An Illustration of the Role of Bias in Machine Learning. In *Proceedings of EMNLP-96*, pages 82–91, Philadelphia, PA.

E. W. Noreen. 1989. *Computer-Intensive Methods for Testing Hypotheses*. John Wiley & Sons.

G. Savova, A. Coden, I. Sominsky, R. Johnson, P. Ogren, C. de Groen, and C. Chute. 2008. Word Sense Disambiguation across Two Domains: Biomedical Literature and Clinical Notes. *Journal of Biomedical Informatics*, 41(6):1088–1100.

M. Stevenson and Y. Guo. 2010. Disambiguation of Ambiguous Biomedical Terms using Examples Generated from the UMLS Metathesaurus. *Journal of Biomedical Informatics*, 43(5):762–773.

M. Stevenson, Y. Guo, R. Gaizauskas, and D. Martinez. 2008. Disambiguation of biomedical text using diverse sources of information. *BMC Bioinformatics*, 9(Suppl 11):S7.

M. Weeber, J. Mork, and A. Aronson. 2001. Developing a Test Collection for Biomedical Word Sense Disambiguation. In *Proceedings of AMIA Symposium*, pages 746–50, Washington, DC.

J. Westbrook, E. Coiera, and A. Gosling. 2005. Do Online Information Retrieval Systems Help Experienced Clinicians Answer Clinical Questions? *Journal of the American Medical Informatics Association*, 12:315–321.

H. Xu, J. Fan, G. Hripcsak, E. Mendonça, Markatou M., and Friedman C. 2007. Gene symbol disambiguation using knowledge-based profiles. *Bioinformatics*, 23(8):1015–22.

A. Jimeno Yepes and A. Aronson. 2011. Self-training and co-training in biomedical word sense disambiguation. In *Proceedings of BioNLP 2011 Workshop*, pages 182–183, Portland, Oregon, USA, June.

Boosting the protein name recognition performance by bootstrapping on selected text

Yue Wang and **Jin-Dong Kim**
Database Center for Life Science,
Research Organization of Information and Systems
2-11-16 Yayoi, Bunkyo-ku, Tokyo, Japan 113-0032
{wang, jdkim}@dbcls.rois.ac.jp

Abstract

When only a small amount of manually annotated data is available, application of a bootstrapping method is often considered to compensate for the lack of sufficient training material for a machine-learning method. The paper reports a series of experimental results of bootstrapping for protein name recognition. The results show that the performance changes significantly according to the choice of text collection where the training samples to bootstrap, and that an improvement can be obtained only with a well chosen text collection.

1 Introduction

While machine learning-based approaches are becoming more and more popular for the development of natural language processing (NLP) systems, corpora with annotation are regarded as a critical resource for the training process. Nonetheless, the creation of corpus annotation is an expensive and time-consuming work (Cohen et al., 2005), and it is often the case that lack of sufficient annotation hinders the development of NLP systems. Bootstrapping method (Becker et al., 2005; Vlachos and Gasperin, 2006) can be considered as a way to automatically inflate the amount of corpus annotation to complement the lack of sufficient annotation.

In this study, we report the experimental results on the effect of bootstrapping for the training of protein name recognizers, particularly in the situation when we have only a small amount of corpus annotations.

In summary, we begin with a small corpus with manual annotation for protein names. A named entity tagger trained on the small corpus is applied to a big collection of text, to obtain more annotation. We hope the newly created annotation to be precise enough so that the training of a protein tagger can benefit from the increased training material.

We assume that the accuracy of a bootstrapping method (Ng, 2004) depends on two factors: the accuracy of the bootstrap tagger itself and the similarity of the text to the original corpus. While accuracy of the bootstrap tagger may be maximized by finding the optimal parameters of the applied machine learning method, the choice of text where the original annotations will bootstrap may also be a critical factor for the success of the bootstrapping method.

Experimental results presented in this paper confirm that we can get a improvement by using a bootstrapping method with a well chosen collection of texts.

The paper is organized as follows. Section 2 introduces the two datasets used in this paper. Following that, in Section 3, we briefly introduce the experiments performed in our research. The experimental results are demonstrated in Section 4. The research is concluded in Section 5 and in the meanwhile, future work is discussed.

2 Datasets

2.1 The cyanobacteria genome database

Cyanobacteria are prokaryotic organisms that have served as important model organisms for studying oxygenic photosynthesis and have played a significant role in the Earthfs history as primary producers of atmospheric oxygen (Nakao et al., 2010).

The cyanobacteria genome database (abbreviated to CyanoBase[1]) includes the annotations to the PubMed text. In total, 39 species of the cyanobacteria are covered in the CyanoBase.

In our cyanobacteria data (henceforth, the Kazusa data for short), 270 abstracts were annotated by two independent annotators. We take the entities, about which both of the annotators agreed with each other. In total, there are 1,101 entities in 2,630 sentences.

The Kazusa data was split equally into three subsets and the subsets were used in turn as the training, development and testing sets in the experiments.

2.2 The BioCreative data

The BioCreative data, which was used for the BioCreative II gene mention task[2], is described as the tagged gene/protein names in the PubMed text. The training set is used in the research, and totally there are 15,000 sentences in the dataset.

Unlike other datasets, the BioCreative data was designed to contain sentences both with and without protein names, in a variety of contexts. Since the collection is made to explicitly compile positive and negative examples for protein recognition, there is a chance that the sample of text is not comprehensive, and gray-zone expressions may be missed.

The reason that we chose the BioCreative data for the bootstrapping is that, the BioCreative data (henceforth, the BC2 data for short) is the collection for the purpose of training and evaluation of protein name taggers.

3 Experiment summary

In the following experiments, the NERSuite[3], a named entity tagger based on Conditional Random Fields (CRFs) (Lafferty et al., 2001; Sutton and McCallum, 2007), is used. The NERSuite is executable open-source and serves as a machine learning system for named entity recognition (NER). The sigma value for the L_2-regularization is optimizable and in our experiments, we tune the sigma value between 10^{-1} to 10^4.

[1]http://genome.kazusa.or.jp/cyanobase

[2]http://www.biocreative.org/

[3]http://nersuite.nlplab.org/

As mentioned in Section 2.1, the three subsets of Kazusa data are used for training, tuning and testing purposes, in turn. We experimented with all the six combinations.

Experiments were performed to compare three different strategies. First, with the *baseline strategy*, the protein tagger is trained only on the Kazusa training set. The sigma value is optimized on the tuning set, and the performance is evaluated on the test set. It is the most typical strategy particularly when it is believed there is a sufficient training material.

Second, with the *bootstrapping strategy*, the Kazusa training set is used as the seed data. A tagger for bootstrapping (bootstrap tagger, hereafter) is trained on the seed data, and applied to the BC2 data to bootstrap the training examples. Another protein tagger (application tagger) is then trained on the bootstrapped BC2 data together with the seed data. The Kazusa tuning set is used to optimize the two sigma values for the two protein taggers, and the performance is evaluated on the test set. With this strategy, we wish the bootstrapped examples complement the lack of sufficient training examples.

Experiment	Seed	BT	BT+SS
E1	368	647	647 (1,103)
E2	368	647	647 (1,103)
E3	366	759	759 (1,200)
E4	366	769	590 (1,056)
E5	367	882	558 (1,068)
E6	367	558	558 (1,068)

Table 1: The number of positive examples used in each experiment. The "BT" column shows the number of positive examples obtained by the bootstrapping in the 15,000 BC2 sentences. In the last column, the figures in parentheses are the number of the selected sentences.

Third, the *bootstrapping with sentence selection strategy* is almost the same with the bootstrapping strategy, except that the second tagger is trained after the non-relevant sentences are filtered out from the BC2 data. Here, non-relevant sentences mean those that are not tagged by the the bootstrap tagger. With this strategy, we wish an improvement with the bootstrapping by removing noisy data. Table 1 shows the number of the seed and bootstrapped examples used for the three strategies. It is observed that the seed

	Training	Tuning	Testing	Baseline	BT	BT+SS
E1	A	B	C	63.7/29.2/40.0 [10^2]	61.3/25.9/36.4 [10^4-10^1]	61.7/38.2/47.1 [10^4-10^4]
E2	A	C	B	65.2/36.9/47.1 [10^3]	67.7/35.0/46.1 [10^4-10^1]	61.7/46.7/53.2 [10^4-10^4]
E3	B	C	A	75.3/36.4/49.1 [10^2]	75.2/31.3/44.2 [10^2-10^1]	67.1/40.0/50.1 [10^2-10^1]
E4	B	A	C	68.5/33.8/45.3 [10^2]	70.2/28.9/40.9 [10^4-10^1]	66.7/36.5/47.2 [10^1-10^2]
E5	C	B	A	77.7/35.1/48.3 [10^1]	71.8/27.7/40.0 [10^4-10^2]	70.9/38.3/49.7 [10^0-10^1]
E6	C	A	B	73.0/39.1/50.9 [10^1]	76.1/32.2/45.3 [10^0-10^2]	67.7/41.8/51.7 [10^0-10^2]

Table 2: Experimental results of using the Kazusa and BC2 data (Precision/Recall/F-score). "BT" and "SS" represent the bootstrapping and sentence selection strategies, respectively. The figures in square brackets are the sigma values optimized in the experiments.

annotation bootstrap only on a small portion of the BC2 data set, e.g., 1,103 vs. 15,000 sentences in the case of E1 (less than 10%), suggesting that a large portion of the data set may be irrelevant to the original data set.

4 Experimental results

The experimental results of all the six combinations are shown in Table 2. The use of the three subsets, denoted by A, B, C, of the Kazusa data set for training, tuning and testing in each experiment is specified in "training", "tuning" and "testing" columns. The results of the baseline strategy that uses only the Kazusa data are shown in the "baseline" column, whereas the results with the bootstrapping methods with and without sentence selection are shown in the last two columns. As explained in Section 3, the sigma values are optimized using the tuning set for each experiment. Note that for bootstrapping, we need two sigma values for the bootstrapping tagger and the application tagger. See section 3.

The performance of named entity recognition is measured in terms of precision, recall and F-score. For matching criterion, in order to avoid underestimation, instead of the exact matching, system performance is evaluated under a soft matching, the overlapping matching criterion. That is, if any part of the annotated protein/gene names is recognized by the NER tagger, we will regard that as a correct answer.

4.1 Results with the bootstrapping strategy

Comparing the two columns, "baseline" and "BT", we observe that the use of bootstrapping may lead to a degradation of the performance. Note that the sigma values are optimized on the development set for each experiment, and the text for bootstrapping is BC2 corpus which is expected to be similar to the Kazusa corpus, but still it is observed that the bootstrapping does not work, suggesting that the text collection may not yet similar enough.

4.2 Results with bootstrapping with sentence selection

Comparing the last column (the "BT+SS" column) to the "baseline" column, we observe that the application of the bootstrapping method with sentence selection consistently improves the performance. The improvement is sometimes significant, e.g., 7.1% of difference in F-score in the case of E1, but sometimes not, e.g., only 0.8% in the case of E6, but the performance is improved in the every experiments. The results confirm our assumption that the choice of text for bootstrapping is important, and that the sentence selection is a stable method for the choice of text.

5 Conclusion and future work

In order to compensate for the lack of sufficient training data for a CRF-based protein name recognizer, the potential of a bootstrapping method has been explored through a series of experiments. The BC2 data was chosen for the bootstrapping as the data set was one collected for protein name recognition.

Our initial experiment showed that the seed annotations bootstrapped only on a very small portion of the BC2 data set, suggesting that a big portion of the data set might be less relevant to the seed corpus. From a series of experiments, it was observed that the performance of protein name recognition was always improved with bootstrapping by selecting only the sentences where the seed annotations bootstrap, and by using them as an additional training data.

The goal was to be able to predict more possible protein mentions (recall) at a relatively satisfactory level of the quality (precision). The experimental results suggest us, in order to achieve the goal, the choice of text collection is important for the success of the use of a bootstrapping method.

For the future work, we would like to take use of the original annotations in the BC2 data. A filtering strategy (Wang, 2010) will be performed. Instead of completely using the output of the Kazusa-trained tagger, we compare the output of the Kazusa-trained tagger with the BioCreative annotations. If the entity is recognized by the tagger and also annotated in the BioCreative data, then the annotation to this entity will be kept. The entity will be regarded as a true positive according to the BioCreative annotations. Otherwise, we will remove the annotation to the entity from the BioCreative annotations.

Further, we also would like to combine the bootstrapping with the filtering. Besides keeping the true positives, we also want to include some false positives from the bootstrapping. Because these false positives helps in improving the recall, when the tagger is applied to the Kazusa testing subset. To discriminate this strategy from the bootstrapping and filtering strategies, different sigma value should be used.

Acknowledgement

We thank Shinobu Okamoto for providing the Kazusa data and for many useful discussion. This work was supported by the "Integrated Database Project" funded by the Ministry of Education, Culture, Sports, Science and Technology (MEXT) of Japan.

References

K. Bretonnel Cohen, Lynne Fox, Philip Ogren and Lawrence Hunter. 2005. Empirical data on corpus design and usage in biomedical natural language processing. *Proceedings of the AMIA Annual Symposium*, 38–45.

Markus Becker, Ben Hachey, Beatrice Alex, Claire Grover. 2005. Optimising Selective Sampling for Bootstrapping Named Entity Recognition. *Proceedings of the Workshop on Learning with Multiple Views*, 5–11.

Andreas Vlachos and Caroline Gasperin. 2006. Bootstrapping and Evaluating Named Entity Recognition in the Biomedical domain. *Proceedings of the BioNLP Workshop*, 138–145.

Andrew Ng. 2004. Feature selection, L_1 vs. L_2 regularization, and rotational invariance. *Proceedings of the 21st International Conference on Machine Learning (ICML).*

Mitsuteru Nakao, Shinobu Okamoto, Mitsuyo Kohara, Tsunakazu Fujishiro, Takatomo Fujisawa, Shusei Sato, Satoshi Tabata, Takakazu Kaneko and Yasukazu Nakamura. 2010. CyanoBase: the cyanobacteria genome database update 2010. *Nucleic Acids Research*, 38:D379–D381.

John Lafferty, Andrew McCallum, and Fernando Pereira. 2001. Conditional Random Fields: Probabilistic Models for Segmenting and Labeling Sequence Data. *Proceedings of the 18th International Conference on Machine Learning*, 282–289.

Charles Sutton and Andrew McCallum. 2007. An Introduction to Conditional Random Fields for Relational Learning. *Introduction to Statistical Relational Learning*, MIT Press.

Yue Wang. 2010. Developing Robust Protein Name Recognizers Based on a Comparative Analysis of Protein Annotations in Different Corpora. *University of Tokyo, Japan*, PhD Thesis.

Author Index